THIS
BLUE
PLANET

Color Illustrations from Time-Life Books

THIS BLUE PLANET

Introduction to Physical Science

CHARLOTTE R. WARD
Auburn University

LITTLE, BROWN and COMPANY Boston

B G S U FIRELANDS LIBRARY

COPYRIGHT © 1972, BY LITTLE, BROWN AND COMPANY (INC.)

ALL RIGHTS RESERVED. NO PART OF THIS BOOK MAY BE REPRODUCED IN ANY FORM OR BY ANY ELECTRONIC OR MECHANICAL MEANS INCLUDING INFORMATION STORAGE AND RETRIEVAL SYSTEMS WITHOUT PERMISSION IN WRITING FROM THE PUBLISHER, EXCEPT BY A REVIEWER WHO MAY QUOTE BRIEF PASSAGES IN A REVIEW.

LIBRARY OF CONGRESS CATALOG CARD NO. 70-179878

FIRST PRINTING

*Published simultaneously in Canada
by Little, Brown & Company (Canada) Limited*

PRINTED IN THE UNITED STATES OF AMERICA

The cover photograph was taken during the flight of Gemini V, August 22, 1965, orbit 14, 1103 Greenwich Mean Time. It shows the Strait of Gibraltar, Morocco, and Spain. Courtesy of NASA.

TO THE STUDENT

```
QC21.2 .W37
Ward, Charlotte Berkley
Reed, 1929-
This blue planet;
introduction to physical
science
```

It is not necessary to have studied science previously to read this book, but we hope it won't be the last science book you ever read. It contains some helps to make your study of science easier. These include:

1. A glossary of terms important to your understanding, but used only a few times. Frequently used terms are defined in the text.

2. A list of symbols and constants, printed inside the back cover where you can find them quickly.

3. A mathematical appendix to refresh your memory about some procedures applicable to scientific thinking. Even though this book doesn't contain mathematical problems, the author could not avoid using graphs and proportions and angles. But they are all explained in the appendix.

4. Brief reading lists at the ends of the chapters. The items listed are "popular" books, on the whole, never advanced texts. Most are books and articles that students like you have found helpful and enjoyable when they were taking a course like yours.

5. Summary statements summarizing each chapter. They aren't all you should learn from the chapter, but they will help you organize your knowledge.

6. Review questions to help you make sure you have the important points in mind. There are also some "thought questions." The book doesn't have the answers to those; sometimes nobody does. But they

will direct your attention to some things we all need to think about.

People become scientists because science is exciting and fascinating to them. This book is aimed at sharing some of this feeling for science with you, and at helping you to feel more at home in a "scientific" world.

PREFACE

There is no room in an Apollo capsule for an untrained person "just along for the ride." This Blue Planet is rapidly approaching the same condition. It is a small world; its resources are not limitless; it is getting noticeably crowded. It has no room for the thoughtless consumer of its resources and space; the careless polluter of its land, water, and air; the person who is unaware of its problems or who feels no personal responsibility for them. How then can we make sure that the people presently riding Spaceship Earth are aware and concerned and take good care of it and bring up their (average of two!) children to do likewise? Undoubtedly, in the last analysis, it is a matter of individual moral decision and commitment. But science also has a part to play, for the overriding, inescapable natural law by which we must live or will perish is the realm of science, and it is science that provides the understanding of the natural working of the world by means of which the personal, societal, and political decisions must be made.

In the twentieth century the attitudes toward science of both the intelligentsia and the man on the street in the western world have been characterized by a strange dichotomy. On the one hand science is extolled for its production of technological miracles ranging from labor-saving machines to drugs and medicines to cosmetics and synthetic fibers. On the other hand science makes bombs and chemical and biological weapons, and is therefore to be feared; it talks in strange symbolic words that few understand, and therefore is somehow alien and beyond hope of common comprehension. The overall result is widespread scientific illiteracy among the highly educated humanists as well as among the uneducated philistines.

Both because of our dependence on that applied branch of science we call *technology* and because of the urgency of finding solutions to environmental problems that are the by-products of our technology, scientific illiteracy is no longer tolerable in any group. College physical science courses are aimed at correcting this condition, at least in those young people who are going to form the well-educated segment of the adult population. Because some of these students will eventually teach a much broader segment of the population in the elementary and secondary schools, they may also be influential in correcting scientific illiteracy on a much larger scale.

This book contains the essentials of scientific principles as a framework into which as much scientific fact may be fitted as becomes available to the student from any sources throughout his life. I hope that the very obviousness of the gaps will remind and encourage the student to go on observing his world and learning about it all his life. If this happens, he will be a scientist in the most important sense of the word: he will be alive, aware, thinking, understanding, acting to improve his environment, within and without, all his life.

The brief reading lists following the chapters of this book consist of books and papers that students of physical science have found readable, interesting, and valuable in providing further knowledge and deeper understanding of the topics discussed in the text. I hope the text will whet the reader's intellectual appetite sufficiently to induce him to try reading some of these.

Acknowledgments

The author would like to acknowledge her debt to the many people whose help and encouragement made this book possible. They include Dr. G. M. Kosolapoff, her colleague in teaching physical science at Auburn University, Dr. Judith C. Damewood, who as graduate teaching assistant contributed much to the development of the course, Dr. Curtis H. Ward, consultant and critic par excellence, and the secretarial staff of the Auburn University Physics Department, especially Mrs. Jo Hawkins and Mrs. Mary Childers, as well as the many excellent teachers, colleagues, and students through the years whose influence is consciously or unconsciously reflected in this book. Finally, the author wishes to express her gratitude to Thomas L. Sears and David W. Lynch of Little, Brown for their untiring efforts on behalf of *This Blue Planet*.

CONTENTS

CHAPTER

1   Scientific Beginnings   1

Magic, religion, and science. Theoretical science. Mathematics. The first scientists. Astronomy. Ptolemy's system. Arab contribution.

2   Motion: The Birth of Classical Physics   19

The revolution in mechanics. Celestial motion. Universal gravitation. Dynamics: the laws of motion. The laws of motion. Circular motion.

3   Energy   45

Mechanical energy. Conservation of energy. Heat and work. Temperature. The kinetic-molecular theory. Avogadro's number. The Carnot cycle. The second law of thermodynamics.

4   Electricity and Magnetism   74

The electrical nature of matter. Magnetic properties. Electrical forces. Producing electrical energy. Electric circuits.

5   Light   97

Measuring light. Geometric optics. Color. Physical optics. The electromagnetic spectrum. Particle nature of light.

CHAPTER

6 Physical Science Circa 1900 — 118

The electromagnetic theory. The Michelson-Morley experiment. Planck's constant. The photoelectric effect. Radiation theories.

7 Relativity — 135

Transformation. Mechanics and electromagnetic problems. Proving mass increase. Time dilation. The fourth dimension. General relativity.

8 Quantum Theory: The Fundamental Structure of Energy and Matter — 154

The structure of matter. Line spectra. Atomic numbers. Wave nature of matter. The uncertainty principle. Diffraction patterns.

9 Chemistry: The Periodic System — 171

Classifying substances. Periodic system of the elements. The guiding principles of chemical reactions. The language of chemistry.

10 Two and One-Half Million Compounds of Carbon — 201

Synthesizing compounds. Unique property of carbon. Types of organic compounds.

11 Molecular Biology: The Physics and Chemistry of Life — 227

The chemical nature of life. The chemistry of heredity. The chemistry of growth. Energy production.

CHAPTER

12 The Atomic Nucleus — 258
Radioactivity. Fission and nuclear power. Radiation uses.

13 Stars and Galaxies — 286
Astronomy. Creation theories. The "steady state" and "big bang" theories. Evolutionary development. Measuring the stars. A model of the universe. Galactic evolution.

14 The Evolving Earth — 323
Solar system formation theories. The earth's crust. Chemical evolution of life. Landscape evolution. Continental drift. Dating the earth's changes. Biological evolution. Geologic time. The age of reptiles. The era of recent life. Weather.

15 Frontiers of Science: Problems and Possibilities — 366
Mysteries of life. Science and moral values.

APPENDIX

A Mathematics — 376

B Scientific Measurement — 389

Glossary — 394

Index — 405

1

Scientific Beginnings

A public service advertisement appearing in magazines in the summer of 1971 to urge financial support of colleges used this opener in large type: "80% of the scientists who have ever lived are alive today." Certainly twentieth-century life is more conditioned by science than human life has ever been, but science is above all human endeavor, and as old as humanity.

Man has always had to cope with his environment somehow. So he has sought to understand it in order to cope more effectively. To understand, he has had to observe and to manipulate his environment. He has of necessity been a *scientist*.

Science has always had two aspects: "pure" or theoretical science, the pursuit of truth about the world for its own sake; and technology, which includes the practical inventions and methods of applying science to every kind of endeavor. The "man in the street" may never meet a theoretical scientist, but technology is all around him; it provides the food he eats, the clothes he wears, his transportation, his amusements, and the tools of his trade, whatever that may be. Nevertheless, technology can advance only as basic scientific knowledge advances, and on the other hand, progress in theoretical science depends on the development of apparatus and techniques that are a part of technology. It isn't really so important to try to separate the two, but on the whole, this book is concerned with the concepts of "pure" science, and in general that is the way science is to be understood here. Science is the open-minded pursuit of truth about one's world, always open-ended, always somewhat tentative about its conclusions, always curious about how things work.

Magic, Religion, and Science

If science begins with curiosity about one's environment, satisfying that curiosity does not always take a form that we would call *scientific*. To begin with, this curiosity is hardly idle. We must seek to understand our world to live in it. We can observe some effect, for example, rain. It becomes obvious that rain is the source of water in the water holes, and that no life is possible on earth without water. We observe that thunder often accompanies rain, or indeed, precedes it. We conclude that thunder produces rain. When rain is needed, therefore, noisemaking in imitation of thunder ought to produce it. So man from earliest times has thought his way from cause to effect, although not always in a rational sequence. Such non-rational cause-and-effect connections lead to a way of coping with the world in terms of what we call magic. Simple laws of probability would require that techniques of magic in influencing nature would occasionally seem to be effective, i.e., it must sometimes rain after a rain dance. Successes are remembered, failures forgotten, and the authority of the shaman is established. He is even more effective in working his magic on persons who die of his curses simply because they believe in them. Anthropological literature abounds in examples.

A second method of coping with the world arose

early in human history: seeking to find the powers or deities behind the happenings of nature and to placate them. It is clear that this primitive approach to religion is still practiced in the present time, just as magic persists in folklore and superstitious practices.

A third method of coping with the world is seeking rational cause-and-effect relations and acting upon them to either control or to adapt to the operations of nature. It may be said at the outset that there is no a priori way of making sure that a cause-and-effect connection is rational rather than magical. In fact it might be better to reserve "magic" for the response to a cause-and-effect connection that depends on subjective actions, as by a shaman or group under the influence of such a person, rather than by objective manipulation of the physical environment. This seems a fairer definition in view of the fact that the best "science" of one age seems utterly fanciful and illogical to a later age that possesses more facts.

So it would seem correct to say that practical science or "technology" is as old as man, because a long-established criterion for classifying remains as human has been evidence of technology: the use of tools, even if only partly shaped pebbles, and of fire. Indeed it may be said that technology was inevitable if humans were to survive at all.

When we think of science, however, we are likely to have in mind "why" questions as well as "how" questions to be answered. The scientist's "why," however, is not the same as the "why" of theology or philosophy. It is not teleological; it does not presume to seek for ultimate meaning or purpose, but rather to find explanation in terms of cause-and-effect: This condition exists because such-and-such a condition preceded it and such-and-such an interaction occurred. Even in those areas of modern physics where strict causality in the classical sense breaks down, the scientist's "why" is aimed at approaching this kind of explanation as nearly as possible. These scientifically ambiguous cases, however, lead to the situation in which modern physics finds itself ever nearer to philosophy.

Theoretical Science

Although practical science, i.e., technology, is as old as mankind, theoretical science in the sense of formulating rational concepts of the universe with powerful predictive properties came later. Cultivating intellectual activity of any kind has certain requisites: curiosity, a sense of wonder about the world, and also leisure in which to indulge it. Those who wonder about the world also must communicate so that in the sharing of insights deeper ones grow, and this communication must be open and honest, unfettered by dogma or preconceived opinions. That observation must be careful and thinking clear almost goes without saying.

Careful observations of natural phenomena were made by prescientific peoples, especially observations of the sequence of the seasons and the stars, sun, and moon. The regular movements of the heavenly bodies enabled the Egyptians as early as 4000 B.C. and the Babylonians by 2000 B.C. to make extremely accurate calendars. The Britons of 1900–1700 B.C. (Plate 1), who are not usually thought of as highly civilized, built the titanic analog computer of Stonehenge to predict such celestial events as solar and lunar eclipses, for reasons we may never know. The Egyptian observers learned early to associate the annual flooding of the Nile with the heliacal (presunrise) rising of Sirius. Because the whole life cycle of the land and people depended on the arrival of the flood waters with their load of silt to renew the land, it is easy to see how those who could predict this crucial event would be held in the highest esteem and elevated to a sort of ruling priesthood, which sometimes oppressed the people on one hand while advancing knowledge on the other. The Babylonians also had their great observers, to whom we trace the Zodiac, the twelve "houses of the sun" – the twelve constellations against which the sun rises in its annual cycle.

Mathematics

Mathematics has been called both the *queen of the sciences* and the *language of science*. (One's choice of terms might be contingent on whether one thought of oneself as a mathematician or as a physical scientist.) Mathematics and physical science have developed together. Mathematical methods have been developed to deal with specific scientific problems in some instances. In others, a mathematical discovery has preceded by years the formulation of physical theory that required it. An example of the former is Newton's development of the calculus (see Chapter 3); an example of the latter would be Einstein's application of the Lorentz transformations to his special theory of relativity (see Chapter 7).

Mathematics in the practical applications of mensuration (measurement) and arithmetic developed in every early civilization. We have evidence of their use, although we are not always able to discern their methods. The Babylonians apparently could solve quadratic equations, but we don't know how they did it. Those great builders, the Egyptians, could use right triangles constructed of string to get length ratios, although they never came up with the theorem of Pythagoras. The Babylonians used a number system based on 60, which has certain advantages over a decimal system. In general, however, mathematics before the Greeks had developed strictly as a practical tool, and simple arithmetic was not really so simple until the Middle Ages because no one before the Arabs of that period devised a simple flexible notation.

The First Scientists

If we think of the prime characteristics of science as the abstraction of coherence and order from the vast and often chaotic observations of the world, then we must consider the Greeks from the sixth century B.C. as the first scientists. The same society that brought philosophy, drama, poetry, and sculpture to its finest flowering in the ancient world did equally well for science. The Greek thinkers of scientific bent produced the first "system of the world," a coherent picture of the universe in terms of observation and inference.

The first name in the annals of Greek science is Thales (640–546 B.C.), a merchant of the Ionian city of Miletus, on the southwestern corner of Asia Minor. Thales seems to embody the spirit of science to perfection. He was a keen observer of the world around him, with an insatiable desire to find relationships among his observations and to test them by prediction. He was able to predict storms as well as celestial events, and put together a "system of the world" that, although it seems fanciful to us, accounted for the facts he had available (Fig. 1.1).

Figure 1.1 Thales' concept of the world.

In sixth- and fifth-century B.C. Greece plane geometry reached its greatest heights. The first great school of mathematical philosophy arose in the Italian city of Croton, where a mystic named Pythagoras listened to the "music of the spheres" and developed the theory of harmonics of strings and of rational numbers. This is the Pythagoras of right triangle fame. His theorem that the square of the hypotenuse of a right triangle is equal to the sum of the squares of the other two sides (usually written $a^2 + b^2 = c^2$) is still one of the most important and useful concepts in trigonometry.

It seems that Pythagoras in considering his great theorem must have been aware that although the theorem is of absolutely general application to plane right triangles, in relatively very few cases are a, b, and c all either integral or rational. Yet his belief in the mystic power of rational numbers led him to reject the idea that pi or π, the ratio of the circumference of a circle to its diameter, is an irrational number, and reportedly even to have executed as a heretic a disciple who dared to insist on the irrationality of π.

This incident illustrates the greatest limiting factor inhibiting the development of science in ancient Greece. Plato's philosophy puts it most clearly: the idea of some perfect reality beyond human grasp, but which can be approached by pure thought more nearly than by observing the objective world. What we normally consider the "real" world was to the Greeks only a shadow of reality. The shadow was, then, less worthy of study than the real thing. This attitude inevitably resulted in neglecting experimentation with or even careful observation of the objective world, and when it came to a choice between observation and logic in finding the truth of a matter, logic won every time. Aristotle's erroneous ideas about nearly everything from the origin of frogs in the spring mud to the centrality of the earth in the universe, which held sway for nearly two thousand years in the western world, are prime examples of the dominance of subjective logic over objective observation and experimentation.

Nevertheless, Greek scientists made some solid achievements, even though some were lost or ignored for centuries. The geometry of Euclid has enjoyed continuous appreciation and application from his own day, around 500 B.C., to the present and on into the foreseeable future, whenever we must deal with plane surfaces. The attempt of Hippocrates (fifth century B.C.) to make medicine scientific instead of magical has gone on, with occasional setbacks, ever since his own time, and twentieth-century doctors still give allegiance to the Hippocratic oath.

Astronomy

The science at which the Greeks excelled is astronomy, a subject that is necessarily, even today, observational instead of experimental. The Greek astronomers, beginning with the disciples of Thales in the fifth century B.C., appreciated the fact that the earth's surface is curved. Anaximander could apparently conceive of a curvature in the east-west direction where he could see ships appearing and disappearing at the horizon, but not in the mountainous north. (Why, we wonder, not in the watery south? Symmetry considerations, perhaps?) He proposed that the earth is a cylinder spinning on a longitudinal axis, unsupported in space. (Anaximander also speculated on the descent of man from fish.)

By the time of Alexander the Great (fourth century B.C.) enough people had traveled far enough — overland with the army to India, and by ship through the Strait of Gibraltar around the British Isles or down the west coast of Africa — to demonstrate the east-west curvature of the earth. Actually a traveller in the north-south direction could observe the curvature of the earth even more easily, because of the constantly changing sets of stars that would come into view overhead or vanish below the horizon and the changing slant of the rays of the sun. Lunar eclipses, observed at different local times in different locations, but observed by all as a curved shadow moving across the moon, gave additional support to the idea of an earth that is spherical. (Plate 2.) Add to this the Greek idea of the sphere as the perfect figure, and you have a spherical earth surrounded by celestial spheres of sun, moon, and the five planets, all enclosed in the crystal sphere containing the fixed stars, surely as ideal and harmonious an arrangement as might ever have produced Pythagoras' celestial music. If this idea was fully developed by Aristotle before 300 B.C., one wonders why people worried about Columbus falling off the edge in A.D. 1492!

However, then even more than now, only a few possessed scientific knowledge, and it seemed clear to the ordinary citizen who never got many miles from home that he lived on a flat earth, with the celestial bodies travelling over it across a vault of heaven most conveniently pictured as an inverted bowl. As to what held it all up, well, something substantial: Atlas, or a large turtle, or an elephant. The idea that the earth is flat has, indeed, persisted to this day, although it is becoming harder and harder to explain satellites, air travel, and men on the moon in terms of a flat earth, and the British Flat Earth Society was reported to suffer a decline in membership after Apollo 8 circled the moon in 1968.

In the last three centuries preceding the Christian era, Greek astronomers proceeded from the idea of a spherical earth to actual measurements and calculations of its size. Aristarchus of Samos, around 270 B.C., devised a method for estimating relative distances and sizes by observing that during a solar eclipse the disk of the moon just covers the sun, and by inferring the necessary arrangement of the sun, moon, and earth on the night of a half moon. It works like this: if the moon under these conditions is half-illuminated by the sun, then the angle earth-moon-sun must be a

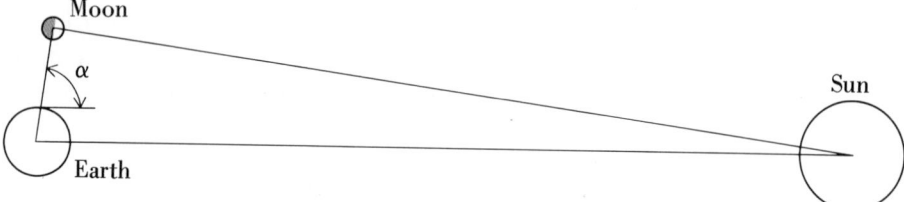

Figure 1.2 Aristarchus' method for estimating the relative distances of the sun and the moon from the earth.

right angle (Fig. 1.2). The angle α, the elevation of the moon observed from the earth, was found to be about 87°. All right triangles having an 87° angle are similar, in Euclid's terms, and all have the same ratio of adjacent side to hypotenuse, 1:19. So, the sun is nineteen times as far away as the moon, by Aristarchus' measurement. His angle was off by 2°50' — the actual value for α is 89°50', giving a moon:sun ratio of 1:395.

However, given a distance ratio, a size ratio can be found as follows (Fig. 1.3): again calling upon the properties of similar triangles because $\triangle ABC$ and $\triangle ADE$ are similar, side BC, the diameter of the moon, must bear the same relation to side DE, the diameter of the sun, as AC, the earth-moon distance, bears to AE, the earth-sun distance. That is, the sun, according to Aristarchus, is 19 times as large in diameter as the moon (Plate 3).

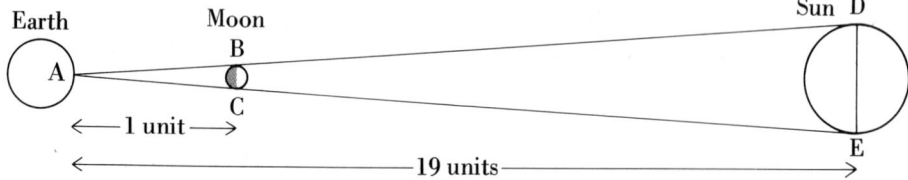

Figure 1.3 Aristarchus' method for estimating the relative sizes of the sun and the moon.

Aristarchus went much further than his contemporaries in another aspect of astronomic theory. He saw that the observed movements of sun, moon, and stars could be explained as well by assuming a moving earth and a stationary sun around which the earth moved. This idea was born 1800 years too soon, however, and it had a very short life. His contemporary Hipparchus of Nicea, whose many detailed observations and correlations of his own work with prior Babylonian and Greek discoveries established him as the greatest astronomer of antiquity, firmly believed in the earth as the center of the universe. After all, he could explain all the celestial motions, too, and it seemed a lot more sensible for the earth to be at the center of things.

After the death of Alexander, the city he had established in the Nile delta grew to be the real intellectual center of the Mediterranean world. An Alexandrian astronomer, Eratosthenes, in the third century B.C. made a very good measurement of the circumference of the earth. He observed that on a day when the noon sun cast no shadow of a gnomon (sundial) at Syene, due south of Alexandria, a gnomon at Alexandria cast a short shadow. If indeed the sun is very far away it can be assumed that the rays of the sun arriving at Syene and at Alexandria are parallel to each other. They make different angles with the gnomons because the earth's surface curves away from the sun between them. The triangle determined by the gnomon and its shadow contains the same angle as the angle subtended by the arc of the earth's surface (Fig. 1.4). Remembering that the angle in radian measure is the

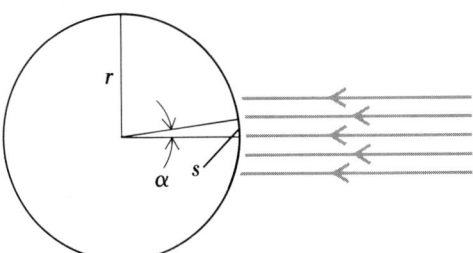

Figure 1.4 Eratosthenes' method for estimating the circumference of the earth.

ratio of the arc length s to the radius r, the radius can be calculated.

Eratosthenes got a remarkably accurate result, especially in view of the fact that Syene is 3° off due south from Alexandria, introducing a small error in s.

Ptolemy's System

The results of centuries of observation were codified by Ptolemy, astronomer and librarian of the great library in Alexandria in the middle of the second century A.D. Ptolemy's system of the world, which held almost undisputed sway until the sixteenth century, placed an unmoving earth firmly in the middle of the universe. The crystal spheres of Aristotle, supposed to hold the sun, moon, planets, and fixed stars in their courses, were dispensed with, but those bodies were still in their ancient places circling the fixed earth. Let it not be supposed that Ptolemy's system was purely a work of imagination. It was based on very precise observations accumulated over centuries of time in many lands. In its final form, it was a complicated scheme worked out with mathematical exactness to accommodate all those observations. The earth was enough off-center to take care of those irregularities we now attribute to the ellipticity of the earth's orbit. Those planets, Mars, for example, that exhibit retrograde or reverse motion at certain seasons were accounted for by assuming they moved on small circles (epicycles) around their circular orbits. To take care of every detail, some of the epicycles even had epicycles, until there were seventy in all.

The complexity of Ptolemy's scheme finally brought about its downfall. An old reliable rule in science, enunciated in the twelfth century by a British scholar named William of Occam and henceforth known as *Occam's Razor,* says that when there are alternative ways of explaining phenomena, that which makes the fewest assumptions is likely to be the best. The Copernican system of the world, which finally displaced the Ptolemaic, in its original form still required forty epicycles to account for the facts. But forty is better than seventy, by Occam's reasoning, and so the idea of a fixed sun and orbiting earth finally displaced the ancient concept of an earth-centered universe. But not without a struggle.

The Greek scientist whose work has best stood the test of time is Archimedes, who lived in Syracuse from 287 to 212 B.C. Philosophically, Archimedes was thoroughly Greek, but his methods seem remarkably modern. His experimentally determined principle of buoyancy (discovered in his bath, as he meditated on how to determine the gold content of the king's crown without destroying it), which states that every body wholly or partly immersed in a fluid is buoyed up by a force equal to the weight of the fluid it displaces, and many other principles of statics, the study of balanced forces, have needed no amendment or correction through two millennia. Archimedes was a theoretical mathematician as well as an engineer. He reportedly lost his life at the hands of an impatient Roman soldier by refusing to move until he finished the mathematical proof he was writing in the sand.

The Romans were engineers, technologists rather than scientists. They preserved Greek thought, but did little to extend it. Lucretius, the poet, latched onto an ancient idea of Democritus about the atomicity of matter and expounded it in his magnum opus *De Rerum Naturae*, but this, too, was an idea born out of due time. In the fourth century much of the great Alexandrian library was burned, and much classic learning was lost to the western world for many centuries; those centuries we call the *Dark Ages*.

In Europe the Dark Ages were enlightened only by the Church. As the concern of the Church lay more with the hereafter than with the here-and-now, the learning it preserved was theological and philosophical rather than scientific. And because nearly all scholars for several centuries were churchmen, the problems to which they addressed themselves were theological rather than related to the physical world.

Arab Contribution

Those innovations in natural science that occurred in the first millennium of the Christian Era were the work of the Moslem Arab culture in North Africa and the Middle East. Considering that these were newly civilized people, it is not surprising that they devoted themselves to the same sciences that occupied new civilizations before them—astronomy, medicine, and mathematics. This is evident in the names of the bright stars. The oldest star names go back to the Greeks—Sirius, Castor, Pollux, and most names of familiar constellations, the pictures in the sky that were so much more apparent to the ancients than to us. Many individual star names in these constellations are of Arabic origin—Deneb, Algol, Betelgeuse.

The most important lasting contributions of the early Arab scholars were in mathematics. Perhaps the greatest deterrent to developing methods of calculation in ancient and medieval times was the absence of a simple system of numerical notation. Anyone who has tried to read the dates on cornerstones of public buildings can imagine trying to do long division with numbers like that! The Arabs invented the simple symbols 0, 1, 2, 3, 4, 5, 6, 7, 8, 9, with which any number can be conveniently expressed, using ten as a base, as we ordinarily do, as well as two or any other base less than ten. Not only is writing processes of arithmetic simplified, but a combination of letter and numerical notation led to another important Arab invention, algebra. It is hard to imagine modern science without that wedding of algebra and geometry called *analytic geometry,* for every scientist loves a graph—a line relating his measured results and interpretable in terms of an algebraic equation that gives him a precise statement of natural law.

It is of course not strictly correct to say that no one thought about natural science in medieval Europe. Robert Grosseteste and his pupil Roger Bacon, whose methods of proof foreshadowed scientific investigation, William of Occam (or Ockham) with his test of simplicity for scientific theories, Jean Buridan and Nicole Oresme, who found flaws in Aristotle's doctrine of motion, Giordano Bruno, who dared to think about the universe outside the context of church dogma, and several others belong in any history of science. After contact with the Moslem world began to be established in the tenth century, the scientific works of Aristotle as well as the new mathematics of the Arabs became available to Europeans. There were disadvantages as well as advantages in this, for a scholarly community accustomed to truth taking the form of dogma, or vice versa, quickly made dogma of Aristotle's scientific pronouncements, most of which, as we have seen, were incorrect.

Meanwhile, halfway around the world in China, technology, if not "pure science," was producing such useful items as gunpowder, rockets, and the magnetic compass, all of which, by the end of the Middle Ages, found their way to Europe.

Historians propose several reasons for the final enlightening of the Dark Ages. That blot on the escutcheon of Christendom, the Crusades, had at least one positive outcome in introducing European nobility to the possibility that life could be comfortable, and undoubtedly importing such Eastern luxuries as silks and spices, fruits and vegetables helped divert some attention to the world of here-and-now. It is just as true that technological advance played a role in ending the darkness—the compass that allowed a ship to venture away from shore even if clouds hid the stars, and the movable type developed by Gutenberg in the fifteenth century that made the widespread dissemination of knowledge possible. By then the Renaissance days of artistic and literary glory were over, but the age of science was approaching its dawn.

SUMMARY STATEMENTS

1. Science is man's attempt to understand the world in rational terms in order to cope with it.

2. The Egyptians and Babylonians were the first to develop science for the very practical purposes of making calendars and measurements for their construction projects.

3. The Greeks were the first people who had leisure enough and inclination to pursue science for its own sake, but because they believed thought to be more reliable than observation in approaching reality, they did not test their thinking by experiment.

4. Nevertheless, some Greek scientists made very good measurements, such as Eratosthenes' determination of the earth's circumference, and Aristarchus' well-planned, if inaccurate, measurement of the relative sizes and distances of the sun and the moon.

5. The geocentric world system put together by Ptolemy in the second century A.D. held sway over astronomical thought for some thirteen or fourteen centuries.

6. During the Dark Ages in Europe, a few scholarly churchmen kept alive some interest in understanding the phenomena of this world, but the greatest scientific progress in those centuries was made by the Arabs, who excelled in mathematics, astronomy, and medicine.

REVIEW QUESTIONS

NOTE: Answers to Review Questions are found in the chapter. Answers to Discussion Questions are not usually found in the text.

1. What three methods have been used by human beings to cope with and understand the world in which they find themselves? What is the essential characteristic of each?

2. For what contributions are these men remembered in the history of science?

Thales	Aristotle	Archimedes
Pythagoras	Aristarchus	Democritus
Euclid	Eratosthenes	Ptolemy

3. What contributions to science did the following Middle Eastern peoples make?

>Ancient Egyptians
>Babylonians
>Arabs of the Middle Ages

4. Name some Europeans who maintained some interest in science through the Dark Ages.

THOUGHT QUESTIONS

1. Astrology is a good example of one generation's science being another generation's magic. Is there any possible scientific basis for astrological beliefs? Can you account scientifically for its current popularity?

2. What do you think of science at this point, before you have begun a formal study of it as a college student?

3. The Greeks were excellent mathematicians and astronomical observers but made little or no progress in biological sciences or in technology (applied sciences). Can this be understood in terms of the temperament, philosophy, and social structure of the Greeks? Do these factors still play a role in determining the way science develops?

4. Why is measurement of such great importance in science?

5. Compare the relative effectiveness of the contemplative method of the Greeks with the experimental method developed by Galileo and his successors in discovering scientific truth.

6. What do you think a scientist means by "scientific truth"? What is the relation between the terms "truth" and "reality"?

7. Considering the actual development of science through human history, what seem to be the essential cultural conditions under which science can develop?

8. What personal qualities do you think scientists need to possess?

READING LIST

Butterfield et al., *A Short History of Science*, Doubleday (Anchor), Garden City, N.Y., 1959. A series of essays on the origins and results of the scientific revolution by British scholars, based on BBC lectures for high school students.

Schneer, Cecil J., *The Evolution of Physical Science*, Grove Press, New York, 1960. A very readable complete history of the development of the physical sciences from early times to the present.

Eiseley, Loren, *The Firmament of Time*, Atheneum, New York, 1970. A beautiful account of the changes in man's understanding of nature and himself.

Young, Louise (ed.), *Exploring the Universe*, Oxford University Press, New York, 1970. Readings on both the scientific and philosophical aspects of the universe, many by the original workers in the various fields.

Snow, C. P., *The Two Cultures and a Second Look*, New American Library (Mentor), New York, 1964. The controversial Rede lecture of 1959 and a follow-up essay written in 1963, examining the lack of understanding between scientists and humanities scholars and its implications for the world.

The following books might be classified as mathematical, but the first two are a kind of science fiction, and the third is a sort of philosophical history of mathematics. They are listed because they are enjoyable and they just might offer some insight into the possibilities of mathematics as a way of understanding the world.

Abbott, E. A., *Flatland*, Barnes and Noble, New York, 1963.

Burger, D., *Sphereland*, Crowell (Apollo Editions), New York, 1965.

Dantzig, T., *Number, the Language of Science*, Doubleday (Anchor), Garden City, N.Y., 1954.

Plate 1 Stonehenge: This huge structure of concentric circles of monoliths and holes, including a "heelstone" that was aligned with the midsummer sunrise, is now thought to be a giant analog computer for predicting lunar and solar eclipses. It was built and rebuilt by otherwise primitive British tribes between 1900 and 1700 B.C.

Color illustrations are from the Life Nature Library or the Life Science Library

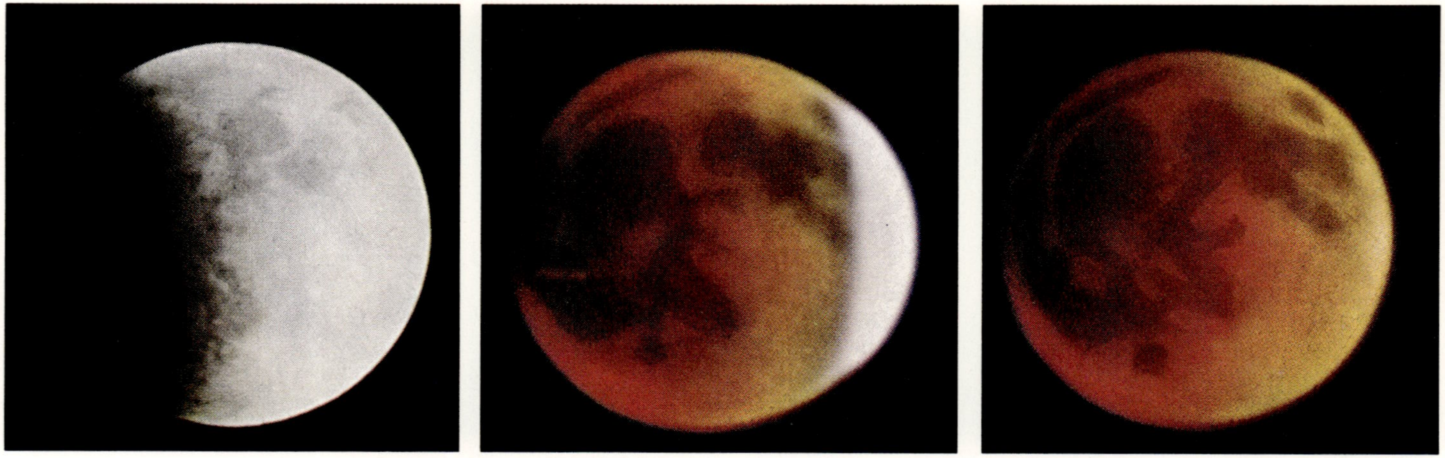

Plate 2 Photographs of a lunar eclipse show that the earth casts a curved shadow.

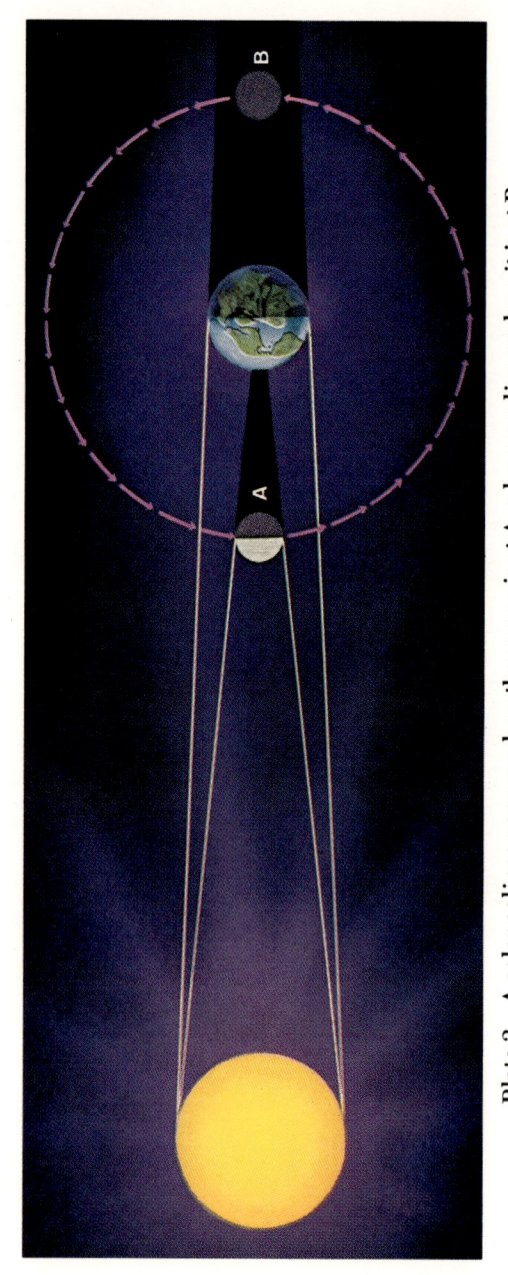

Plate 3 A solar eclipse occurs when the moon is at A, a lunar eclipse when it is at B.

Aphelion

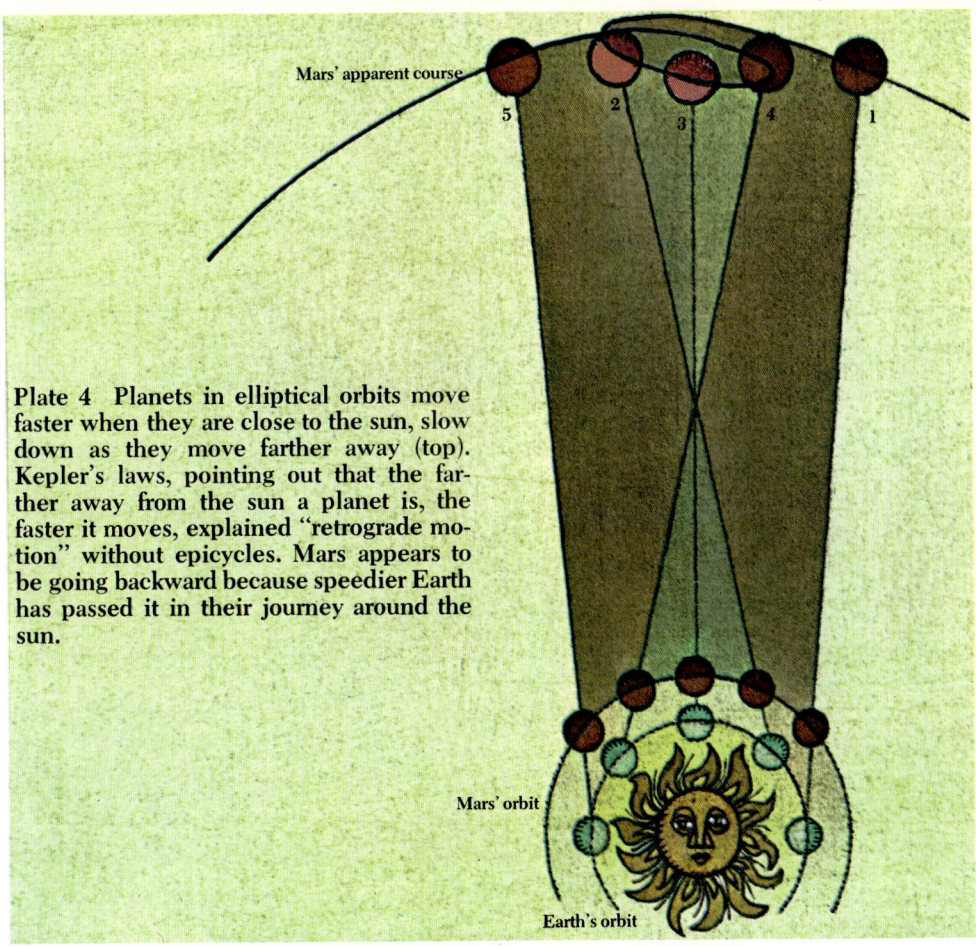

Perihelion

Mars' apparent course

Mars' orbit

Earth's orbit

Plate 4 Planets in elliptical orbits move faster when they are close to the sun, slow down as they move farther away (top). Kepler's laws, pointing out that the farther away from the sun a planet is, the faster it moves, explained "retrograde motion" without epicycles. Mars appears to be going backward because speedier Earth has passed it in their journey around the sun.

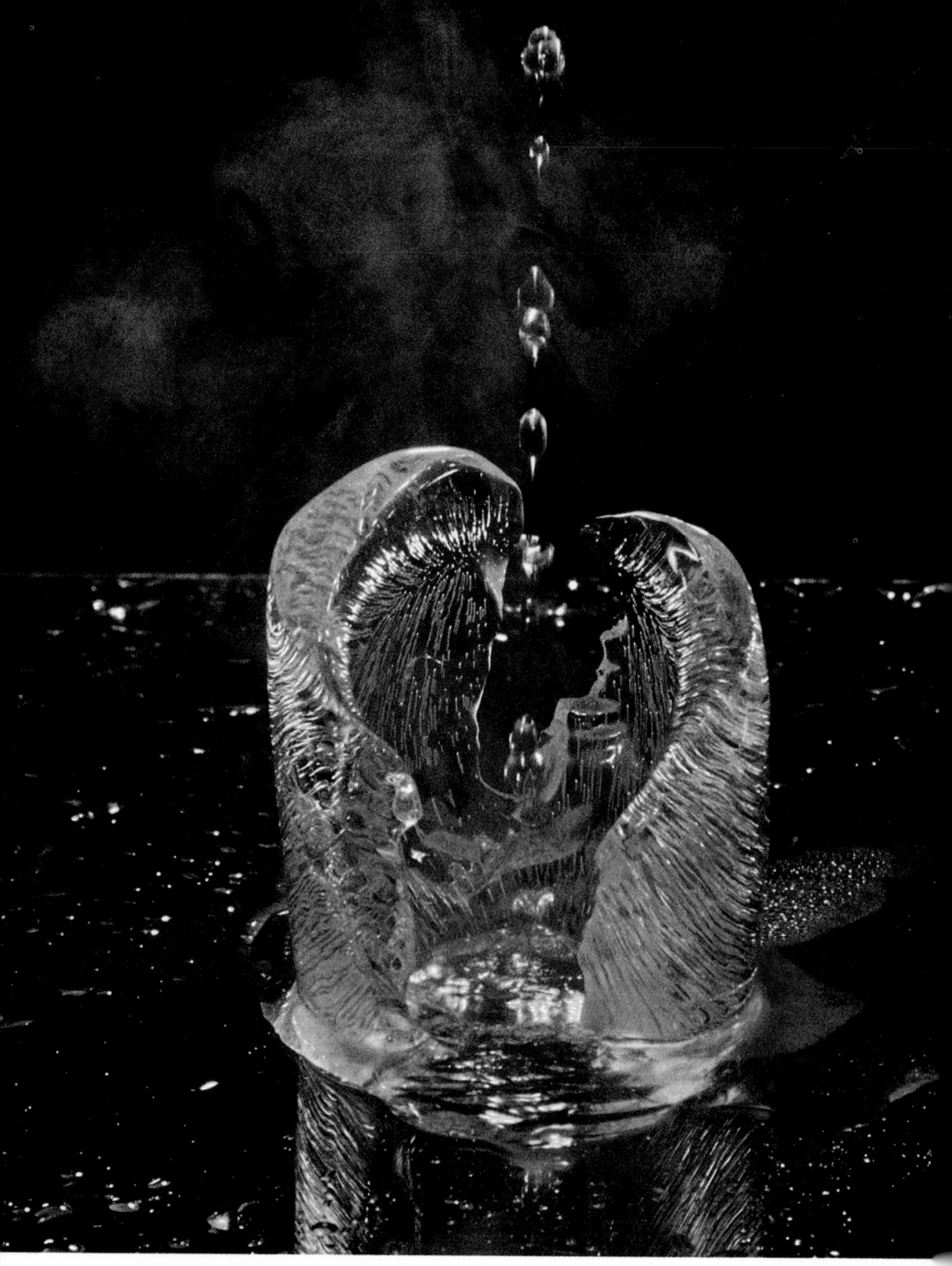

Plate 5 Water is remarkable for being solid, liquid, and gas within a narrow temperature range.

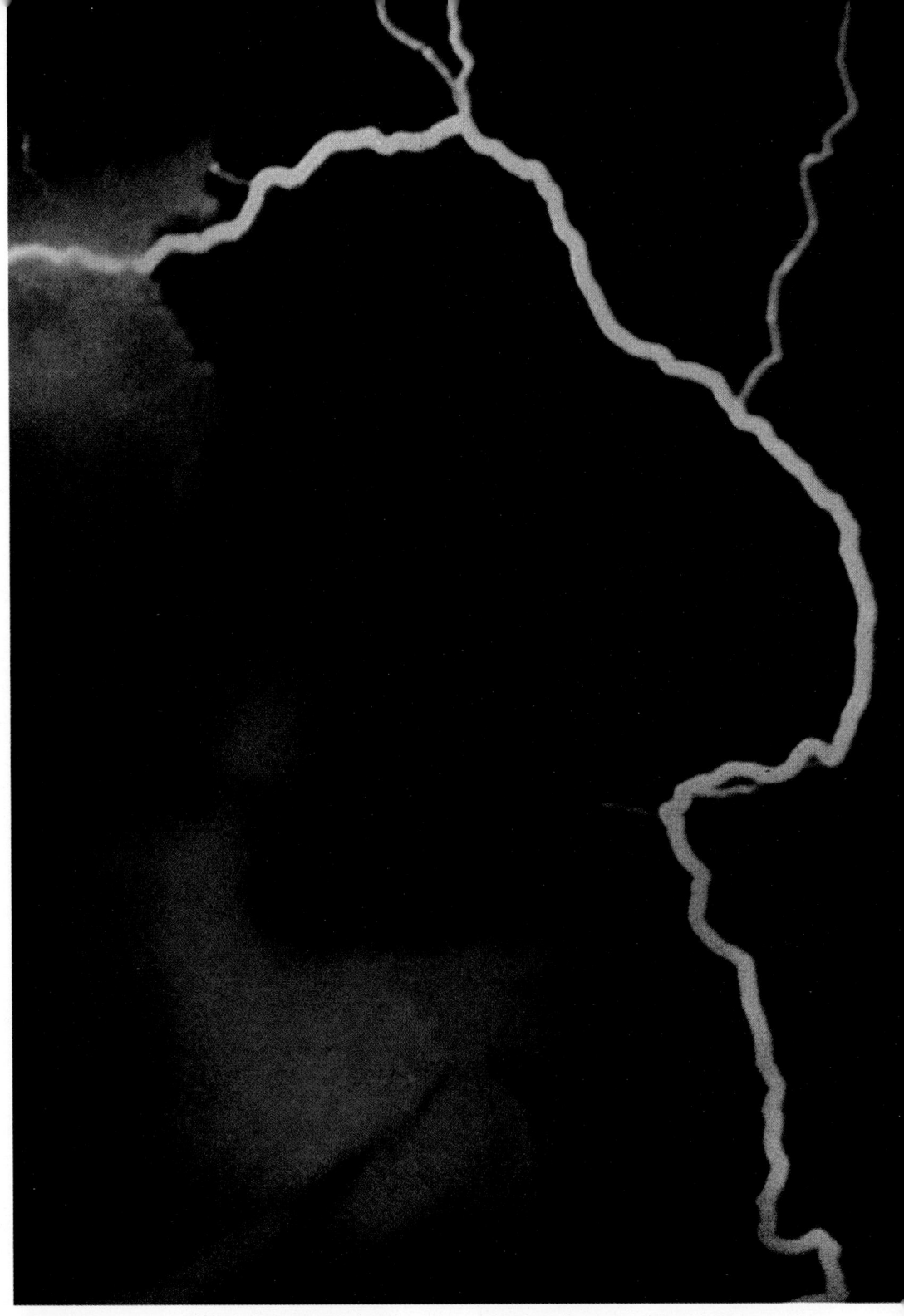

Plate 6 A streak of lightning is the visible path of an electric discharge in the atmosphere.

Plate 7(a)　A continuous spectrum produced by a white light source.

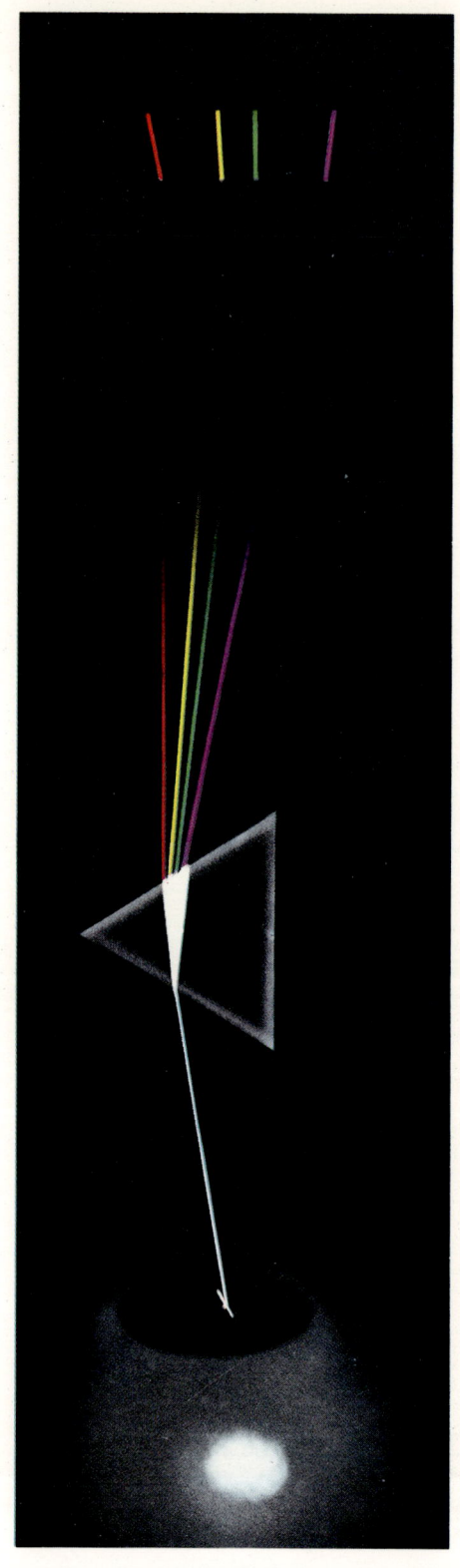

Plate 7(b)　A bright line spectrum emitted by "excited" atoms of a single element.

Plate 7(c) A dark line spectrum showing wavelengths absorbed by the atoms in becoming "excited."

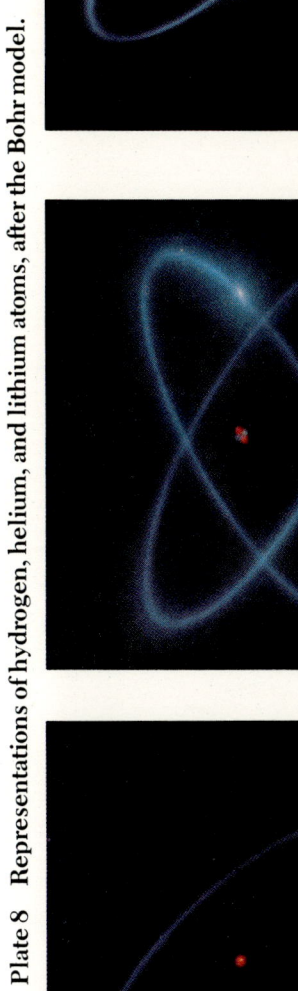

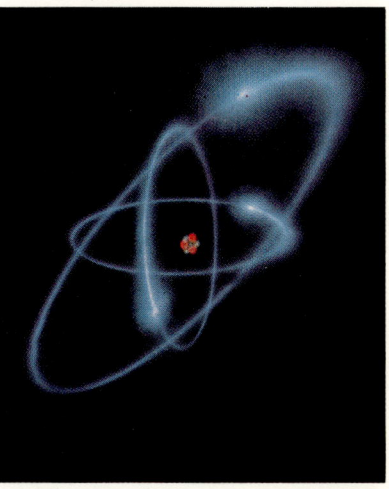

Plate 8 Representations of hydrogen, helium, and lithium atoms, after the Bohr model.

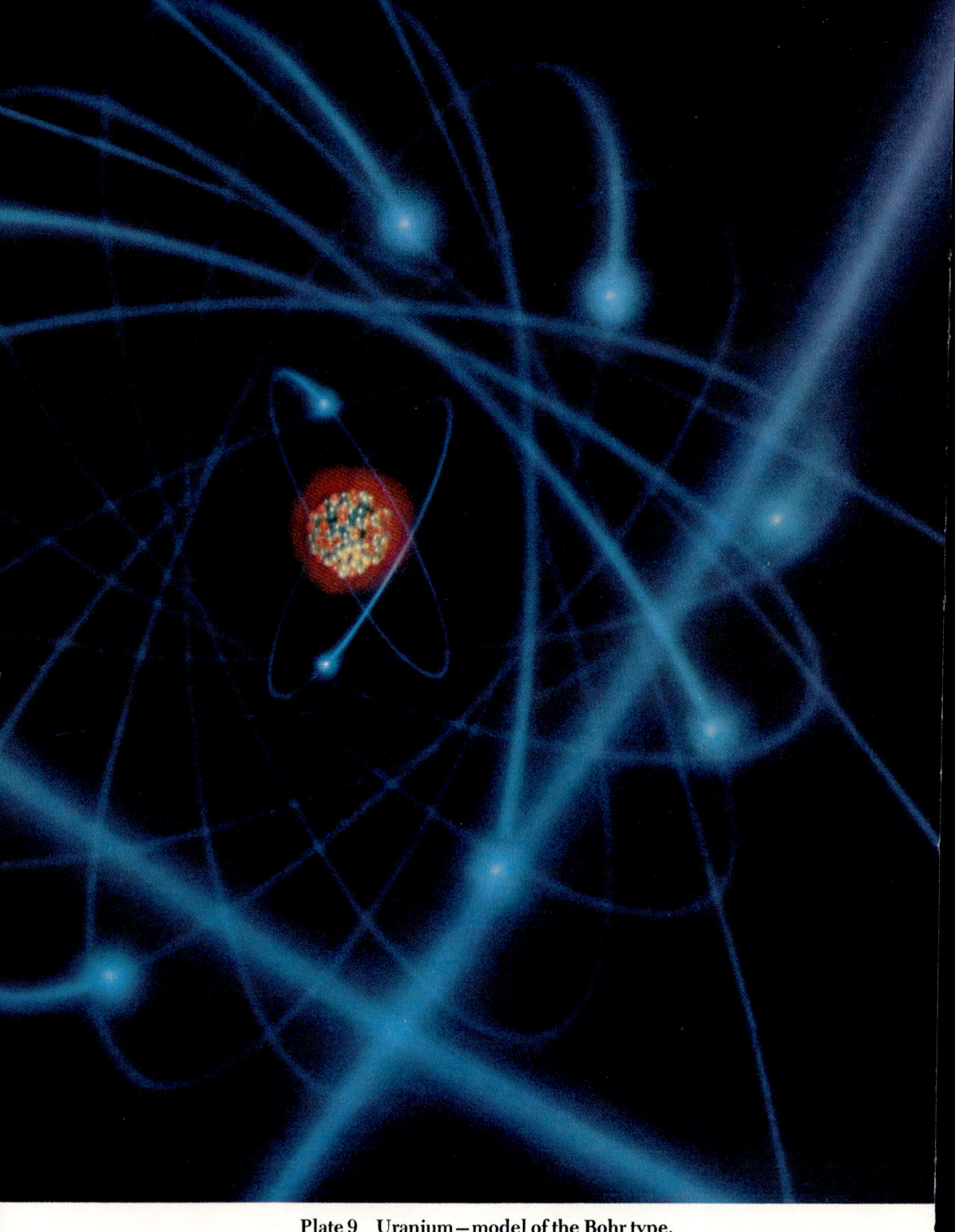

Plate 9 Uranium—model of the Bohr type.

2

Motion:
The Birth
of Classical Physics

It turns out to be a very risky business to try to predict how a science will develop, mainly because it is always easier to say after the fact what really crucial questions should have been asked and answered—in actuality one can ask at any time only questions provoked by what one already knows. So it was that the topic that occupied the greatest minds of the fifteenth, sixteenth, and seventeenth centuries does not at first glance seem to the twentieth-century mind to be truly fundamental. The topic was motion.

The universe is full of motion. In fact, motion rather than rest seems to be its natural state. But how is motion to be understood? Is it continuous or does it progress by finite, separate—that is, discontinuous—stages? This is an ancient question, indeed. Recall the paradox of Achilles and the tortoise, formulated by Zeno in the fifth century B.C. If Achilles can go twice as fast as the tortoise, and so gives the tortoise a head start of half the distance to the finish line, the tortoise will always win, because in the time it takes Achilles to get half way, the tortoise will go three-fourths of the way, and so on, with the athlete getting closer and closer, but never passing his slowpoke competition. Now in real life, everybody knows that Achilles will win. What is wrong with Zeno's logic? The trouble lies in the fact that motion is continuous, not discontinuous, as Zeno's race course is assumed to be.

Another question of long standing concerned the cause of motion. It was observed that an object generally moved if you pushed (or pulled) on it, but that it sooner or later stopped if you quit pushing. Aristotle thus acknowledged this universal observation when he stated that force (push or pull) was necessary to produce and to maintain motion. This led to a hard question: what pushes the celestial spheres in their orbits? No answer was apparent.

Perhaps the longest known kind of motion of inanimate objects is that which we call *falling*. Why does everything fall down? Or does everything fall down? Aristotle's answer: Things seek their own natures. Heavy objects fall toward the solid earth. Smoke and vapors go up into the lighter air. Clearly, then, the heavier an object the faster it falls — or does it?

The Revolution in Mechanics

Before we judge Aristotle too harshly, let us remember two things. First, because the Greeks extolled reason and the ideal reality was not susceptible to observation, they did not perform experiments on real systems. Second, moving objects do stop if you don't keep pushing; a rock does fall faster than a feather, in air. The new developments that set the stage for a revolution in mechanics, the science of material bodies and their motions, were, first, an experimental approach, and second, the idea of extrapolating from real experimental conditions to ideal if unreachable conditions.

The man whose name is associated with the first

experiments designed to answer the questions of motion and to test Aristotle's answers to them was Galileo Galilei, who lived in Italy from 1564 to 1642. Galileo investigated the motions of bodies and of light, which was supposed by many in his day also to be corpuscular. His attempts to measure the speed of light, which he assumed to be finite, were unsuccessful, and he conceded that light might be transmitted instantaneously after all. The actual answer to this problem was another century in coming. Although Galileo was never able to slow down light and measure its velocity, he was able to slow down falling motion by using inclined planes so that it could be studied. The gentler the incline, the slower a sphere's speed as it rolled down, so that the velocities of bodies of various sizes could be compared. The results were not in accord with the doctrines of Aristotle.

Although a feather and a rock certainly fall with unequal speeds, a big rock and a little rock (whether or not Galileo ever really dropped them from the Tower of Pisa), released simultaneously, do land together if they fall through distances up to some tens of feet. However, any falling object if it falls far enough will reach a "terminal velocity," a constant speed at which it falls the rest of the way down. The new thing that Galileo saw in this situation was that the terminal velocity depended on the effect of air resistance on the falling body. The falling of a feather, which has a very large surface in comparison to its weight, is much more determined by air resistance than by whatever is pulling the feather down. Indeed, dust and smoke, rather than going up indefinitely, also come down eventually, although at first the air bears them up. Galileo was able to extrapolate in thought, if not experimentally, to a state of no air resistance, or vacuum, in which he felt confident that a rock and a feather would fall together. (For many years, until recently, nearly every basic physics course included a demonstration of a coin and a feather falling together through a tall glass tube from which nearly all air had been pumped.)

During his experiments on the motion of spheres on inclined planes, Galileo was led to another important extrapolation (Fig. 2.1). He observed that a ball rolled down one plane would roll up another to nearly the height at which it started. What would happen as the second plane became more and more nearly horizontal?

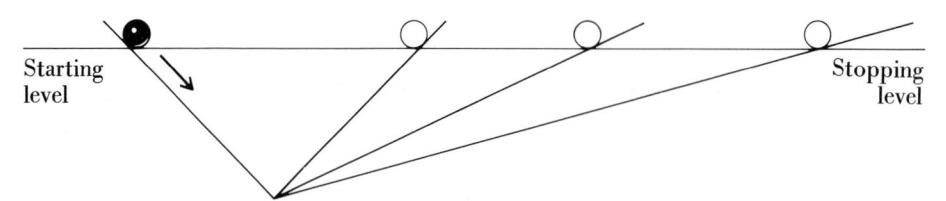

Figure 2.1 Galileo's inclined plane experiment.

Clearly if the ball returns to its initial level, it will travel farther and farther in the horizontal direction each time. And if the second plane were horizontal, the ball would have to go on forever, seeking its original level unsuccessfully.

Of course, this doesn't happen. The ball, however smoothly spherical, on a plane however slick and smooth, does eventually stop. But we have now reversed our question about its motion. We can see its continued motion as its "natural" state, and so we ask, "What stops it?" Thus arose the concept of inertia as a property of matter that makes it tend to resist changes in its state of motion, whether that state be rest or *uniform linear motion*—motion at a constant rate of speed in a straight line. But since all things in our common experience do stop, we must recognize an ever-present stopping force, which we call *friction*. Friction is the force that opposes motion. It is the force we call "air resistance" when we see a falling body reach its terminal velocity. It stops the sphere on the smooth plane by interactions of molecular forces we are only beginning to understand. But whatever its form, it is always present in the real world.

In 1642, the year that Galileo died, the man was born in England who later arranged in a simple and elegant form the principles by which motion can be understood. His name was Isaac Newton, and the problem that primarily occupied him was understanding motion on the grand scale of the solar system.

Celestial Motion

In 1543, a Polish churchman, Nicholas Copernicus, gave permission on his deathbed for the publication of a manuscript that was to ensure the immortality of his name. The manuscript, like its author's name, was in Latin, and it was titled "De Orbium Coelestium Revolutionibus" – "Concerning the Revolutions of the Celestial Orbs." Osiander, the friend to whom its publication was entrusted, immediately appreciated its subversive and revolutionary potential, and hastily added a preface intended to protect his teacher's good name after the man himself was beyond reach. Osiander insisted that Copernicus's book was only a suggestion of mathematical possibilities and in no way claimed to be an actual description of the real world.

What was the incendiary idea in "De Revolutionibus"? Simply the suggestion that the observed motions of sun, moon, and planets could be explained much more simply by assuming the sun, not the earth, to be the center of the system, with the earth and its attendant moon revolving about the central sun along with the five other known planets, Mercury, Venus, Mars, Jupiter, and Saturn. And why was this such a dangerous idea that its author dare not promulgate it during his lifetime? The answer lies in the domination of culture and learning by a religious hierarchy that in fifteen centuries had evolved from the free, even radical, movement that was the early Christian church to a ponderous structure of dogma and ritual that conceived of itself as eternal and unchanging, possessed of all essential knowledge in all fields.

Thus a new viewpoint on almost anything, and more especially a theory that dethroned the Earth, the abode of God's favored creature Man, from the center of the universe, was obviously heresy and an unconscionable attack on revealed truth.

It was, again, Galileo who saw the possibilities in the theory of Copernicus, and performed the experiments that helped to substantiate it. Galileo's evidence, today, seems very indirect. It was simply the view that struck Galileo's eye when he turned his newly invented telescope toward the heavens: the craters and (as he named them) *maria* of the moon and the four principal satellites of Jupiter orbiting the parent planet. It was evidence that the celestial orbs were not qualitatively different from earth—not perfect crystal spheres or divine lights but made of rock and (as the name *maria*, seas, suggested) water like earth. And if little moons went around Jupiter as our moon goes around the earth why shouldn't both Earth and Jupiter revolve around the sun? Of course, Galileo couldn't get his Inquisitors even to look through his telescope, and in his old age he was forced to recant and deny the truth of the Copernican concept of a heliocentric "system of the world." One may hope he really did mutter under his breath, "Nevertheless, it does move." But the power of an "idea come to its time" is well demonstrated by the survival of the Copernican theory and its development and correction in Protestant northern Europe where the Inquisition held no sway.

The Copernican system of the world needed correction. Just as Ptolemy had found that the orbits of the heavenly bodies could not be absolutely centered on the earth, but required a complicated system of epicycles and eccentrics, so also Copernicus was obliged to accommodate the observed facts by having his stationary sun a little off the center of his circular planetary orbits and by using more than half as many epicycles as Ptolemy had required. Still, by Occam's

Razor, forty epicycles are probably nearer the truth than seventy, and so is the symmetry of the idea of planets orbiting and reflecting the light of the unique member of the family, the central sun, than the idea of a stationary earth.

The man who put the finishing touches on our picture of the system of the world was Johannes Kepler, who was born in the German province of Württemberg in 1571, and who received his training in astronomy from one of the greatest astronomical observers of all time, the Dane, Tycho Brahe. Tycho never had a telescope, but from his Castle of the Sky, Uraniborg, he watched and charted the sky for many years, building up such a body of accurate data on the motion of Mars that Kepler, the mathematician, was not willing to accept a calculated path for Mars that differed from Tycho's observations by eight seconds of arc.[1] This small discrepancy led Kepler to discard the idea of circular orbits for planets. If God intended to make circular orbits, reasoned the devout Kepler, he wouldn't make an error of eight seconds. Kepler then concluded that the orbits must be ellipses, and found that his calculations confirmed that hypothesis exactly. So the basic laws of celestial mechanics, the science of the motions of the heavenly bodies, were codified in the sixteenth century as Kepler's Laws of Planetary Motion:

1. *The orbits of the planets are ellipses with the sun at one focus.*
2. *A line joining the sun and any planet sweeps out equal areas in equal intervals of time.*
3. *The square of the orbital period of a planet is proportional to the cube of its mean orbital radius.*

These three statements sum up our picture of the solar system. The sun is central in the system, yet not

[1] (A second of arc is 1/60 of a minute of arc, which is 1/60 of a degree, which is 1/360 of a circle. Eight seconds of arc is a very small error.)

at the center, for an ellipse has two foci. The second law implies the earth's, or any planet's, orbital speed varies, for in order for that imaginary line to sweep out equal areas in equal time intervals the planet must move faster the closer it is to the sun. And third, the Harmonic Law relates distances to orbital velocities via orbital radii—the ellipses that constitute the planets' orbits are so nearly circles that we often make that approximation (Plate 4).

Universal Gravitation

So the motions of the members of the solar system were described, but not explained. Here was where Newton came in. As the famous story goes, the young Newton, home on an enforced vacation from Cambridge in 1666 because of an outbreak of the plague, found all the pieces to his celestial puzzle falling into place as an apple fell from a tree. The problem, as Newton stated it, was one of central forces. The great insight he had was that the central force pulling falling objects faster and faster to the earth and the force holding the moon in its orbit around the earth are the same. He drew his analogy thus: Suppose you simply drop an object. It falls straight down, along a line directed toward the center of the earth. But if you throw the object forward it will travel some horizontal distance before landing, although if it is released at the same height in each case, it will take exactly the same length of time to reach the ground, assuming that the horizontal distance is small enough for the earth to be considered flat over that range. If the forward speed given the object, or projectile, is so great that its range is large enough for the curvature of the earth to be noticeable, the projectile will be seen to be falling around the earth until its path intersects the earth's surface (Fig. 2.2). A satellite such as the moon is simply a body whose horizontal velocity is so great that, although the same central force is acting

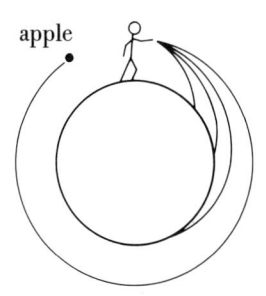

Figure 2.2 Falling motion and orbital motion. The apple falls around the earth if its horizontal velocity is sufficiently high to prevent its path from intersecting the earth.

to pull it toward the center of the earth, its path never intersects the earth's surface, so that it falls around the earth forever.

What is the nature of the central force? Here again, Newton exercised great insight in synthesizing the ideas that came to be stated as the Law of Universal Gravitation, which proposes that:

> *An attractive force exists between every pair of material bodies in the universe.*

The earth pulls on the apple, but so does every other apple. The apple falls to earth because the earth is so much bigger than any apple. So the attractive force must be directly proportional to the mass of the object, i.e., the greater the amount of material in an object, the greater the force on it, and the greater the force it exerts.

Another characteristic of this attractive force is that it diminishes as the distance between the bodies increases. That is, force and distance are somehow inversely proportional. To determine an exact mathematical relationship, one begins with the simplest possible function and goes on to more complicated ones until the relationship that best agrees with experimental observation is found. The simplest function of a number that gets smaller as the number gets larger is its reciprocal. So we guess, first, that force is proportional to the reciprocal of distance, or, $F \propto 1/d$. This does not correspond to observation, however, for F diminishes much faster than $1/d$. The next function to try, then, is $1/d^2$, which clearly gets smaller (as d increases) much faster than $1/d$ does. This turns out to be the correct choice: the observed force between two bodies does indeed vary inversely as the square of the distance between them.

We may now state the Law of Universal Gravitation, as Newton did:

There exists between every two material bodies in the universe a force of attraction which is directly proportional to the masses of the two bodies and inversely proportional to the square of the distance between them.

If we recall that when any quantity is proportional to two or more other quantities, it is proportional to the product of those quantities, we can write the Law of Gravitation in mathematical form:

$$f \propto m_1 \qquad F \propto m_2 \qquad F \propto 1/d^2$$

$$F \propto \frac{m_1 \times m_2}{d_{1,2}^2}$$

where $m_1 =$ mass of one body, $m_2 =$ mass of second body, $d_{1,2} =$ distance between m_1 and m_2.

Mathematically, a proportionality expression is not very useful for making actual calculations. However, the idea of proportionality implies that if two quantities are proportional, their ratio is constant, or

$$\frac{F}{\frac{m_1 \times m_2}{d_{1,2}^2}} = \text{constant.}$$

We generally call this proportionality constant "big G" and write the resultant equation thus:

$$F = G \frac{m_1 m_2}{d_{1,2}^2}.$$

As Newton, who was a cautious man and waited twenty years to publish his results, said after checking his equation against observations of the moon's motion, they "answered pretty nearly." The small error turned out to be due to a poor value for the earth-moon distance, not to the law. So the law turned out to be valid to explain both the motion of falling bodies on earth and the motion of planets and moons in their orbits in space. Calculations showed that a central force following such an inverse square law would naturally produce motion in elliptical orbits with orbital speed being dependent on the distance between the two bodies attracting each other. Kepler's third law also followed from the Law of Gravitation, as we shall see later.

Dynamics: The Laws of Motion

Along with the Law of Universal Gravitation, Newton developed three general laws of motion which form the foundation of classical mechanics, the science of motion of rigid bodies. A measure of the success of classical mechanics may be suggested by a remark made in 1968, when the Apollo 8 spacecraft circled the moon and returned to earth: "It was a great success for Isaac Newton."

Classical mechanics deals with six quantities: mass, distance, force, velocity, acceleration, and time. The mathematically expressed relationships among them constitute our scientific description of motion.

Mass (m): The quantity of matter in a body. The measure of the inertia of a body (because inertia, or resistance to change in its state of motion, is the fundamental property of matter).

Distance or displacement (s): The length of a straight line between two points.

Time (t): The measure of the interval between two events.

Velocity (v): The rate of change of displacement with time $v = \Delta s/\Delta t$ where Δ indicates a change in the quantity following it.

Acceleration (a): The rate of change of velocity with time

$$a = \frac{\Delta v}{\Delta t} \text{ or } \frac{v - v_0}{t - t_0}$$

where the subscript $_0$ refers to initial conditions.

Force (F): That which, when unbalanced, produces a change in the state of motion of a body.

The first three quantities differ from the others in that they are *fundamental properties* of the material world and are in some sense directly measurable. The latter three are *derived quantities*, that is, they are defined in terms of relationships among the fundamental three.

The quantities of mechanics fall into two groups: *scalar* and *vector*. Scalar quantities have only size, or magnitude, whereas vector quantities are characterized by a direction as well as a magnitude. Mass, length, and time are scalars. Displacement, force, velocity, and acceleration are directed or vector quantities. To see the distinction between vectors and scalars, consider this example: an inexperienced pilot, taking off from Chicago for New York, climbs above the dense cloud cover and inadvertently heads west instead of east, only to land at Nowhere, Nebraska, instead of the Big Town after an hour and a half of flying time. The direction of the 600-mile displacement is critical.

The Laws of Motion

The laws of motion are based on the extrapolation to perfect conditions in which a force is not required to maintain motion, only to change it. They thus represent situations much simpler than those actually encountered in the real world, but because they deal with the ideal or limiting cases we can understand real situations in terms of small corrections. For example, a marble rolling on a long smooth table does stop, eventually, but we understand that another force, friction, has acted to stop it, and if we want to keep it going, we must push to overcome friction. The laws themselves are usually stated this way:

1. *Law of Inertia: A body at rest will remain at rest, or a body in motion will remain in uniform linear motion, unless an unbalanced force acts upon it.*
2. *Law of Acceleration: The acceleration of a body is directly proportional to the unbalanced force acting upon it and inversely proportional to its mass.*
3. *Law of Reaction: If a body A exerts a force on a second body B, then B exerts an equal force on A but in the opposite direction.*

The first law simply expresses the property of inertia whereby matter resists changes in its state of motion, so that an unbalanced force is required to accelerate it. The word *unbalanced* is necessary, for a body may remain at rest while being acted on by a number of forces, if these forces balance or cancel each other. A book lying on a table is undoubtedly acted upon by the earth and pulled downward, but it doesn't move downward because the table exerts an upward force on it of equal magnitude. The tag on the rope doesn't move if the two teams in a tug-of-war game are equally matched, and pull equally in opposite directions. An airplane may fly at a constant speed in a straight line although four separate forces act upon it if their sum, or resultant, is zero (Fig. 2.3). A body that is subject to no unbalanced force is said to be in equilibrium. The equilibrium may be *static* as

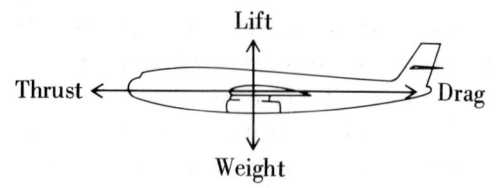

Figure 2.3 A system in dynamic equilibrium.

in the case of the book on the table, or it may be *dynamic*, as in the case of the airplane flying at a constant velocity. Remember, an unbalanced force is not required to maintain motion, but only to change it.

The second law of motion is an operational definition of force, through the term *operational definition* gained currency some two centuries after Newton. It means simply a definition in terms of observed or measured quantities. A force, in this kind of definition, is simply what, when unbalanced, produces an acceleration. However, as the second law points out, the magnitude of an acceleration depends on mass as well as force. It is rather obvious that it takes a bigger push to accelerate a piano than a ping-pong ball. For a given force the acceleration of a body is inversely proportional to its mass. In mathematical shorthand, $a \propto F$ and $a \propto 1/m$, so that $a \propto F/m$, or, crossmultiplying, $f \propto ma$. It is convenient, whenever possible, to have the proportionality constant simply be 1, which is achieved in the second law by an appropriate definition of the units[2] in which the quantities are meas-

---

[2] Using clearly defined and generally agreed upon units is basic to the operation of science. The system of units most widely used by scientists is the metric system. The metric unit of length was originally intended to be one ten-millionth of one-fourth of the earth's circumference, but is now defined by the wavelength of green light from a krypton lamp. The meter is equal to 39.37 inches, and is divided into tenths, hundredths, thousandths, etc. One one-hundredth of a meter is a centimeter (cm), which is the basic length unit in the cgs (centimeter-gram-second) system, used by chemists and by physicists dealing with small scale systems. The "g" in cgs stands for *gram*, the unit of mass defined as the mass of a cube of water measuring one centimeter on each edge, that is, a volume of water of one cubic centimeter. One thousand grams, called a *kilogram*, is on a scale comparable to a meter, and these two units form the basis of the

ured. F (measured in dynes) $= ma$ when mass is measured in grams and acceleration is the change in velocity (measured in centimeters per second per second). A dyne is the amount of force required to accelerate a mass of one gram by one centimeter per second per second. A pound, similarly, is the amount of force required to accelerate a mass of one slug by one foot per second per second.

Table 2.1   Equivalents

Force = mass × acceleration	cgs units	mks units	British units
$F = ma$	dyne = $\dfrac{\text{gram-cm}}{\text{sec}^2}$	newton = $\dfrac{\text{kg-m}}{\text{sec}^2}$	pound = $\dfrac{\text{slug-ft}}{\text{sec}^2}$

Newton did not state his second law of motion in terms of force and acceleration at all. He put it like this: the change in momentum of a body is proportional to the impulse acting on it. Momentum is a quantity describing a moving body in terms of its mass and its velocity. Momentum = mass × velocity, or

$$p = mv.$$

Impulse is equal to the product of a force and the interval of time during which it acts: impulse $= F\Delta t$. In mathematical shorthand, choosing units so that the proportionality constant is 1, we have

$$F\Delta t = \Delta mv$$

where Δ indicates change. If we recall that $a = \Delta v/\Delta t$ then we see that

$$F\Delta t = \Delta mv \text{ and } F = ma$$

are equivalent statements of the second law of motion.

---

mks (meter-kilogram-second) system of units. In both cgs and mks systems, the "s" stands for *second*, the basic unit of time, and is defined as 1/86,400 of an average day in the year 1900. Day length does vary minutely for various reasons. All other units in mechanics are based on these.

The third law of motion is one of the most useful, most misquoted, and most misunderstood laws of science. Its popular statement is "for every action there is an equal and opposite reaction," and everyone knows that's how rockets work. One of the best illustrations of the third law is the functioning of gravitation. Consider the earth and Newton's apple, and let d_{Ea} be the distance between their centers. Then the force exerted by the earth on the apple is

$$F = G \frac{m_E m_a}{d_{Ea}^2}$$

and its direction is toward the center of the earth. But the Law of Gravitation says "every two bodies attract each other," so we can write the expression for the force of the apple on the earth as

$$F = G \frac{m_E m_a}{d_{Ea}^2};$$

its direction is toward the center of the apple.

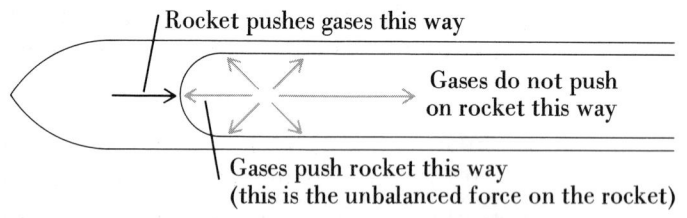

Figure 2.4 Rocket propulsion as an illustration of Newton's Third Law.

Rockets are indeed reaction engines, and they work like this: Fuel burns inside the rocket, producing gases that expand and push outward in all directions (Fig. 2.4). The rocket pushes back in all directions but one, so the exhaust gases go out the back, and their unbalanced forward force on the rocket pushes it forward.

A major point of confusion seems to be between action-reaction pairs and forces in equilibrium. To get these straight, consider a simple system[3] of a book lying on a table on the earth (Fig. 2.5). Consider first the forces acting on the book: F_g, the gravitational attraction of the earth, and F_t, the force of the table, supporting the book. Because the book is not accelerating, we must conclude that the forces acting on it are balanced. F_g and F_t, having the same magnitude but opposite directions, and both acting on the same object, are an example of two forces in equilibrium or balanced forces.

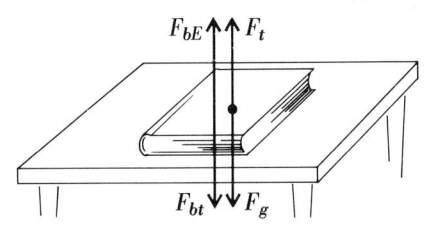

Figure 2.5 Third Law forces and forces in equilibrium: F_{bE} and F_g are a third law pair, as are F_t and F_{bt}. The forces F_t and F_g, acting on the book, are in equilibrium.

Now consider the forces exerted by the book,

Force exerted by the book on the table: F_{bt} = force due to weight of book: directed downward.
Force exerted by the book on the earth: F_{bE} = equal and opposite to earth's pull on book: directed upward.

The application of the third law always involves two bodies exerting forces on each other. In our system we have two "third law pairs" of forces acting:

1. F_g and F_{bE}, the force of the earth on the book and the equal and opposite force of the book on the earth.
2. F_t and F_{bt}, the force of the table on the book and the equal and opposite force of the book on the table.

---

[3] "System" is any part of the universe isolated for consideration.

We could of course add a third pair, the force of the earth on the table and of the table on the earth.

We can now put together the laws of motion and of gravitation to answer some remaining questions about motion. First, why do all things in the absence of air resistance fall with the same acceleration, a quantity called "little g"—the acceleration due to gravity? A look at the mathematical formulation of the laws should help. For any accelerated motion of a mass m, $F = ma$. The second law of motion is universal in its application. If the body is falling to earth, the acceleration will have the value g. So we can describe the force on a falling body as $F = mg$. But the force that causes the falling motion is gravity, which we write as

$$F = G\frac{mm_E}{d^2},$$

where m is the mass of the falling body, m_E is the mass of the earth, and d is for practical purposes the radius of the earth, since the falling body normally starts its fall rather near the earth's surface.[4] Thus we have

$$F = G\frac{mm_E}{R_E^2} = mg,$$

whence

$$g = \frac{Gm_E}{R_E^2}.$$

Because the mass and radius of the earth are essentially constant and G, determined numerically by Henry Cavendish some fifty years after Newton, is a universal constant, it is clear that g is essentially constant for all bodies falling near the earth.

The expression $F = mg$ defines the concept of weight. The weight of any body on earth is simply

---

[4] That is to say, if the apple begins to fall from the tree 10 ft above ground level at such an elevation that the distance to the center of the earth is 4000 mi, then the distance between their centers, 4000 mi + 10 ft, is essentially 4000 mi.

the force of the earth's gravity pulling on it. One weighs less on the moon than on earth because "little g" for the moon is less than "little g" for earth, due to the fact that the moon's mass and radius are smaller than earth's.

Because we are small bodies living on a comparatively very large body, the earth, gravity seems to us the dominant force in our lives. Investigations of the effects of "weightlessness" in space—that condition in which a body is falling freely in orbit around the earth or moon or whatever and so is not subject to the normal resistance to gravitational force offered by the earth's solid crust that makes him aware of his weight—indicate that adaptation to our particular value of "little g" is a very fundamental characteristic of the development of earth-dwelling organisms. However much it may dominate the earthly scene, gravity is actually a very weak force. The proportionality constant G, which relates the magnitude of the gravitational force to masses and distances for all matter in the universe, is a very small number. In 1798 Henry Cavendish, by allowing the force of gravity between pairs of massive lead spheres to twist a wire with a force that could be determined by the angle through which the wire turned, found G to be 6.67×10^{-11} newton-meters^2 per kilogram^2. The mass of the earth is nearly 6×10^{24} kilograms. Yet it exerts on the average adult human a force of less than 900 newtons, or less than 200 pounds. (A newton is the force required to accelerate a mass of one kilogram one meter per second per second.)

In Cavendish's experiment, the attraction between known masses M_1 and m_1, M_2 and m_2, and also between M_1 and m_2 and M_2 and m_1, which would twist the wire was observed and the distances between the paired masses measured (Fig. 2.6). The amount of force required to twist the wire through a known angle was previously measured. The angle could be measured by the reflection of an angular scale (a) in a

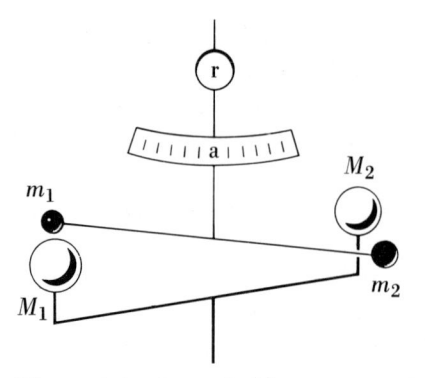

Figure 2.6 Cavendish's apparatus for determining G.

mirror (r) rigidly fixed to the wire. In the gravitation equation, only G would be unknown and could be calculated. Once G was known, given the weight of any mass, the mass of the earth could be determined, so Cavendish's experiment came to be known as "weighing the world."

Circular Motion

Newton's laws also help us answer the question of circular motion. Motion in a circular path is always accelerated motion, because although the magnitude of the velocity remains constant, its direction continually changes, so anything that is moving in a circular (or elliptical) path is subject to an unbalanced force. A force that acts to accelerate a body toward the center of a circle is called a *centripetal force*. The commonly used illustration of its action is the behavior of a stone whirled at the end of a string.

If one whirls a stone on the end of a string, the central force of one's hand sets the weight in circular motion. If the string is suddenly released, the central force ceases to act and the stone flies off in a straight line tangent to the circular path at the point of release. This straight line motion is exactly what the first law of motion leads us to expect when no force is acting.

Circular Motion

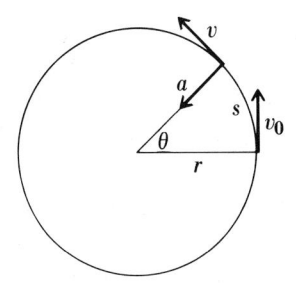

Figure 2.7 Velocity and acceleration vectors in circular motion.

The central force, then, had been acting to bend the straight line path into a circle, that is, to accelerate the stone toward the center of the circle (Fig. 2.7). A little elementary geometry, plus our definition of acceleration, will give us a mathematical definition of our central or centripetal force.

Acceleration = change in velocity with time

$$a = \frac{v - v_0}{t} \quad \text{or} \quad at = v - v_0$$

But we recall that a and v are vector quantities and vector arithmetic is played by somewhat different rules than the ordinary arithmetic of scalars. When we subtract, we have to consider directions as well as magnitudes, as in Figure 2.8(a). Now since the velocity vectors are tangent to the circle, and thus perpendicular to the radii, the triangle in Figure 2.8(b) is similar to the triangle in Figure 2.8(c). We have called both the arc of the circle and the chord s, but

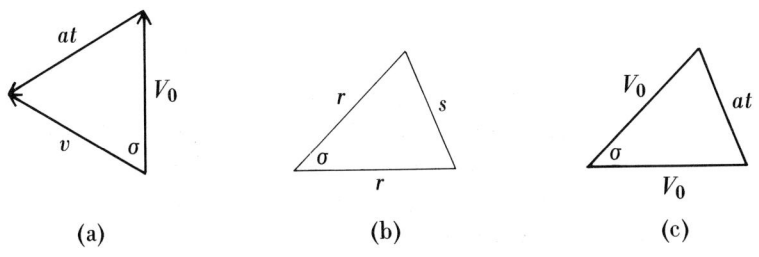

(a)   (b)   (c)

Figure 2.8 Similar triangles showing vector subtraction.

if we choose θ to be a very small angle, the difference will be insignificant. Then we could say that $s/r = at/v$, because corresponding sides of similar triangles are in the same ratio to each other. Then a little algebra enables us to solve for a: $a = s/t \times v/r = v^2/r$ because $v = s/t$ by definition. Centripetal acceleration thus is given by v^2/r and centripetal force by mv^2/r.

A term one often hears is *centrifugal force,* which means "center-fleeing." Centrifugal force is the "third law partner" of centripetal force. The hand holding the string exerts a centripetal force on the stone; the stone pulls back on the hand with centrifugal force, equal and opposite, and exerted on the body that exerts the centripetal force.

Gravity is the centripetal force that accelerates planets in nearly circular paths around the massive central sun. That is to say,

$$G\frac{m_s m_p}{R^2} = \frac{m_p v^2}{R}$$

where m_s is the mass of the sun, m_p is the mass of the planet, v_p is the planet's orbital velocity, and R is the average (mean) orbital radius of the planet's orbit around the sun. Because velocity is distance covered per unit time we can express it thus:

$$v = \frac{2\pi R}{T}$$

because $2\pi R$, the circumference of the orbit, is the distance covered in the time denoted by the orbital period T. Making this substitution, we get

$$T^2 = \frac{4\pi^2 R^3}{G m_s}$$

As $4\pi^2/G m_s$ is a constant, we see that this expression relating T^2 to R^3 is just Kepler's third law.[5]

Another kind of motion that is frequently encountered was also first studied scientifically by Galileo, who timed the swinging of a chandelier in church by

---

[5] See Appendix A, Mathematics, for details.

Circular Motion

his own pulse, and discovered that its period of vibration depended only on the length of the chain by which it was suspended.

Such motion as the swinging of the chandelier is called *periodic motion* because it repeats itself at regular intervals of time or *periods*. The time required for one round trip, cycle, vibration, or oscillation is a period. The inverse of period is *frequency*, or cycles per unit time, usually cycles per second. A swinging pendulum and a weight suspended from a vibrating spring are examples of periodic motion. Such a motion is characterized not only by a changing velocity but by a constantly changing acceleration, which of course means that a constantly changing unbalanced force is acting on the moving body. In the case of the pendulum the force is the resultant or sum of the pull of the string holding the weight and the pull of gravity, as in Figure 2.9(a). In the case of the weight on the spring, the resultant force at any time depends on gravity and on the restoring force of the spring, as in Figure 2.9(b).

A study of periodic motion enables us to understand other important concepts in mechanics. We will return to it in the next chapter.

Resultant force F_R produces acceleration toward equilibrium point, B. One cycle is motion from A to C and back again.

(a)

$F_r =$ restoring force

Restoring force in stretched spring

(b)

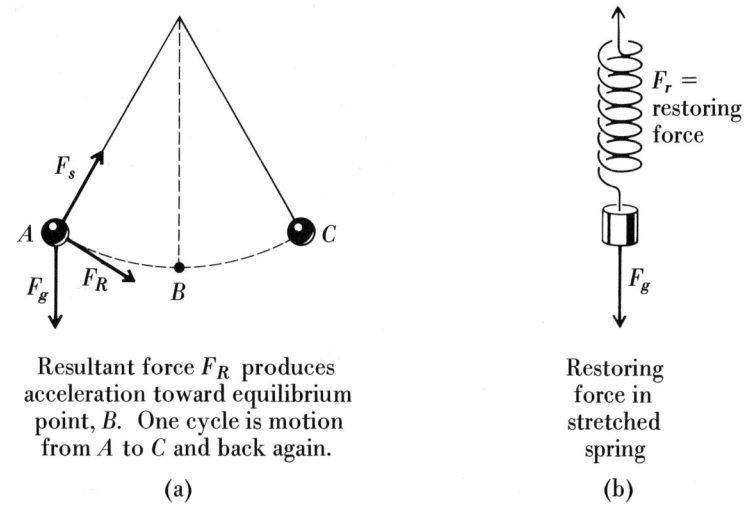

Figure 2.9 Periodic motion of pendulum and spring.

SUMMARY STATEMENTS

1. Experimental science began with Galileo in the sixteenth century. Galileo dealt mainly with questions concerning the motion of bodies, especially falling bodies.

2. Galileo's experiments pointed to the conclusion that force is not required to maintain motion, only to change it.

3. Johannes Kepler, Galileo's contemporary, formulated three laws of planetary motion:
   a. The orbits of planets are ellipses with the sun at one focus.
   b. A line joining any planet to the sun sweeps out equal areas in equal time intervals.
   c. The square of a planet's orbital period is proportional to the cube of its mean orbital radius.

4. Isaac Newton, in the seventeenth century, formulated four laws that described motion accurately and defined the central force that makes bodies fall to earth and holds moons and planets in their orbits.
   a. Law of Inertia: A body does not change its state of motion unless an unbalanced force acts on it.
   b. Law of Acceleration: The acceleration of a body is directly proportional to the unbalanced force acting on it and inversely proportional to its mass.
   c. Law of Reaction: If one body exerts a force on a second body, then the second exerts an equal and opposite force on the first.
   d. Law of Universal Gravitation: There exists between every two masses in the universe a force of attraction which is directly proportional to their masses and inversely proportional to the square of the distance between them.

5. A body that has no unbalanced forces acting on it is said to be "in equilibrium." It may be at rest or moving at a constant velocity.

6. Motion that repeats itself in a definite interval or *period* of time is called *periodic motion*. It is characterized by a constantly changing acceleration due to a constantly changing force acting on the body performing the periodic motion.

REVIEW QUESTIONS

1. What were some aspects of motion that scientists from early times tried to explain?

Motion: The Birth of Classical Physics

2. Whose method, Aristotle's or Galileo's, produced the more correct explanations? Why?

3. Why was Copernicus' theory considered heretical?

4. On what evidence did Kepler establish his laws of planetary motion?

5. State Kepler's laws.

6. State Newton's Laws of Motion.

7. State in words and symbols the Law of Universal Gravitation.

8. Define the following terms:

inertia	momentum	weight
mass	impulse	centripetal
force	scalar	periodic motion
acceleration	vector	"third law pair" of forces
velocity	equilibrium	

9. Identify these men:

Copernicus   Galileo
Kepler   Newton
Tycho Brahe   Cavendish

10. What are the dimensions, in terms of mass, length, and time, in the British, mks, and cgs systems of the following quantities?

acceleration force
velocity   momentum

THOUGHT QUESTIONS

1. The Greeks observed the world and developed theories about it by pure thought; Galileo experimented with real systems and then extrapolated to ideal conditions. If both methods lead to idealized theories, why is Galileo's method preferred today?

2. During the flight of Apollo 8 around the moon it was remarked that this flight was a great triumph for Isaac Newton. In what sense was this true?

3. In what other realms besides motion does the question of continuity-discontinuity arise?

4. When astronauts are in orbit around the earth they are said to be "weightless." From what you know about forces acting on orbiting bodies, deduce the actual meaning of "weightlessness." Can you think of a situation in which almost anyone might find himself that would make him weightless for a few seconds?

5. State brief explanations for the following statements in terms of the laws of Kepler and Newton:
    a. The earth goes faster along its orbital path in January than in July.
    b. Most children would rather live on Mercury than on Saturn (all other things being equal) because "Christmas comes but once a year."
    c. Copernicus still needed a bunch of epicycles to make his sun-centered system "work."
    d. A baseball thrown horizontally from 4 ft above the ground will land simultaneously with a baseball dropped from the same height, if they are released simultaneously.
    e. If Newton's apple weighed a quarter of a pound, it exerted a force of a quarter of a pound on the earth.
    f. If you weigh yourself on a spring scale (such as bathroom scales) on the moon, you will read a different result than on earth, but if you stand on one side of an equal arm balance it will take exactly the same counterweights to balance you both places.
    g. The moon is not likely to fall on the earth, although it is always falling toward earth's center.

6. Given a one kilogram mass, a value for "little g" of 9.80 m/sec^2, $G = 6.67 \times 10^{-11}$ newton m^2/kg^2, and 6400 km as the radius of the earth, calculate the earth's mass. (Answer: about 6×10^{24} kg.)

7. You are performing EVA (extra-vehicular activity) in earth orbit when your tether breaks at the space ship. What will be the best procedure to follow in order to regain your ship?

READING LIST

Cohen, I. Bernard, *The Birth of a New Physics*, Doubleday (Anchor), Garden City, N.Y., 1960. A historical development of Newtonian physics with some solid scientific material, too.

Andrade, E. N. da C., *Sir Isaac Newton*, Doubleday (Anchor), Garden City, N.Y., 1954. A biography by a distinguished modern physicist.

Arons and Bork (eds.), *Science and Ideas*, Prentice-Hall, Englewood Cliffs, N.J., 1964. Readings primarily concerned with classical physics, but including some good essays on twentieth-century philosophy of science.

3

Energy

Styles change in scientific expression as well as in everything else. In Galileo's time the word for the producer of motion was *impetus*, a term that has wide literary usage still, but is no longer heard in the sciences. With Newton's formulation of classical mechanics the key word became *force*. In some physics circles "force" now seems to be on the way out, and the in term is *energy*.

Energy is what makes things go. It is what produces change in matter, whatever the nature of the change may be. Its traditional definition is "the ability to do work." Energy is perhaps as hard to define as force, but it seems easier to grasp energy as a "real thing" than to give such a reality to force.

The idea of energy as the ability to do work is useful in enabling us to recognize the forms that energy can take. To do work, to run things, we use forms of energy: heat, electricity, muscular activity, motions of bodies, light, and chemical reactions. Energy can be stored, too, for future use, as in water behind a dam or a battery on a shelf. This leads to the concept of *potential energy*.

Mechanical Energy

Work is a familiar concept. Everyone knows what work is, but each person's definition is likely to be highly subjective. The objective mechanical definition of work is nothing like that. It is simply, mathematically, the product of a force and the distance through which it acts. We write it as

$$W = F \cdot s.$$

The dot used as a multiplication sign is a way of saying that, although force and displacement are vector quantities, work is a *scalar*. Work, thus, is what is done when a force is exerted through some distance.

When work is done, energy is changed in form. Since this is not immediately obvious, let us look at an example. If you lift a 2 lb book from the ground to a table three feet above the ground, you will do work in the amount of 2 lb times 3 ft or 6 ft-lb of work. (The force exerted was that required to lift the book against gravity, i.e., its weight.) The book has now acquired potential energy equal in amount to the work done on it, because the book could fall from the table and do some useful work such as squash a bug or operate the foot pedal that turns on the water at the drinking fountain. In the process of falling, the stored energy would be changed into the energy of motion called *kinetic energy*. When the book lands, it and the ground and air are warmed slightly by the energy of motion becoming heat. This heat is so diffused we can't use it to run a heat engine or warm a house, so we say that the energy has been degraded, and has become unavailable to do further work.

Conservation of Energy

Implicit in this discussion is the idea that energy persists through many changes. This is so well substantiated that it forms the basis of perhaps the most important principle in science, the law of conservation of energy, which may be simply stated as:

Energy can be neither created nor destroyed, only changed in form.

Because this statement purports to be universal it sets a certain limit on the universe: the total amount of energy in the universe is fixed and presumably finite; it cannot be changed. It is rather obvious that this law cannot be proved, because we cannot observe every part of the universe, even indirectly, to keep tabs on all the energy there. Yet almost the whole structure of science is built upon this foundation principle. It is a prime example of what we mean by a fundamental law in science: it is essentially unprovable; it is based on vast numbers of observations and experiences; and no exceptions have ever been found. Of course, if a genuine exception were to be found, we would have to junk the conservation law and start all over again to build a firmer foundation.

To illustrate change and conservation of energy let us return to our system in periodic motion, a simple pendulum. A pendulum is a small heavy object or bob suspended on a string of negligible mass. It is about as near as we can come to that ideal body, the point mass. (All the laws of mechanics assume that all rigid bodies can be theoretically reduced to point masses, and the so-called center of mass of a body is the point at which all forces act.) A pendulum hanging still is a system in equilibrium. The tension in the supporting string exactly balances the force of gravity on the bob, its weight. To set a pendulum in motion, work must be done to lift the bob above its equilibirum position. The amount of work $(F \cdot s)$ is mgh, where mg is the force required to lift the weight against gravity, and h is the distance through which the force acts (Fig. 3.1). (Yes, we did move the bob sideways, too, but if we neglect the minute force of air resistance, we over-

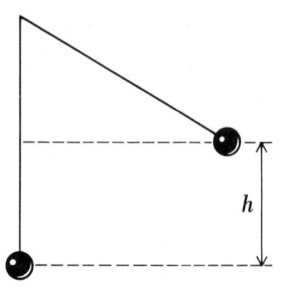

Figure 3.1 Pendulum with potential energy mgh from work done against gravity.

came no opposing force in the horizontal direction, and hence did no work in that direction.) The pendulum, by virtue of the work done in lifting it, has now acquired an amount of potential energy equal to the work done: mgh. That's conservation.

When the pendulum is released and begins to fall, h, and consequently mgh diminishes, but now the bob is moving, and thus has kinetic energy or energy of motion (Fig. 3.2). How much? At any point, its kinetic energy will be the difference between the amount of potential energy mgh it had to begin with less the amount mgh' it still has. When the bob gets back to its equilibrium position, its kinetic energy will equal mgh, because it will have none of its original potential energy left. That's conservation. (This tells us something about potential energy: we usually arbitrarily define some zero level of potential energy, here the

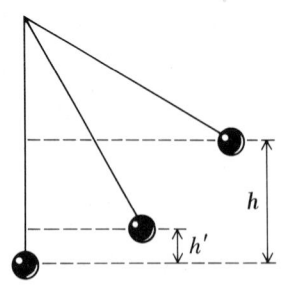

Figure 3.2 Pendulum falling, converting potential energy to kinetic energy.

equilibrium position, and measure potential energy from that zero level.) The bob does not stop at the equilibrium position; it goes on swinging until it has reached the same height on the other side of the equilibrium position. (Reminds one of Galileo's spheres rolling up the inclined planes; could there be a connection?) When the pendulum has reached its maximum height, all its kinetic energy has been converted back to potential energy again.

Because kinetic energy is a property of moving bodies its mathematical definition is given in terms of its velocity. We can arrive at this definition by starting with work, the original source of the energy of the moving body:

Given: $KE = W = F \cdot s$
But, always, $F = ma$, so $W = mas$
Acceleration, $a = \dfrac{v - v_0}{t}$ by definition.

Also, from our definition of velocity, $s = vt$. However, this last equation implies that v is unchanging during the time interval, t, which clearly it is not if there is acceleration. As long as the acceleration is constant, however, we can replace constant by average v, which is just $(v + v_0)/2$. This gives us $W = m(v - v_0)/t$ $(v + v_0)/2t = \tfrac{1}{2}m(v^2 - v_0^2)$. In most cases we are concerned with the kinetic energy of bodies that were originally at rest, i.e., for which $v_0 = 0$. So kinetic energy[1] is defined as

$$KE = \tfrac{1}{2}mv^2.$$

As long as a pendulum performs its periodic motion there is a continual interconversion between potential and kinetic energy. However, every pendulum eventually stops. Because stopping involves a

---

[1] It will be seen by inserting the usual units for mechanical quantities that the dimensions of energy are newton meters (called *joules*) in the mks system; dyne centimeters (ergs) in the cgs system, and foot pounds in the British system.

change in velocity, a stopping force must be acting. This force is friction, acting between the string and its support and appearing also as air resistance. Eventually the energy of the pendulum is "used up" in overcoming friction and so the bob comes to rest at its equilibrium position (Fig. 3.3).

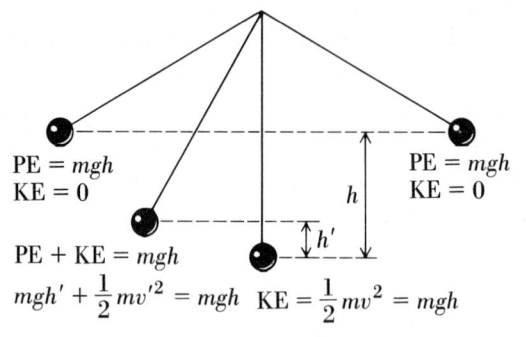

$PE = mgh$
$KE = 0$

$PE = mgh$
$KE = 0$

$PE + KE = mgh$
$mgh' + \frac{1}{2}mv'^2 = mgh$ $\quad KE = \frac{1}{2}mv^2 = mgh$

Figure 3.3 Conservation of energy in a pendulum system.

Heat and Work

If the pendulum no longer has the energy it started with—$F \cdot s$—and if energy is conserved, where has it gone? The form energy always takes when it used to overcome friction is heat, as anyone knows who has rubbed his hands together vigorously on a cold morning.

Heat is something a body acquires when it is put into a fire or near something hotter than itself. Heat flows from hotter bodies to colder ones. As late as the eighteenth century, heat was popularly supposed to be a fluid, a material substance that was weightless (a contradiction in modern terms) and invisible that flowed from one body to another. The idea that heat is energy rather than material began with a renegade Yankee named Benjamin Thompson, who is remembered in the history of science as Count Rumford of the Holy Roman Empire.

Thompson from an early age proved to be long on scientific ingenuity and short on moral sense. He married a widow half again his age to achieve financial independence with which to pursue his experiments, and left her without, apparently, a second thought when the Continental Army got wind of the fact that he was spying for the British. He departed hastily for England, and finally wound up at the court of the Duke of Bavaria, who was also Holy Roman Emperor. Thompson made several useful contributions to the welfare of the Bavarian army, including designing better fabrics for uniforms and developing a recipe for a nutritious potato soup. His next assignment was to supervise the boring of a cannon. Here he observed that as the hole was bored in the metal, the metal got hot. This happened every time; the supply of heat, or "caloric fluid," as it was called, never seemed to run out. That in itself was peculiar behavior for a material substance. How could the supply of it be inexhaustible? Perhaps instead heat was a form of energy produced from the mechanical work done in boring the cannon. Thompson's experiments were inconclusive but persuasive. So many more observations of the behavior of heat could be explained in terms of energy than in terms of a caloric fluid that the fluid model was gradually abandoned. But as the concept of heat energy developed, a whole new picture of matter had to evolve, too.

The continuity-discontinuity argument that characterized early discussions of motion raged for a much longer time with respect to the nature of matter. The first glimmerings of the idea of the particulate nature of matter go back to Democritus, who lived in Greece in the fifth century B.C. Democritus conceived of matter as being made of ultimate, indivisible particles he called *atoms* (meaning indivisible), rather than being infinitely divisible. His idea was promulgated a few centuries later by Lucretius, and was never entirely forgotten, but for a long time it just didn't seem necessary.

Modern atomic theory is usually dated from the

writings of John Dalton, a British schoolmaster of the early nineteenth century who provoked mirth, sarcasm, and considerable skepticism by insisting that the various simple chemical substances called *elements*, more and more of which were being identified, were composed of discrete ultimate particles or atoms that were all alike for a given element, but differed from one element to another. Dalton's atoms enabled a great many phenomena of the new science of chemistry to be understood. A chemical compound, resulting from the combination of two or more elements, must be made of small ultimate particles, too, but these particles must consist of several atoms held together by whatever it is that we call a *chemical bond*. These smallest particles of compounds came to be called *molecules*. Over the years *molecule* has come to mean the smallest independently existing bit of matter, whether it is a simple atom of an element or a group of chemically combined atoms of a compound. In our discussion of heat energy we shall use molecule in this latter, most general sense.

What happens when heat is applied to materials? The mercury in the thermometer rises; ice melts; water boils. In every case motion is observed. That implies kinetic energy, but not in the sense that a massive body moves from one place to another. The movement is internal. The ice cube doesn't move, intact, to another place on the table. It falls apart, where it is, into a liquid, which spreads out over a larger area of the table, possibly even over the edge. Heat energy, then, manifests itself as the kinetic energy of molecules. When we observe a heated metal, whether liquid mercury or solid iron, expanding, we understand that the molecules of the metal are moving faster and taking up more space.

Notice that increased kinetic energy manifests itself in increased velocity. Mass may be assumed unchanging in ordinary mechanical and thermal systems. (The absolute validity of that assumption will be examined in Chapter 7. It is valid here.) So when by heating we increase the kinetic energy of molecules we do not change either the mass or the volume of molecules, only their velocity.

To illustrate the relation of expansion of a material to increased molecular velocities, let us imagine a room with elastic walls containing ten Cub Scouts. Under any imaginable circumstances the Cub Scouts will be in some degree of motion. But if they are suddenly provided with the stimulus of a football, an observer outside the room will observe its walls to stretch, i.e., the room will expand. Now the Cubs didn't grow; we can even assume the ball was lying there all along, unnoticed. The thing that has changed is the velocity of the Cubs now that they are running around rapidly, passing and catching the ball, shoving and tackling each other, and frequently bumping into and stretching the walls.

We are accustomed to specifying how hot or cold a thing is in terms of *temperature*, which is a measure of the intensity of heat energy. Temperature does not measure the amount of heat. To express the quantity of heat energy involved in a process we need to consider not only the intensity of the energy change, that is, the change in temperature, but also the amount and nature of the material that is heated. The quantity of heat required to heat a pot of water clearly depends on the quantity (mass) of water and on the change in temperature, but it also depends on the fact that we are heating water, not mercury or alcohol or iron or salad oil. We can sum it up like this:

$$Q = mc\Delta T$$

where Q is the amount of heat,
m is the mass of the water in the pot,
c is a characteristic property of water called specific heat,
and ΔT is the temperature change.

The quantity c, the specific heat or heat capacity per

unit mass, is the amount of heat required to change the temperature of a unit mass of a substance by one degree, and depends on several factors connected with the nature of the material.

The unit in which heat is measured when cgs units are used is the calorie. It was defined originally as the amount of heat required to change the temperature of one gram of water one degree on the centigrade scale of temperature. In the mks system (or if you are dieting) the heat unit is the kilocalorie or Calorie. In British units, the British thermal unit (Btu) measures heat. It is the amount of heat required to raise the temperature of a pound of water one degree Fahrenheit, and is equal to about 252 calories.

Temperature

Temperature is measured according to various more or less arbitrarily established scales. Galileo, as usual, was the first to devise a thermometric device, a sort of gas thermometer that had the disadvantage of responding to pressure changes as well as temperature changes. Some two centuries later, a German physicist named Fahrenheit developed a mercury thermometer and a scale with small divisions running from zero as the coldest temperature recorded in his native city of Danzig in 1720 to the boiling point of water at 212°. The freezing point of water on this scale is 32°. The particular numbers arose from the doubling and then redoubling the number of degrees between the freezing and boiling points of water on a scale devised about 1701 by Isaac Newton, who had chosen to call body temperature 12. In the nineteenth century a Swedish scientist named Celsius suggested that because water is the universally used standard substance, its melting point (freezing point) be called zero and its boiling point be 100°. The scale of 100 divisions for a long time was called the *centigrade scale*, but now the term *Celsius* is coming into use (Plate 5).

We have defined heat as the kinetic energy of molecules. We can now define *temperature* as a measure of the average kinetic energy of a collection of molecules. This definition arises out of the new model of matter that arose in the mid-nineteenth century, called the Kinetic Theory of Gases.

The Kinetic Theory of Gases is based on the properties of a highly idealized (and quite nonexistent) substance called an ideal or perfect gas. A perfect gas has these properties:

1. The molecules of a perfect gas are point particles. Because a point has no volume, the volume of a gas is the space in which its molecules move.

2. The molecules of a perfect gas exert no forces, either of attraction or repulsion, on each other.

3. The collisions of perfect gas molecules are perfectly elastic, so that when two molecules collide no kinetic energy is lost.

4. The molecules of a perfect gas are in constant, random motion, so that a gas always "fills" or occupies all the space in which it is contained. (That is, you can have a jar half full of water, but not half full of air.)

The Kinetic-Molecular Theory

When the kinetic energy of molecules increases, the property which increases is the velocity. Because energy is continually being exchanged among colliding molecules, some will occasionally acquire a very high velocity and others will go very slowly. But the vast majority will have velocities clustering about an average intermediate velocity in what we call a "normal distribution." This average kinetic energy is measured in terms of temperature. The mathematical relationship is written as

$$\tfrac{1}{2}mv^2 = \tfrac{3}{2}kT$$

where v^2 is the average value of the square of the velocity and k is a proportionality constant (Fig. 3.4).

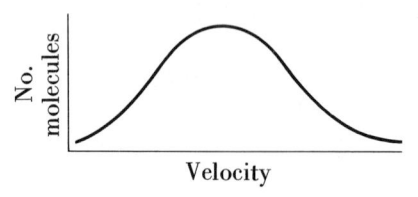

Figure 3.4 Distribution of velocities in a collection of gas molecules.

If there are no ideal gases in the real world one might wonder why the kinetic theory is of any use or importance. It is useful in the same way that Newton's first law is useful—although it specifies conditions never fully met in the real world, it defines limits to which reality can be extrapolated and by which it can be simplified. The real world is nearly always too complex to be understood directly. Our models enable us to cope with simple situations that approximate reality and in terms of which reality can be understood. Perhaps the basic task of science is to make models, either physical or mathematical, and to test them against the observation of the real world, discarding those models that do not correspond to reality and refining and improving those that do.

Although real gases differ from the ideal in rather obvious ways (molecules are not points; they do exert forces on each other), we can understand their behavior from the model. It becomes clear why air in a room doesn't settle out on the floor due to gravity and have to be stirred up every so often, and why gases are compressible over a wide range of volumes. Pressure is understandable as the force of molecules colliding with an area of surface, so we can define pressure either as the force per unit area or the kinetic energy per unit volume (kinetic energy density). We can see why gas pressure is the same in all directions and why a pressure exerted on one part of a gas is transmitted to all parts of it.

The kinetic theory can be extended to help us understand the other states or phases in which matter occurs: liquid and solid. All gases will become liquids under some conditions, and the essential condition is cooling, although compression will generally help, too. The essential difference between the gas state and the liquid state would seem to be one of kinetic energy. *Cooling* means reducing the average kinetic energy of the molecules, slowing them down to where the forces of attraction between the molecules become effective. (An ideal gas could never become a liquid but all real gases can.) This is fine, except that we find that a phase change, such as the liquefaction of a gas, or conversely, the vaporization (boiling) of a liquid, is a constant temperature process. That is, if we put a thermometer into a pot of water and begin to heat it, we will observe the temperature to rise steadily to 100°C (or 212°F) and then to stay there until all the water is boiled away. If your pot is a pressure cooker and we can continue to measure the temperature of the water vapor after all the liquid is gone we will see that the temperature rises again.

This led an eighteenth-century Scottish scientist, Joseph Black, to apply the term *latent heat* to that amount of heat that must be applied to break apart the liquid molecules and free them to behave as gases. Latent heat, because it does not raise the temperature, can be looked upon as potential rather than kinetic energy.

Similarly, a solid differs from a liquid mainly in that its molecules have less freedom of movement and less velocity than those of a liquid. We cannot apply our definition of temperature as the measure of average kinetic energy strictly to molecules in the liquid and solid states, and here the forces between molecules play an overwhelmingly important role, but the rough idea of difference in kinetic energy content being the essential difference between solids and liquids is still useful. We find that the phase transition we call *melting* (or, conversely, *freezing*) is also a constant temperature process and that it also involves a latent heat.

Table 3.1 Thermal Data for Water and Some Other Liquids

Substance	Specific heat	mp	Latent heat of fusion L_f	bp	Latent heat of vaporization L_v
Water	1.00 cal/g°C	0°C	79.9 cal/g	100°C	540 cal/g
Mercury	0.033	−39.9	2.82	356	65
Ethyl Alcohol	0.45	−117.3	24.9	78.5	204
Ammonia	1.05	−77.7	108.1	−33.4	327.1

If heat energy is added to a substance, either of two things, with attendant effects, happens: the temperature rises, with an accompanying expansion in most cases; or a phase transition occurs. Let us start with a piece of ice somewhat below its freezing point and change it into steam somewhat above its boiling point, and add the amounts of heat required for the steps of this process.

Overall transition:
    1000g ice @ −20°C → 1000g steam @ 120°C

Steps involved:

    Ice @ −20° to ice @ 0° $mc\Delta T$
        = 1000g × 0.5 cal/g°C × 20°C = 10,000 cal.
    Ice @ 0° to water @ 0° mL_f
        = 1000g × 80 cal/g   = 80,000 cal.
    Water @ 0° to water @ 100° $mc\Delta T$
        = 1000g × 1 cal/g°C × 100° = 100,000 cal.
    Water @ 100° to steam @ 100° mL_v
        = 1000g × 540 cal/g   = 540,000 cal.
    Steam @ 100° to steam @ 120° $mc\Delta T$
        = 1000g × 0.5 cal/g°C × 20°C = 10,000 cal.
                     Total   760,000 cal.

To carry out these transitions on a little over two pounds of water requires 760 kilocalories—about the calorie content of two large hamburgers with all the trimmings. Water is an unusual liquid in that its specific heat and its latent heats of phase transition are very much larger than one would expect for the small molecule H₂O (two atoms of hydrogen combined with one atom of oxygen), and its melting and boiling points are quite abnormally high. In addition, ice is less dense than water so that ice floats and bodies of water freeze from the top down. Consequently ice acts as insulation on top of deep lakes, keeping the water in the bottom above the freezing point. Both the climate of earth and the development of protoplasm as a water-based system are due to these "abnormal" properties of water.

Let us return to our consideration of the simplest phase of matter, gas, to understand some additional aspects of heat energy. We shall assume our gas to be ideal, and consider its changes under different applications of heat energy.

Because a gas completely fills any container in which it is held, we can specify the volume of the gas by specifying the volume of the container. If a given mass of gas — which is to say, a given number of molecules — is placed in a container, the molecules in constant random motion will keep bumping into the sides of the container (Fig. 3.5). Because molecules are incredibly small (2g of hydrogen gas contains

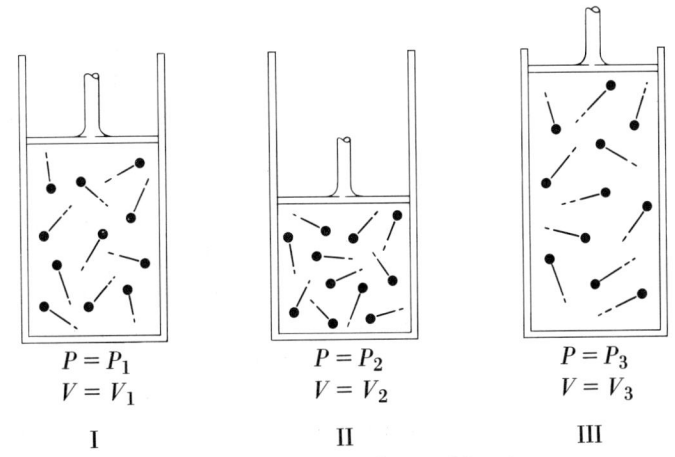

$P = P_1$
$V = V_1$

I

$P = P_2$
$V = V_2$

II

$P = P_3$
$V = V_3$

III

Figure 3.5 Pressure-volume relationships in gases at constant temperature.

roughly 6×10^{23}, or 6 followed by 23 zeros, molecules), and because their motions are entirely random, each side of the container will be pounded uniformly by the molecules. On a macroscopic scale, we can then observe the pressure of the gas to be the same in all directions on all surfaces. Suppose now that the container is a cylinder closed by a movable piston, so that the volume may be reduced by pushing in the piston. The molecules are now squeezed closer together. They have less room in which to move, but because we have not reduced the temperature, they are still moving with the same average speed. This means they will bump into the walls more often, and the overall pressure will be observed to increase. Conversely if we pull the piston out so as to increase the volume, encounters with the walls will become rarer, and we will observe the pressure to drop. We may sum up this pressure-volume relationship as follows: At a given temperature, the pressure of an ideal gas is inversely proportional to its volume. Our point of view has been to consider the pressure to be exerted by the gas. However, Newton's Third Law assures us that the pressure exerted by the piston on the gas is equal (and opposite) to the pressure of the gas on the piston. Thus, the volume of a gas varies inversely as the pressure upon it.

In this form, we have the principle discovered in the seventeenth century by Robert Boyle, known as Boyle's Law. In mathematical shorthand we can write $V \propto 1/p$ or $V = k(1/p)$ or $PV = k$.

When two quantities are inversely proportional their product is a constant. This brings us back to the conservation of energy. Let us see how. If we keep the temperature constant, we are not putting any energy into or removing any energy from the system. So its energy must remain constant. If we consider pressure as "kinetic energy density," kinetic energy per unit volume of the gas, we see that the statement

$PV = k$ is the same as $KE/V \times V = k$ or, KE is constant for our system.

Suppose next we fix the piston firmly in the cylinder so that the volume of our gas cannot change. Now if we add heat energy, the molecules' kinetic energy will be increased, their velocity increases, or, that is to say, the temperature rises. What macroscopic effect will be observed? Clearly the faster-moving molecules will collide with the walls more often, which constitutes an increase in pressure. Conversely, cooling the gas means slowing the molecules and reducing the pressure. So we see that there is a direct relationship between pressure and temperature, which we can write as $P \propto T$ or $P = kT$ or $P/T = k$.

Finally, let us allow the piston to move freely up and down in the cylinder so that as the gas is heated the pressure stays constant. As we have seen before, the hotter the molecules the faster they move, and the more space they occupy. This principle is enunciated (after eighteenth-century French scientist Jacques Alexandre Charles) as Charles' Law:

The volume of a gas varies directly as its temperature.

Mathematically, this is $V \propto T$ or $V = kT$ or $V/T = k$. (This k and the k in Boyle's Law do not have the same numerical value. We are using the symbol k to represent any constant.)

The variation of gas volume with pressure and with temperature leads us to some very fundamental ideas about heat energy. A graph of Boyle's Law and one of Charles' Law (Fig. 3.6) will suggest some of them, to begin with. The $PV = k_B$ curve is a hyperbola. The curve never crosses either axis. As the pressure becomes very large, the volume becomes very small, but neither quantity ever becomes zero. On the other hand, $V = k_c T$ represents a straight line of which k_c is the slope. The general equation of a straight line is

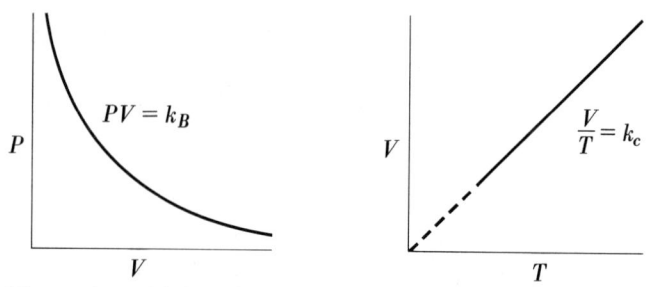

Figure 3.6 (a) Graph of Boyles' Law. (b) Graph of Charles' Law

$y = mx + b$, where m is the slope and b is the y-intercept, or the value of y when x is zero. It will be noted that in the Charles' Law equation $b = 0$. That is, there is (theoretically, at least) a temperature at which an ideal gas has no volume. Remember our model of an ideal gas states that the molecules are points, and a point has no property except position; it takes up no space. So we can conceive of an ideal gas having no volume; it means only that the molecules are absolutely still. On the other hand, it is impossible to attach meaning to the term *negative volume*, so it makes no sense to ask what happens if we reduce the temperature still further. Moreover, it is hard to see how a motionless point could slow down any more. So we arrive at the concept that an absolute zero point of temperature exists, below which it is theoretically impossible to go.

Because we arrived at the idea of absolute zero by using a highly simplified model, the ideal gas, we had better ask whether this concept has any meaning in the "real" world. Let us begin by checking the correspondence of our model to reality, as compared in the list on page 63 (opposite).

So we may accept the idea of an absolute zero of temperature as objectively true. What about the other end of the scale? Is there an absolute maximum of temperature? A geometric answer from the Charles' Law graph would be "no," because a straight line is infinite in extent. However, temperature is a measure

Ideal Gas Model

1. Molecules are point particles.

2. Molecules exert no forces of attraction on each other.

3. Collisions of molecules are elastic, and do not slow them down, on the average.

4. Molecules are in constant random motion, occupying all the space in their container.

Real Substances

At high temperatures and low pressures, real gas molecules *have* so little volume compared to the space they *occupy* that they may be considered points, and they are so far apart and so fast moving the forces between them are negligible. *But,* at low temperatures and high pressures, the molecular volumes and the intermolecular forces are not negligible, and the forces eventually become strong enough as the molecules slow down and clump together for the gas to condense into a liquid, and eventually to solidify into a rigid crystal structure. No real substance remains gaseous down to absolute zero, although helium comes close, condensing at about 4 centigrade degrees above absolute zero.

The molecules of liquids and solids have some motion, but it is greatly restricted by the attractive forces between neighboring molecules.

of kinetic energy, and the Conservation Law suggests that the total amount of energy is finite. So one might argue for a theoretical maximum temperature, but we do not know how much energy there is nor could we easily agree on a distribution of it that would enable us to speak meaningfully about a maximum temperature.

Avogadro's Number

We can now put together our two laws about gas behavior and obtain a general gas law, or an "equation of state" descriptive within limits suggested above of all gases and their behavior. Remember that if $V \propto T$ and $V \propto 1/P$ then $V \propto T/P$ or $PV/T =$ constant. This constant is usually written as R, when the particular quantity of gas under consideration is one mole. This unit was developed in chemistry, where it was observed that certain definite weights (or masses) of chemical elements were always involved in chemical combination. It early became apparent that elemental units, i.e., atoms, had different weights relative to each other, and a convenient scale of these weights was constructed. Then an Italian named Avogadro made the interesting observation that at a given temperature, equal volumes of all gases behaved as if they all contained the same number of particles. So if we define a *mole* of a substance as the number of grams of that substance equal to its relative molecular weight,[2] we have a quantity of material containing a definite number of molecules, which if it is an ideal gas, has a perfectly definite volume at any given pressure and temperature, regardless of its composition. So we can say that if we have one mole of gas,

---

[2] For example, 1 mole of water, chemical formula H_2O, is 18 grams, because the atomic weight of H (hydrogen) is 1 and of O (oxygen) is 16 ($2 \times 1 + 1 \times 16 = 18$). If it were possible to have 18g H_2O in the vapor state at 0°C and 1 atmosphere pressure, it would occupy 22,400 cm³.

containing 6.023×10^{23} (Avogadro's Number) molecules, the following law holds:

$$PV = RT$$
For n moles, $PV = nRT$.

For gases that cannot be considered ideal, this law must be corrected. One widely used corrected form, the van der Waals equation, accounts for molecular volume and intermolecular attraction in this way:

$$(P + a/V^2)(V - b) = RT$$

where a/V^2 accounts for intermolecular (van der Waals) forces at high pressures and b accounts for molecular volumes. The constants a and b are experimentally determined for individual gases.

It will be seen from the ideal gas equation that T must be so defined that $V = 0$ when $T = 0$. So our arbitrary temperature scales, the Fahrenheit and Celsius scales, cannot be used to specify temperature in these equations. We must instead use the absolute scale of temperature (Fig. 3.7).

The Kelvin scale, an absolute temperature scale

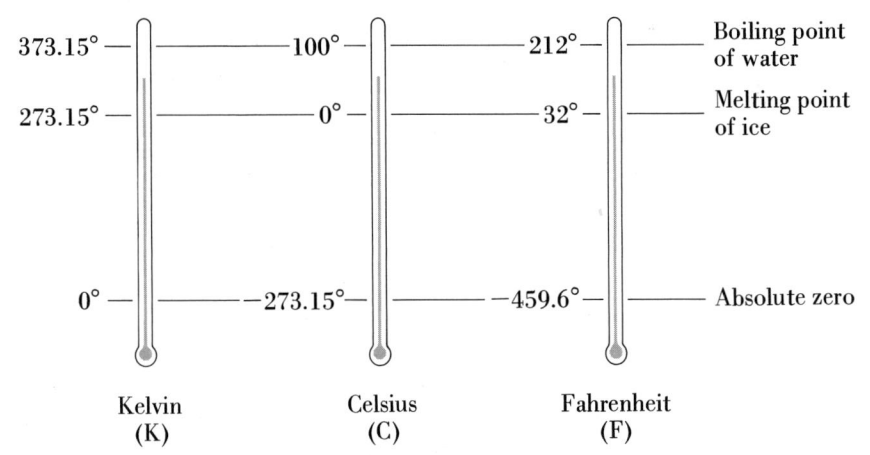

Kelvin (K)	Celsius (C)	Fahrenheit (F)	
373.15°	100°	212°	Boiling point of water
273.15°	0°	32°	Melting point of ice
0°	−273.15°	−459.6°	Absolute zero

Figure 3.7 Temperature scales. The calibration points are the freezing and boiling points of water.

with Centigrade-sized degrees, is the scale most widely used in science, for nearly every temperature-dependent property of matter is a function of its absolute temperature.

We seem to have wandered a long way from the subject of energy, but actually we have circled back, because we can now see the connections between heat and work that led scientists to two of the most fundamental and important laws of the universe.

The Carnot Cycle

Count Rumford was able to demonstrate that heat is a product of mechanical work against friction. Because of the property of reversibility and symmetry in nature, we would expect that heat could be used to do work. This, of course, was well demonstrated in Rumford's time by the invention of steam engines by Thomas Newcomen and James Watt, its application to water travel by Fulton and others, and the use of the Newcomen engine in mines. Steam engines, in which the expansion of steam against a piston produces mechanical motion, are very inefficient. Mechanical efficiency of a heat engine is measured by the ratio of the work output to the heat input. The more heat that can be turned into work, the greater the efficiency of the heat engine. In the 1820's, a young French engineer, Sadi Carnot, set out to discover how to increase the efficiency of heat engines.

Although Carnot still thought in terms of a caloric fluid, he saw the analogy between heat and mechanical energy: both have a capacity factor and an intensity factor, the difference in temperature being analogous to the difference in height in mechanical potential energy. Carnot reasoned that if one could construct two steam engines of differing efficiencies, one could use the more efficient engine to drive the less efficient one in reverse, thus lifting the same

The Carnot Cycle

amount of "caloric" to a higher temperature each cycle. The cycle would thus produce work and accumulate energy, becoming self-sustaining without further heat input. This is "perpetual motion of the second kind," and is just as impossible, Carnot realized, as "perpetual motion of the first kind," which is based on the possibility of creating energy in violation of the Conservation Law. He further showed that the maximum efficiency available from any engine depends only on the two temperatures, T_h (for *hot*) and T_c (for *cold*), between which it operates.

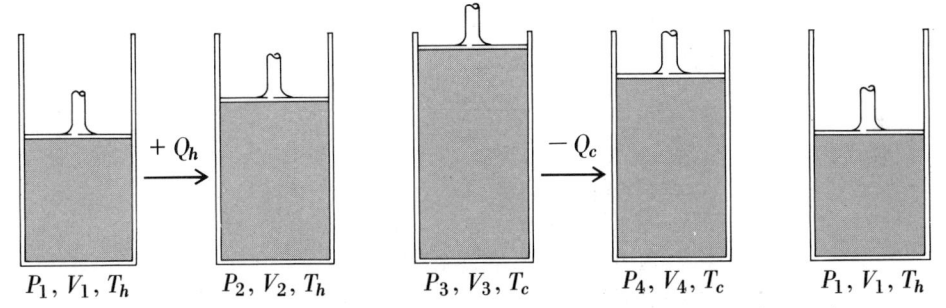

Figure 3.8 Pressure, volume, and temperature relationships in an ideal engine, representing the Carnot Cycle.

The Carnot Cycle is the name applied to the operating cycle of the most perfect heat engine imaginable. The working substance is an ideal gas, the material that expands when heated and pushes on the piston. So no energy is wasted overcoming intermolecular forces. The piston in this ideal engine is weightless and moves up and down in its cylinder without any friction. (Unfortunately, no one has been able to manufacture this engine!) So the only possible source of inefficiency in this engine is the loss of some heat by removing it when the gas is recompressed, rather than turning it all into work. The cycle goes like this (see Fig. 3.8):

Step I: Isothermal expansion—a mass of gas in a cylinder occupies volume V_1 at pressure P_1 and a relatively high temperature, T_h. The gas is allowed to expand to a new volume V_2 at which its pressure is P_2, but its temperature is maintained at T_h by adding an amount of heat Q_h. This is necessary, because otherwise the work of expansion that pushes the piston and thereby does some work outside the system would be done at the expense of the internal energy of the gas and its temperature would fall.

Step II: Adiabatic expansion — the gas is allowed to expand again, but this time it is insulated so that no heat can be exchanged with its surroundings. This adiabatic expansion will thus result in lowering the gas temperature to T_c as the volume increases to V_3 and the pressure falls to P_3. (V is no longer simply inversely proportional to P, but increases slower than P decreases under adiabatic conditions.)

Step III: Isothermal compression — the compression half of the cycle starts with an isothermal compression of the gas to V_4 and P_4, with the temperature maintained at T_c by removing the energy of compression as an amount of heat Q_c.

Step IV: Adiabatic compression — the initial conditions V_1, P_1 and T_h are restored by compressing the gas adiabatically.

William Thomson (later Lord Kelvin), who continued to study heat engines for a long lifetime after Carnot's early death by smallpox, saw, as Carnot had seen, that the efficiency of the ideal engine depends only on the difference between the two temperatures between which the cycle operates. Thomson also saw that the only way it would be possible to achieve an efficiency of 100% would be to reduce T_c to absolute zero. This became the basis for the absolute temperature scale that is named for him, and which was arrived at by reasoning like this:

$$\text{Efficiency} = \frac{\text{work output}}{\text{heat input}}$$

But work output = heat input at T_h − heat removed at T_c. So,

$$\text{efficiency} = \frac{Q_h - Q_c}{Q_h} = 1 \text{ (or 100\%) only if } Q_c = 0.$$

Kelvin proposed a temperature scale so defined that $Q \propto T$ at every point so that

$$\frac{Q_h}{T_h} = \frac{Q_c}{T_c}, \text{ etc.}$$

Then efficiency equals

$$\frac{T_h - T_c}{T_h}$$

where T is in degrees Kelvin, and 100% efficiency is possible only at $T_c = 0°K$.

The question now comes: Can we reach absolute zero? The answer is: No. Let us see why. First of all, our common experience tells us that heat always flows spontaneously from a hotter to a colder body, never in the opposite direction, and that if we wish to do the nonspontaneous job of moving heat from a cooler to a hotter body (as in a refrigerator) an expenditure of energy (as in the refrigerator motor) is required. The lower the temperature of a system, the more work will be required to lower it still further, until we must conclude that an infinite amount of work would be required to remove the last bit of heat. In the late 1960's, experimenters in both the Soviet Union and, a little later, in the United States achieved temperatures of $10^{-6}°K$ (one millionth of a degree above absolute zero), but we have good theoretical reason to believe that we will never get to $0°K$.

The Second Law of Thermodynamics

Again our consideration of energy has led us up against a stop sign in nature. Just as the Conservation

Law reminds us that our energy supply is finite, our investigation of heat engines has suggested that a limitation exists on the availability of energy. We summarize this set of ideas as the Second Law of Thermodynamics, the Conservation Law being the First Law, and "Thermodynamics" the name of the science of heat and work. Unlike the First Law, the Second has no simple single statement. In fact, at first glance, its various statements don't even seem to have much connection with each other. Here are some of them:

1. *Heat never spontaneously flows from a cold body to a hot one.* (This is simply the observation of our common experience.)
2. *No heat engine, however perfect, can be 100% efficient.* (This is, of course, due to the impossibility of reaching 0°K.)
3. *Any system will become disordered spontaneously; order is produced only by the expenditure of energy.* (This is the law of science every housewife knows, whether she knows she knows it or not!)
4. *The most random distribution or arrangement of a system is always most probable.*
5. We define a quantity called *entropy*, which is a measure of the randomness of a system or the unavailability of its energy, and can summarize the Second Law as did the German physicist, Clausius: *The entropy of the universe strives toward a maximum.*

The Second Law is rather like a "one-way street" marker in the universe. It suggests that the universe is set on an irreversible course, spontaneously tending toward disorder, randomness, and unavailability of energy to do useful work, even though all the energy is still there. Nineteenth-century philosophers, especially in Germany, extrapolated these ideas to predict a final *Wärmetod* or "heat death" for the universe, a final state of equilibrium at some uniform temperature at which, there being no temperature gradient, no work can be done.

A rather obvious "exception" to the Second Law has probably already suggested itself to the reader. What about living systems? Do they not represent increasing degrees of order along the evolutionary scale as well as in the course of growth of a single organism? They do, indeed. But the Second Law talks about the increase in entropy of the *universe*, and does not preclude some systems becoming increasingly ordered as long as the increased disorder in their environment is also taken into account. Twentieth-century civilization would seem to ideally illustrate the Second Law: our highly ordered, energy-consuming society has produced an alarming amount of disorder called *pollution* in its environment. The Second Law also suggests that the situation is not completely reversible, but rather than taking this to be license to despair and do nothing, it might better urge us to be careful and to slow down, at least, those processes we cannot fully reverse.

From a scientific point of view, then, we might say that energy makes the world go around. However much our picture of the universe may appear to have changed from the beginning to the end of this book, the two laws of energy are our fundamental concepts for understanding the universe.

SUMMARY STATEMENTS

1. Energy is what produces changes in matter.
2. Energy is conserved; that is, it can neither be created nor destroyed. This principle, the Law of Conservation of Energy (or the First Law of Thermodynamics) is *empirical*, that is, it is based on experience, but is unprovable.
3. Mechanical energy may be categorized as *kinetic energy*, the energy of moving bodies, or *potential energy*, energy due to place or position.
4. Heat is the kinetic energy of molecules.
5. Matter is made up of minute particles called *molecules*

that move relatively independently of each other, and randomly, in the gas state, but that are held more or less rigidly by intermolecular forces in the liquid and solid states.

6. Temperature is the measure of (1) the intensity of heat energy; (2) the average kinetic energy of gas molecules; (3) the degree of randomness of a system.

7. The "one-way-ness" of nature, its spontaneous tendency toward the most random distribution, and the impossibility of constructing a 100% efficient heat engine are summarized as the Second Law of Thermodynamics.

8. The laws of thermodynamics are *laws of impotence;* they set limits on what can be done in the world because energy is both conserved and less and less available to do useful work.

REVIEW QUESTIONS

1. What is energy? How could you prove it is a real thing?
2. What is kinetic energy?
3. What is potential energy?
4. Consider a pendulum in motion, as shown. What kind(s) of energy does it have at points A, B, C, and D? At what point(s) is its velocity a maximum? a minimum?

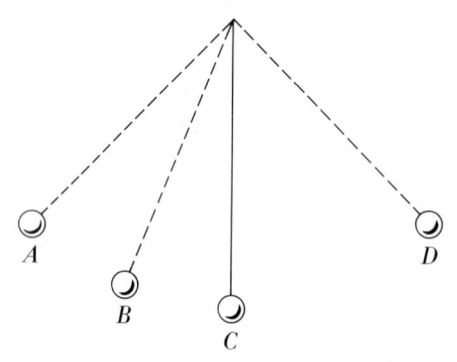

5. Why was the caloric theory of heat unsatisfactory?
6. How did the assumption that heat is energy affect our understanding of the nature of matter?
7. If you kept heating some system (in a closed container out of sight) and observed that for a period of several minutes the temperature did not rise, what would you assume was going on inside the container?
8. How do you go about setting up a temperature scale?

9. Define temperature in three different ways.

10. What is the difference between the solid, liquid, and gaseous states of a substance?

11. Mention some of water's important "abnormal" properties.

12. State and illustrate the operation of (1) Boyle's Law; (2) Charles' Law; (3) the general gas law.

13. State the First and Second Laws of Thermodynamics.

THOUGHT QUESTIONS

1. Name all the energy transformations involved in getting you from your bed to your first class of the day.

2. The laws of thermodynamics are called "laws of impotence" because they set certain limits on what can be done in the world. Would an understanding of these laws be useful in tackling and solving present environmental problems? How about such economic problems as development of industry or use of natural resources?

3. Watch for everyday illustrations of the Second Law. Keep a list of them for one day.

4. If energy is strictly conserved, why should anyone worry about the earth's civilizations running out of energy?

5. What is the use of such impossible ideas as ideal gases, weightless and frictionless pistons, and absolute zero?

6. Why will no automobile manufacturer ever offer for sale a car with a 100% efficient engine?

7. If you saw a movie in which a red solution changed into a colorless solution with a pellet of red Easter egg dye rising to the top, gathering all the color to it as it rose, what would you conclude? Does this have anything to do with Arthur Eddington's definition of *entropy* as the "arrow of time"?

READING LIST

Brown, Sanborn C., *Count Rumford, Physicist Extraordinary*, Doubleday (Anchor), Garden City, N.Y., 1962. A fascinating biography of a truly extraordinary character.

MacDonald, D. K. C., *Near Zero*, Doubleday (Anchor), Garden City, N.Y., 1961. A well-written account of both the idea of absolute zero and the strange phenomena occurring in that neighborhood.

4

Electricity and Magnetism

Matter and energy: the components of the physical world. Surely anyone could tell them apart! But the further we pursue the ultimate nature of each the fuzzier the distinction between them becomes, until we come to the startling idea that matter is made of electricity, or rather of electrically charged particles; electrons, and protons.

We are used to thinking of electricity as a form of energy, because it runs so many things in our world, but our forebears in generations not more remote than the nineteenth century just as easily thought of electricity as a material fluid, like the "caloric fluid," and having similar properties. Certainly that which was identified as a charge of electricity moved with varying degrees of ease through material bodies and could be "condensed" or collected in containers just as steam could be condensed into water in a cold flask.

The production of electric charges was known to the Greeks, from whom the word *electricity* is derived, because *elektron*, or amber, rubbed with fur, was the best source of electric charge they knew. The whole matter of electric charge, however, was simply a novelty until near the end of the eighteenth century when

Benjamin Franklin and others began to investigate electric phenomena in an orderly and scientific fashion. It was Franklin, of course, who performed the extremely foolhardy experiment with the kite and key in a thunderstorm and miraculously lived to tell the world that lightning and electricity are one and the same thing (Plate 6).

The Electrical Nature of Matter

Franklin, too, recognized that electricity came in two varieties with apparently opposite properties, which he designated positive (+) and negative (−). These designations were completely arbitrary, but once the labels were assigned, the world was stuck with them, and every twentieth-century engineering student has had occasion to wish that Franklin had named his charges the other way around.

Around 1800, these facts were known about electricity:

1. *A charge could be produced on many objects by rubbing them with fur or cloth. All these various materials, however, acquired a charge that was either like that acquired by amber rubbed with fur (−) or like that of glass rubbed with silk (+).*
2. *Objects having unlike charges exert a force of attraction on each other, whereas objects having like charges repel each other.*
3. *Metals do not hold an electric charge, but rather conduct electricity, i.e., allow an electric current (thought of as a fluid) to flow through them.*
4. *The electrical fluids (+ and − varieties) could be condensed and stored in such "condensers" as a Leyden jar, a glass jar lined inside and out with a thin metal coating. Charges were conducted into the jar by a metal chain connected to a small metal sphere on which the charge was generated by "friction," i.e., by rubbing.*

Magnetic Properties

In the early years of the nineteenth century, a number of the leading natural philosophers of the Western world studied electrical phenomena. Many of these men observed a close parallel between the behaviors of static or stationary electric charges and of magnetic poles, which had been investigated extensively by William Gilbert more than a century before. The parallels are indeed striking. Compare the following list of magnetic properties with the preceding list of electrical properties.

1. *Some substances can be magnetized (i.e., given magnetic properties) by rubbing them with another magnetic substance. The two ends of a magnetic substance acquire opposite magnetic properties, distinguished as north and south poles.*
2. *Unlike poles attract; like poles repel each other.*
3. *Magnetic forces act through many media, but only a few substances can be magnetized. In addition, it was recognized that the earth has magnetic properties, as if there were a bar magnet through the globe, skewed slightly with respect to the rotational axis, so that the earth's magnetic poles and geographic poles do not quite coincide. The magnetic compass had been introduced from China and had come into common use in fifteenth-century Europe, making possible the great age of discovery that opened up the new world.*

Among the scientists who tried to find some connection between electricity and magnetism was the Danish physician Hans Christian Oersted, who finally gave up medicine entirely for "natural philosophy," as physics was called in his day. Oersted's success in establishing the connection between electricity and magnetism was due to one of those happy accidents in science that, as someone remarked, only happen to those who deserve them. As the story goes, Oersted's search for a connection had been unsuccessful for so long that he had begun to change his lectures to point out that in spite of the similarity no connection existed. Then someone staying behind after a lecture to ask a question knocked the demonstration apparatus out of alignment, and presto! the connection appeared (Fig. 4.1).

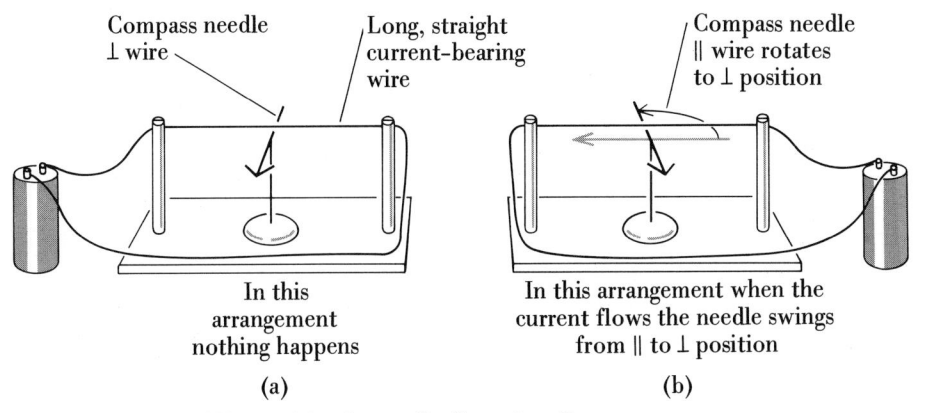

Figure 4.1 Oersted's "accident."

Because magnets line up parallel to each other, it had been expected that, if a current-carrying wire had magnetic properties, it would line up parallel to a compass needle. In fact, the magnetic field, which the wire indeed does have, is perpendicular to the wire.

The way was now open to explore the exact relationship between the two phenomena, electricity and magnetism. A French scientist, Charles Augustine de Coulomb, determined experimentally the laws describing both electrostatic and magnetic forces by using apparatus analogous to that used by Cavendish when he "weighed the world," because these laws are analogous to the law of gravitation (Table 4.1, Fig. 4.2).

The laws of electrostatic and magnetic attraction thus became known fairly early, before the nature of either phenomenon was understood. In fact, nearly all the groundwork was laid for the practical use of *electromagnetism,* as it came to be called, before any insight into the nature of charge and magnetism was gained. As the concept of energy developed, however, the notion of electrical fluid fell into disuse and it was possible to deal with electric (and magnetic) phenomena as forces, energies, potentials, and fields.

Table 4.1

Law	Mathematical relationship	Definition
Gravitation	$F = G(m_1 m_2 / d_{1,2}^2)$ (newtons)	m = mass in kilograms d = distance in meters G = universal gravitational constant (in new m²/kg²)
Electrostatic force	$F = k(q_1 q_2 / d_{1,2}^2)$ (newtons)	q = charge in coulombs[a] d = distance in meters k = proportionality constant (in new m²/coul²)
Magnetic force	$F = \kappa(p_1 p_2 / d_{1,2}^2)$	p = pole strength in ampere-meters d = distance in meters κ = proportionality constant, in new m²/(amp m)² or new/amp²

[a]Because the q's have a sign (+ or −) as well as a magnitude, F will be − when the force is attractive and + when it is repulsive. The unit of q, the coulomb, is the quantity of charge in mks units. It was chosen simply to be consistent in magnitude with other mks units. One coulomb equals 1.6×10^{19} electronic charges.

Electrical Forces

Let us consider a little further the laws describing electric forces. The proportionality constant k, which

Electrical Forces

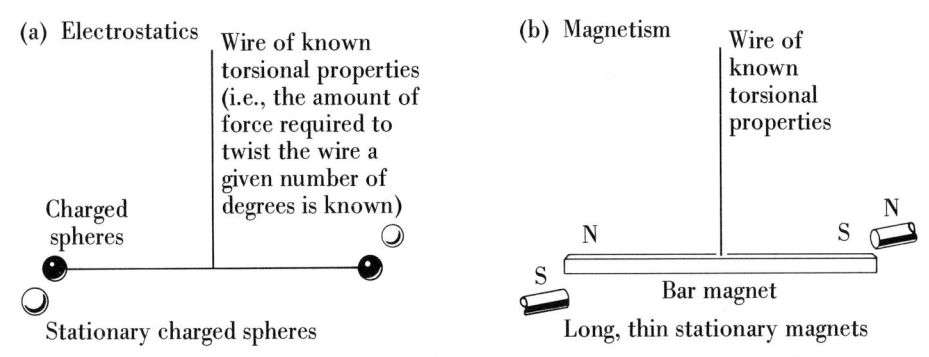

(a) Electrostatics — Wire of known torsional properties (i.e., the amount of force required to twist the wire a given number of degrees is known); Charged spheres; Stationary charged spheres

(b) Magnetism — Wire of known torsional properties; Bar magnet; Long, thin stationary magnets

Figure 4.2 Coulomb's apparatus for determining electrostatic and magnetic forces.

corresponds to G in the law of gravitation, is actually made up of several factors:

$$k = \frac{1}{4\pi\epsilon_0}.$$

The 4π factor is related to the fact that the force field around a charge is spherical. The factor ϵ_0 is called the *electric permittivity* of the medium in which the forces act. Unlike gravity, electrical force is altered by the medium in which it acts. Illustrations of the law of electrostatic attraction usually assume air to be the medium, but many technological applications of electricity involve other media. *Dielectric constant* is the ratio of the permittivity of a given medium to that of vacuum. In general, a material of high dielectric constant is a poor conductor of electricity, and therefore is a good insulator.

The operation of electrical forces presents the same difficult question as does gravity—the problem of "forces at a distance." Everyone knows that if you want to move a piano you have to put your hands on it and shove. It won't go far if you stand across the room and wave at it. Yet the dominant forces in the universe act between entities that are not in contact—masses, charges, magnetic poles, perhaps also the fundamental particles of the nucleus of the atom. Partly to avoid the difficulty of understanding the action of forces at a distance, many physicists have come to prefer the concept of fields to that of forces.

A field may be defined as the sphere of influence of a force, or whatever exerts a force. "Sphere of influence" is used in very much the same way in physics as in history and political science. A nation's sphere of influence includes all those nations whose internal and external affairs are to some extent determined by that strong nation. So a charge, because it exerts a force on all other charges in its vicinity, to some extent determines the behavior of those other charges said to be in its field. Similarly the earth is held in orbit because it is in the gravitational field of the sun. If the sun suddenly disappeared, the earth's path would become a straight line (slightly influenced by the other planets) instead of an ellipse.

Field

Field as a mathematical concept is due largely to one of the greatest scientific minds of the nineteenth century, James Clerk Maxwell. In his relatively short life of forty-eight years he made fundamental contributions to thermodynamics and brought together the ideas of electricity and magnetism into a coherent, comprehensive, and mathematically elegant theory that has needed little correction.

The mathematical definition of a field is:

The magnitude of a field is the force on a unit of whatever quantity responds to that force.

Field is a vector quantity. The strength or magnitude of the electric field at any point at a distance r from the charge q that causes the field is

$$E = k\frac{q}{r^2}, \text{ which is seen to be } \frac{F}{q'}$$

where q' represents a unit change. We can similarly express magnetic field, B as $B = \kappa p/r^2$ and gravitational field I as $I = Gm/r^2$, where p is the pole strength of the pole causing the magnetic field and m is the magnitude of the mass causing the gravitational field. It is clear that I is simply "little g" when m and r are the mass and radius of the earth.

Michael Faraday, one of the greatest experimentalists who ever lived, was no mathematician, but he was a great model maker. He envisioned a field as consisting of lines of force, the intensity of the field being indicated by the number of lines. We still find this picture useful (Fig. 4.3).

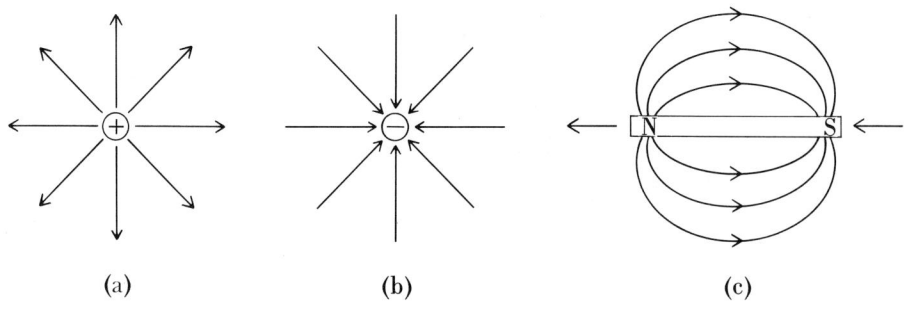

(a)   (b)   (c)

Figure 4.3 Faraday's model of electric and magnetic fields. The arrow indicates the direction in which a positive charge or a north pole would move in the electric or magnetic field, respectively.

One rather striking contrast between electric charges and their fields and magnetic poles and their fields is that although we may have isolated charges, either positive or negative, it is impossible to observe isolated poles, either north or south. It was not possible to explain this observation until a rather complete picture of the structure of matter was available, and we therefore will defer a theory of magnetism until Chapter 9.

Potential

Potential is another extremely useful concept in electricity. It bears the same relationship to energy as field does to force. We are accustomed to thinking of electricity as a form of energy; let us describe it in words analogous to those we used in describing mechanical and heat energy.

We begin with work: $W = F \cdot s$. Because charges respond to Coulomb forces, we must write

$$W = \frac{(kq_1q_2)}{r^2} \cdot s.$$

Now r is the distance between the charges, but the work being done *on* one of the charges to move it the distance s is being done by the other charge, and the distance r is not going to stay the same, because the force acts along the line joining the charges and they move either toward or away from each other. Newton's calculus was invented to take care of just such constantly changing situations, and its application produces this definition of work:

$$W = \frac{kq_1q_2}{r}.$$

Potential is defined as *work per unit charge*. The potential at a point in an electric field is the amount of work that must be done to put a charge at that point. If there were no electric fields in the area, then no electrical work would be done at all. If the field is due to a positive charge work will have to be done at the expense of some external energy source if a positive charge is to be brought into it. The same field would itself do work on a negative charge to bring it nearer the positive charge. Potential, like work, is a scalar quantity, but it will be positive or negative depending on the sign of the charge to which the field, and therefore the potential, is due. The unit of potential is the *volt*, and its symbol is V:

$$V = \frac{W}{q'} = \frac{kqq'}{\frac{r}{q'}} = \frac{kq}{r} = \frac{\text{joules}}{\text{coulomb}} = \text{volt}.$$

In this expression q' represents the unit + charge in terms of which potential is defined, and r is the distance from q to the point at which V is being specified.

The concept of potential at a point is rather less useful than the concept of potential difference, which is exactly what its name suggests: the difference in potential between two points. Recalling our Conservation Law, we recognize that if work is done to place a charge in an electric field or to move it from one point to another in a field that work must be converted to potential energy of the charge, just as mechanical potential energy is given to a mass if it is lifted to a higher level against gravity. So it is useful to think of potential difference as the potential gained or lost by a unit charge as it moves between the two points between which the potential difference is specified. Just as a mass slides downhill, from a point of high potential energy to a point of lower potential energy, so electric charges flow from a point of high potential to a point of lower potential. This flow of charge is what we call an *electric current*. Of course, if the force of friction is great enough, the mass won't slide. If the electrical friction we call *resistance* is great enough, current will not flow. In that case, electric charges can be held in storage, as it were, between two points of different potential.

A practical device for storing electric charge in this way consists of two parallel metal plates separated by a nonconducting material (that is, a material of very high resistance). When a charge is put on one plate it stays there, even though a potential difference has been created between the two plates. Such a device is called a *capacitor* today. It is no different in principle from the Leyden jar "condensers" known to Benjamin Franklin. A radio or television set is tuned to a station or channel by varying the charge-storing ability (capacitance) of the tuning capacitor of the set.

In most electrical circuits we use, the idea is to promote the flow of current rather than to stop it. To return to our mechanical analogy, the speed of our mass sliding down the hill will depend on two things: the steepness of the hill and the frictional resistance it encounters. Similarly the magnitude of electric current will depend on the steepness of the electric hill, which is measured by the potential difference, and the resistance offered by the conductors constituting the circuit. A German physicist, Georg Ohm, was the first to formulate this relationship precisely, so we call his formulation *Ohm's Law:*

$$I \propto V \quad \text{and} \quad I \propto \frac{1}{R}, \quad \text{so} \quad I \propto \frac{V}{R}.$$

If we measure I, the current, defined as the time rate of flow of charge, in coulombs per second, or amperes, and V, the potential difference in volts, then we can choose a unit of resistance such that the proportionality constant is 1, and write Ohm's Law as $I = V/R$. The resistance unit is the *ohm*.

The list of electrical units is a roll call of physicists who laid the foundations of electricity and magnetism as a branch of physics (Table 4.2).

Producing Electrical Energy

For electric current to flow, two conditions are necessary. (1) There must be a potential difference—a "hill" down which electrons can flow. So there must be a device to provide this potential difference. (2) There must be a complete circuit. Unless the conducting path is continuous, few electrons will move at all. The first condition is, on a closer look, simply a result of the Conservation Law. There must be an energy source to lift the electrons to the top of the hill. In general, two kinds of energy sources are applied to electrical circuits. Each necessarily makes electrical energy from some other kind of energy, because you can't make energy from nothing.

The first kind of energy source we will consider is the *voltaic cell*, a device in which the energy of a chemical reaction is used to push electrons around the circuit. In the early nineteenth century, an Italian

Table 4.2

Quantity	Name of unit	Namesake
q charge	coulomb	discovered laws of electric and magnetic force.
I current	ampere	early student of currents and their associated fields.
V potential; potential difference	volt (Volta)	invented the electrochemical (voltaic) cell for production of electric energy from chemical energy.
R resistance	ohm	discovered law of current and resistance.
B magnetic field intensity	$\dfrac{\text{webers}}{m^2}$	Weber invented the galvanometer (with Gauss) and measured the proportionality factor between electric and magnetic units which turned out to be the velocity of light.
H magnetic field	gauss	German mathematician.
C capacitance (of a capacitor)	$\dfrac{\text{coulombs}}{\text{volt}} = \text{farad}$	Faraday discovered electromagnetic induction, electrochemistry, and much more.
L inductance (related to magnetic properties of coils)	henry	co-discovered electromagnetic induction.
W work (any kind)	joule	made many contributions to thermodynamics; measurer of the "mechanical equivalent of heat," i.e., relation of heat and work units.

physician, Galvani, observed the twitching of a frog leg (severed from the frog) when it was touched by two wires, and the discovery was called animal

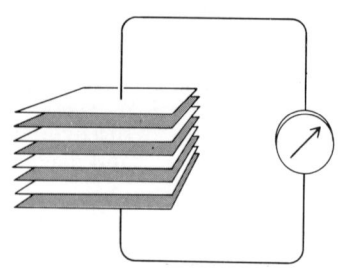

Figure 4.4 Diagram of Volta's cell.

electricity. The key to its production seemed to be the conductivity of the fluids in the animal tissue. Another Italian, Alessandro Volta, pursued this phenomenon further, and found that animal tissue could be replaced by any salt solution, and that if two different metals were put in electric contact via a conducting solution a potential difference could be measured externally between the metals. The earliest Voltaic cells consisted of blotting paper soaked in a salt solution, stacked alternately with sheets of unlike metals (Fig. 4.4). The most familiar Voltaic cells are the dry cell, which is a primary cell that we discard when it stops putting out energy, and the secondary or storage cell (such as the lead storage battery or the Edison battery), which can be recharged when it runs down. The chemical reactions in these three cells are not simple, so let us take a simpler, though less useful, example of a Voltaic cell for our example. The essential elements of any such cell are: two metals or other chemical substances that differ in their degree of chemical activity, and a fluid that conducts electricity. Your suspicion that chemical activity is connected to electricity is entirely correct, and will be pursued at length in Chapter 9. For now, let us accept the fact that "chemical reaction" involves shifts of electrons, and let us consider a Voltaic cell in which the simplest kind of shift takes place (Fig. 4.5). The two metals we choose are copper (chemical symbol Cu), which is a

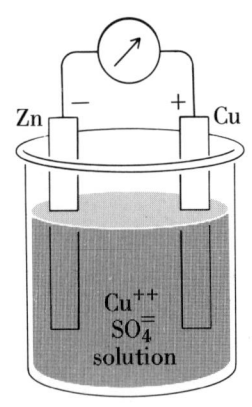

Figure 4.5 Simple voltaic cell.

rather inactive metal, and zinc (Zn), which is a rather active metal. We can judge the activity of a metal by the uses to which we put it. We make wire and pots and pans of copper, confident that it will last a long time without corroding (corrosion is a chemical reaction), although we do not generally use pure zinc for structures or objects. The strips of these two metals, or *electrodes*, and the conducting fluid, or *electrolyte*, are immersed in a solution of the blue salt called copper sulfate (chemical formula $CuSO_4$). If we connect a voltmeter by wires between the two metals (outside the solution) we will observe that it registers a difference in potential energy between the electrons at one electrode and those at the other, and it also indicates that an electric current is flowing, because it takes current to make the meter work. Where did the energy—and the flowing electrons—come from? This is what happens: The active metal, zinc, contains electrons that it releases rather easily, thus: $Zn \rightarrow Zn^{++} + 2e^-$. This equation represents a half-cell reaction. It says that an uncharged atom of zinc releases two negatively charged electrons, leaving a positively charged fragment called a zinc ion. If there is someplace for electrons to go (after all, they do repel each other, so a bunch of them are not going to hang around together), this reaction keeps on happening because the Zn^{++} ion represents a lower state of energy than the Zn atom, and the Second Law of Thermodynamics assures us that energy will always flow "downhill." Where does the energy go? It pushes the electrons around the circuit to the other electrode, where the other half of the cell reaction uses up the electrons in this way: copper ions in the copper sulfate solution absorb electrons to become copper atoms that "plate out" or coat the copper electrode. The half-cell reaction is written thus: $Cu^{++} + 2e^- \rightarrow Cu$. Current is conducted through the solution by the movement of $SO_4^=$ (sulfate) ions toward the Zn electrode and by the migration of Cu^{++} and Zn^{++} ions toward the Cu electrode. The $Cu^{++} + 2e^- \rightarrow Cu$ reaction also produces energy, and the total energy produced by the whole cell reaction, expressed in energy per unit charge (volts) is the "electromotive force," or emf, of the cell. A dry cell, no matter what size, has an emf of about 1.5 volts; a lead storage cell produces about 2 volts (Fig. 4.6). An ordinary 12-volt battery consists of six cells connected so that the current flows through each in turn. This kind of connection is called a *series connection*.

Because the cell reactions go only one way by themselves, a voltaic cell always has its electrons going in one direction. Such a one-directional flow of electrons is called *direct current*.

The current we commonly use in household circuits is *alternating current*, which flows first in one direction and then the opposite. Ordinary "60 cycle a.c." reverses its direction 120 times per second. (That means the lights go out 120 times per second, but our eyes don't respond fast enough to notice it.) Alternating current is produced by generators, devices in which mechanical energy expended to move electric charges in a magnetic field is transformed into electric energy. If a wire, which, like all matter, con-

Producing Electrical Energy

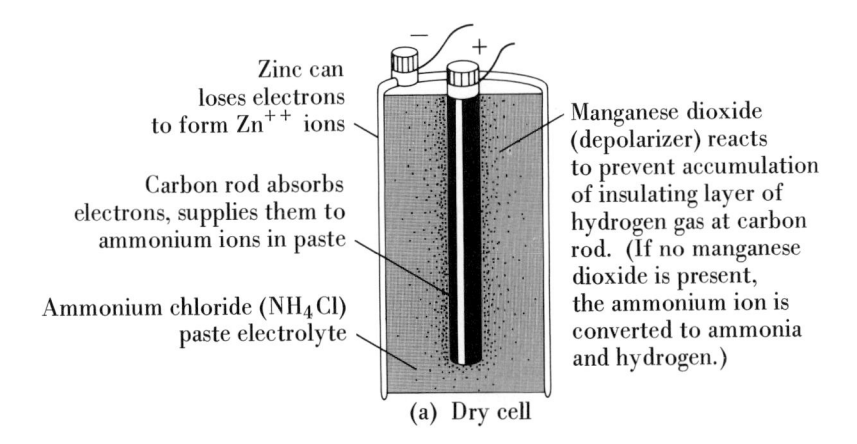

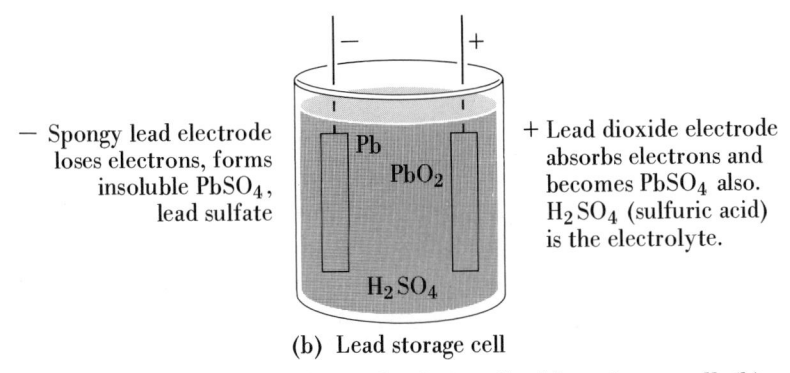

Zinc can loses electrons to form Zn^{++} ions

Carbon rod absorbs electrons, supplies them to ammonium ions in paste

Ammonium chloride (NH_4Cl) paste electrolyte

Manganese dioxide (depolarizer) reacts to prevent accumulation of insulating layer of hydrogen gas at carbon rod. (If no manganese dioxide is present, the ammonium ion is converted to ammonia and hydrogen.)

(a) Dry cell

− Spongy lead electrode loses electrons, forms insoluble $PbSO_4$, lead sulfate

Pb
PbO_2
H_2SO_4

+ Lead dioxide electrode absorbs electrons and becomes $PbSO_4$ also. H_2SO_4 (sulfuric acid) is the electrolyte.

(b) Lead storage cell

Figure 4.6 Commonly used voltaic cells: (a) a primary cell; (b) a secondary cell.

tains electrons, is pulled through the space between the poles of a horseshoe magnet, the magnetic field will exert a force on the electrons and displace them toward one end of the wire (Fig. 4.7a). So a potential difference will exist between the two ends of the wire, and if another wire were connected across its two ends, electrons would flow through the completed circuit (Fig. 4.7b). If the wire is pulled back in the other direction, the electrons are displaced in the opposite direction. If the wire is pulled back and forth, the direction of the current alternates.

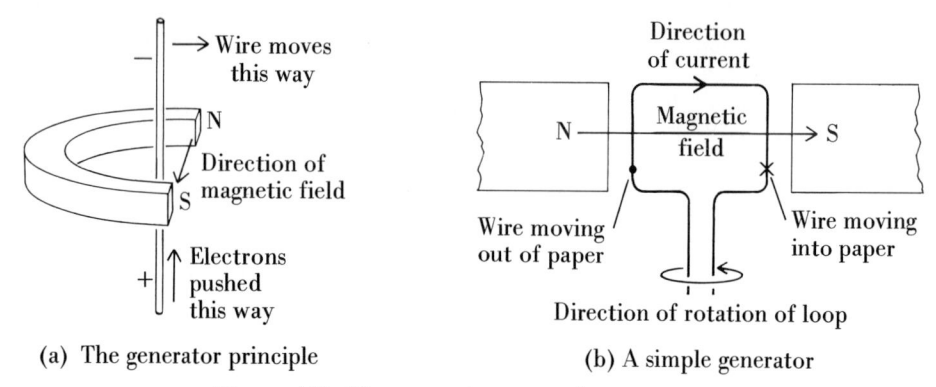

Figure 4.7 How a generator works.

(a) The generator principle

(b) A simple generator

Obviously, one couldn't light a city by pulling wires between the poles of horseshoe magnets. But the giant dynamos run by steam or water or nuclear power do essentially the same thing.

A practical generator uses steam, produced by burning coal or oil or heated in a nuclear reactor, or falling water to turn a turbine that turns coils of wire in the space between the pole pieces of powerful magnets. Obviously, if a coil of wire keeps turning, the wires connected to it directly will get all twisted up. So the connection is not made directly but through a system of "brushes" and "slip rings" (Fig. 4.8).

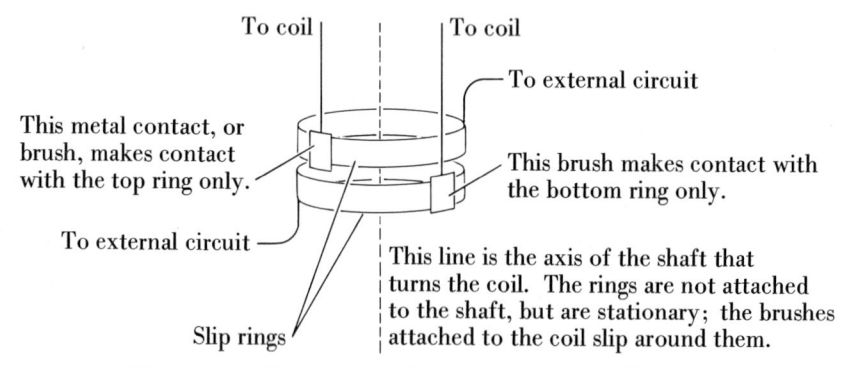

Figure 4.8 How current from the armature coil of the generator is conducted to the external circuit.

Producing Electrical Energy

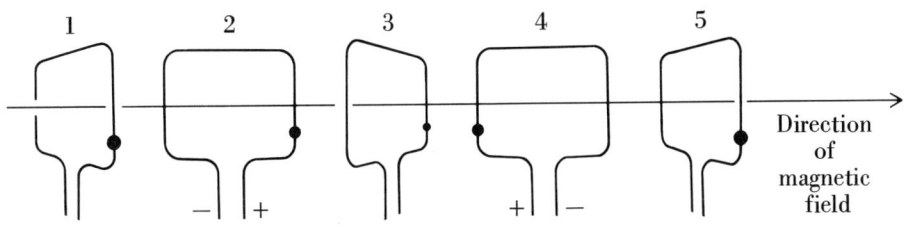

Figure 4.9 Successive positions of the rotating coil (armature) during one cycle of alternating current generation.

A closer look at the arrangement (Fig. 4.9) will indicate why the current in the external circuit alternates. Note how the marked side of the loop moves. At positions 1, 3, and 5, it is moving parallel to the magnetic field, and its electrons are not disturbed. At position 2 it is cutting across the field in a direction into this textbook page; at position 4 it is cutting across the field in the opposite direction. In between these positions, the moving loop cuts across the field at continuously varying angles, and the more nearly perpendicular the moving sides of the loop are to the field, the greater the force on their electrons. If we were to graph the magnitude of the force as indicated by the amount of potential difference generated relative to the position of the loop, and if we remember that the electrons move in a direction at position 4 that is opposite to that at position 2, we get a wave-shaped curve that represents one cycle of alternating current (Fig. 4.10).

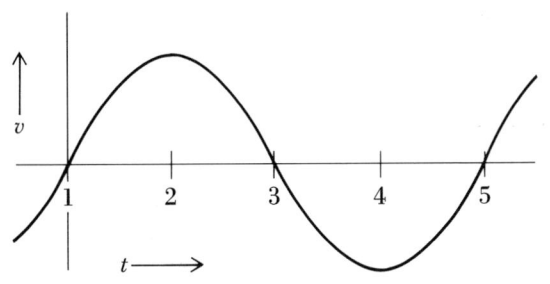

Figure 4.10 Variation of emf with time during one cycle of alternating current.

Electric Circuits

What else may a circuit contain besides "electron pumps"—the sources of electric energy in a circuit? First, it must consist of a complete conducting path. Charges lose energy as they travel through any circuit. In fact, they lose in an "external circuit" exactly as much energy as they gain in the "electron pump." The law requires it—the law of conservation of energy. (If they kept working their way around the circuit, but stashing away a little energy at the end of each trip, they would in essence be acquiring energy from nowhere, and that never happens.) The energy lost by the flowing electrons turns up as some other kind of energy. If it is lost in getting through a large resistance, such as an electric iron, it turns up as heat. If the current flows through a coil, the coil acquires a magnetic field, and electric energy becomes magnetic energy. If a capacitor is put into the circuit, the energy will be stored there as the charges are stored.

Circuit elements may be connected in a circuit in two ways. If they are connected consecutively, so that all the current flows through each element in turn, we have a *series circuit*. If the connections provide alter-

Symbols:

| voltaic cell | resistance | capacitance | inductance | a.c. generator |

d.c. circuit with resistances in series

d.c. circuit with resistances in parallel

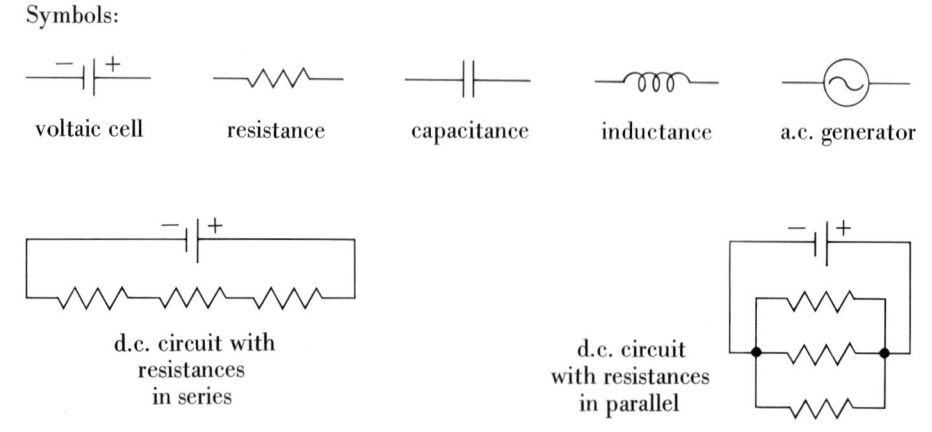

Figure 4.11 Circuit symbols; series and parallel connections in d.c. circuits.

Electric Circuits

Grid: the potential difference between cathode and grid can control electron flow

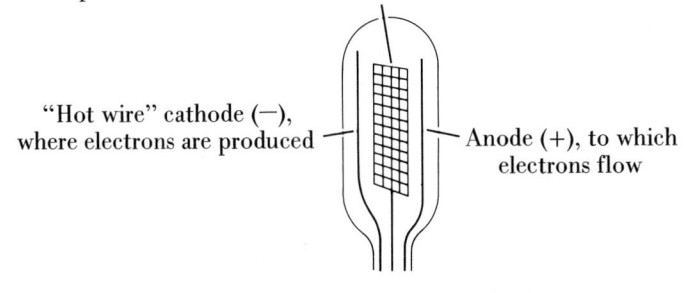

"Hot wire" cathode (−), where electrons are produced

Anode (+), to which electrons flow

(a) Vacuum tube or "valve"

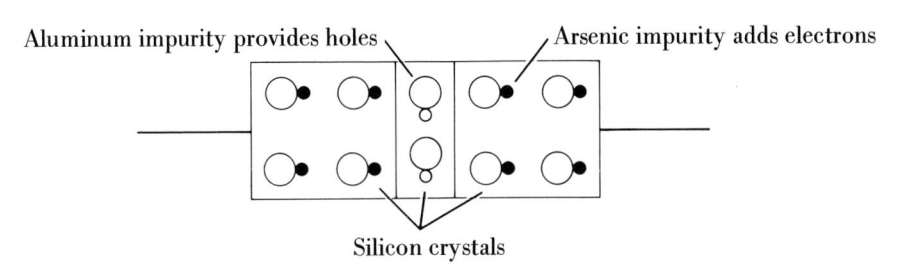

Aluminum impurity provides holes

Arsenic impurity adds electrons

Silicon crystals

"Semiconductors" contain impurities that provide either extra electrons or "holes," representing a deficiency of electrons. This is an N-P-N transistor, with a relatively positive layer between two negative layers. Electrons will flow from left to right. By maintaining appropriate potential differences across the P-N junctions a transistor can serve as a rectifier or amplifier.

(b) transistor

Figure 4.12 Electronic devices: (a) a simple triode; (b) diagrammatic representation of an N-P-N transistor.

nate paths to the current, it is a *parallel circuit*. Scientists and electricians use standard circuit symbols to show how the elements are connected. The complete picture is a circuit diagram (Fig. 4.11). Remembering that a circuit must be complete, or closed, if current is to flow at all, do you think that the various outlets, light fixtures, etc., that comprise your room's circuit are connected to each other in series or in parallel?

Since the English physicist J. J. Thomson identified electrons with negative particles torn out of metals by applying high potential differences across them, it has been understood that current in a wire is indeed a flow of electrons. (Ampere assumed that positive charges flowed, and electrical engineers still sometimes speak of "conventional current," which is opposite in every way to electron current.) But if the potential difference is high enough, electrons can be made to flow through space or through certain kinds of solid materials that are not normally conductors. The technology based on this phenomenon is known as *electronics* (Fig. 4.12).

SUMMARY STATEMENTS

1. Electric charge is a property of the fundamental units of matter, *electrons* and *protons*, which have opposite kinds of charge, designated *negative* and *positive*, respectively.

2. Like charges exert mutually repulsive forces; unlike charges attract each other.

3. Magnetic phenomena are very similar to electrical; magnetic poles, north and south, interact much as do electric charges.

4. Both electric and magnetic forces follow force laws analogous to the gravitational force law for masses.

5. The space in which an electric charge exerts force is called an *electric field*. The intensity, or strength, of the electric field is defined as the force it exerts on a unit positive charge.

6. The space in which magnetic forces act is called a *magnetic field*. Its strength is defined as the force it exerts on a hypothetical unit north pole.

7. Unit poles are hypothetical because it is impossible to separate the north and south poles of a magnet. The tiniest possible fragment of a magnet still has both poles.

8. A magnetic field always accompanies the flow of electric current, the direction of the field being perpendicular to the direction of the current.

9. Moving a conductor in an electric field induces a current in the conductor (generator principle); conversely, a current-carrying loop will rotate in a changing magnetic field (motor principle). Generators produce alternating currents; a motor runs when alternating current is supplied to its armature coil.

10. Direct current (d.c.) is produced by voltaic cells. A voltaic cell consists of two electrodes of different chemical

activity immersed in a conducting solution called an electrolyte. The emf, or energy per unit charge, supplied by the cell depends on the cell's chemical reaction.

11. In order for current, usually understood as moving electrons, to flow in a circuit, there must be (1) a source of energy, and (2) a closed path.

12. The opposition to flow of current is called *resistance*. It is defined in terms of Ohm's Law: $R = V/I$, where V is the potential difference, i.e., difference in potential energy per unit charge, between the two ends of the resistance and I is the amount of current (charge per unit time) flowing through it.

13. Circuit elements, such as resistances, may be connected in series or in parallel. Current flows in all elements in a series circuit, or none; current may flow through each element of a parallel circuit independently.

14. Electronics is the science of electron flow without wires or conducting solutions, i.e., through space (in vacuum tubes) or semiconductors (in transistors).

REVIEW QUESTIONS

1. What happens when: like electrical charges are brought near each other? unlike charges are brought near each other?

2. How is Coulomb's law for electric forces like Coulomb's law for magnetic forces and like the Law of Gravity? In what ways do these three kinds of forces differ?

3. What is a field? Give examples of three kinds of physical fields.

4. The following terms have very similar meanings. What is the common meaning and how is each used?

    potential   voltage
    potential difference   IR drop
    emf

5. Define and name the unit in which each is measured:

    electric charge   resistance
    current   capacitance
    potential difference

6. What are the two requirements that must be met if current is to flow in a circuit?

7. What parts or substances must you have to make a voltaic cell?

8. What elements must a generator contain?
9. Explain the use of slip rings.
10. Distinguish between an electric circuit and an electronic circuit.

THOUGHT QUESTIONS

1. Why did the introduction of steam and electric power have such a drastic effect on the structure of Western society? Trace the ultimate effects of the discoveries in the scientific areas of heat, electricity, and magnetism on the following:

family and home life	distribution of wealth
education	"woman's place"
growth of cities	children
communications	government
manufacturing and commerce	transportation
	international affairs

2. An earlier "revolution," the great age of exploration and discovery beginning in the fifteenth century, also owed its existence to a scientific discovery. What was it? What methods did it replace? Why was it superior?

3. The relation between electricity and magnetism illustrates the principle of symmetry in science. That is, if A interacts in some way to produce or affect B, then it might be expected that some interaction of B would produce or affect A. Specifically: because electric charges and magnetic poles obey identical (in form) force laws, it was expected that charges might produce magnetic effects. Then reasoning from symmetry would suggest that magnetic fields would affect charges. Describe how both of these symmetric effects can be demonstrated. Have you encountered any other symmetric laws or relationships in science? Watch for them.

READING LIST

Bitter, Francis, *Magnets*, Doubleday (Anchor), Garden City, N.Y., 1959. Not only magnetism but the making of a scientist is developed.

MacDonald, D. K. C., *Faraday, Maxwell, and Kelvin*, Doubleday (Anchor), Garden City, N.Y., 1964. Biographies of three of the greatest British scientists of the nineteenth century, on whose work rests the structure of electromagnetism and thermodynamics.

5

Light

As organisms evolving on a planet near enough to its parent star to receive from it great quantities of radiation of wavelengths between 0.00004 and 0.00007 centimeter, we have developed special organs for sensing this abundant form of energy, which we call *light*. Man, birds, and many other organisms receive most of their information about their environment via the sense of sight. Inevitably, human beings have wondered about the nature of the light by which we see.

Early theories of light and sight varied from the idea that the eye sends out emanations, possibly particles, which are bounced back to the eye from the object that is thus seen, to the suggestion that objects themselves send out emanations, possibly particles, which produce sight by falling on the eye.

Optics, the study of light phenomena, attracted the attention of many great minds of sixteenth- and seventeenth-century science. Galileo, as we have seen, tried to measure the velocity of light, and reluctantly conceded it to be infinite when he failed to measure any lapse of time between the uncovering of his assistant's lantern as a signal that his signal had been received and the sending of his own signal, though their distance apart in space was considerable. The first indication that the speed of light, though finite, is far too great for Galileo's experiment to succeed came through the astronomical observations of Olaus Roemer, but perhaps Galileo deserves some credit, too, for it was he who first turned a telescope toward the heavens.

Interest in optics and the invention of optical instruments probably exemplifies one of those chicken-or-egg relationships. A Dutch spectacle-maker named Hans Lippershey had discovered that looking through two lenses separated from each other made distant things appear nearer, much nearer than through one lens, only. Galileo put the two lenses at either end of a dark tube and discovered a new universe in a night.

Measuring Light

Roemer in 1675 was studying the eclipsing of Jupiter's moons by the planet. He observed that the variation in time of reappearance of an eclipsed moon could be as great as 16 minutes over an observational period of six months, and correctly deduced that the variation was due to the differences in distance between Earth and Jupiter at different times. Because Jupiter takes about twelve times as long to orbit the sun as Earth does, the maximum difference in the separation of the

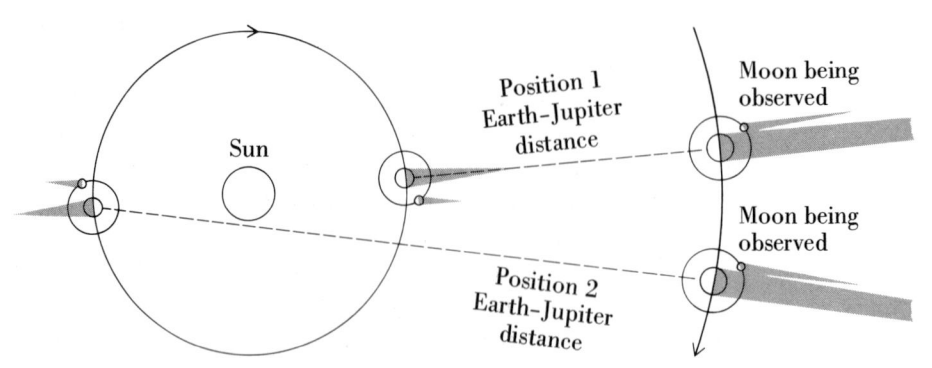

Figure 5.1 Roemer's method for determining the velocity of light.

Measuring Light

two planets is roughly the diameter of the earth's orbit (Fig. 5.1). Dividing that distance, which was not very accurately known in Roemer's time, by the 16-minute time difference gave the almost unbelievable figure of 200,000,000 meters per second for the velocity of light—and further work has shown Roemer's figure to be too small!

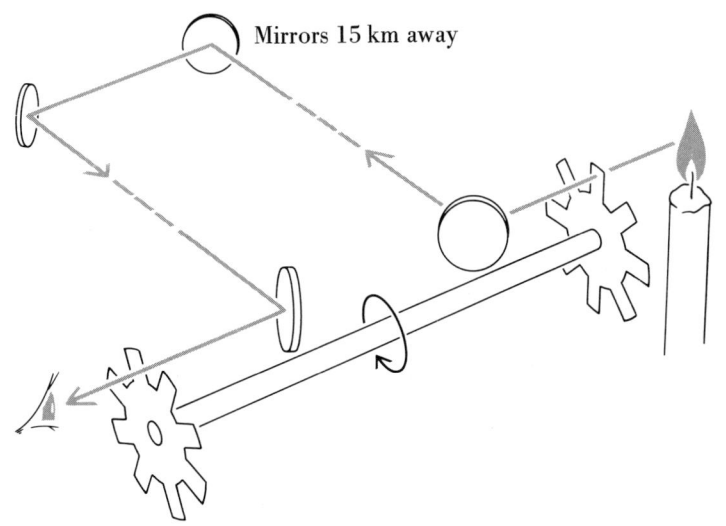

Figure 5.2 Fizeau's method for determining the velocity of light.

Light velocity was first measured under "laboratory" conditions by H. L. Fizeau in Paris in 1849. He used a set of toothed wheels (cogs or gears) in rotation as "light choppers" (Fig. 5.2). A pulse of light from the source must go between two teeth of the first wheel to reach the mirrors that would reflect it back to the observer, who could see it only if it managed to get between the teeth of the second cogwheel. By using mirrors a path of some 30 kilometers was achieved, and a speed of revolution of the wheels could be found that would enable a single pulse to be timed.

The most accurate work on the velocity of light was done by A. A. Michelson, the first American to win the Nobel prize in physics. He used a rotating mirror as his light chopper, and the carefully measured distance between Mt. Wilson and Mt. San Antonio in California for his light path, and calculated the speed of light to be some 2.9979×10^8 meters per second. Further refinements such as confining the entire light path to an evacuated tube and lengthening it by multiple reflecting mirrors at each end have confirmed this figure, which for convenience we shall round off to 3×10^8 meters/second.

This very great speed might have argued early against a particle nature of light had not this theory been favored by no less a scientific personage than Isaac Newton. In his first published paper Newton revealed his discovery of the dispersion of white light into all the colors of the rainbow by a prism. White light must then be composed of parts—not exactly particles—of different colors. These parts, or rays, Newton conceived to be associated with the bodies seen rather than to be mere motions (i.e., waves) in the medium transmitting them, as his contemporaries Robert Hooke and Christian Huygens believed them to be. Thus was born the great argument over the nature of light: wave or particle? As Newton's fame grew, the wave theory waned in popularity; correctly or not, his name came to be associated with a corpuscular or particle picture of light.

Geometric Optics

Light rays, whether they are waves or beams of particles, exhibit certain kinds of behavior that can be described by straight lines. This is the realm of geometric optics, and it can be understood without recourse to any particular model or theory of the nature of light, either particle or wave.

It is our common observation that light rays traveling through a homogeneous, transparent medium such as air at a constant temperature follow a straight-line path. That is, in learned polysyllables, light is rectilinearly propagated. We are all so firmly con-

Geometric Optics

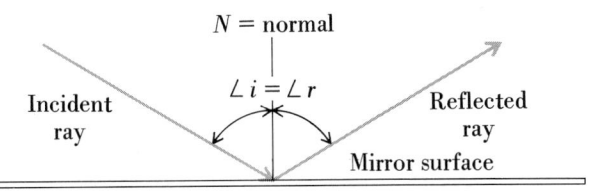

Figure 5.3 The laws of reflection.

vinced of this, in fact, that we interpret the path of every light ray that reaches our eyes as having been a straight line, whatever may have happened to deviate it. The baby or the puppy upon first seeing its reflection in a mirror tries to get behind the mirror where the "other" baby or puppy seems to be.

In addition to their straight-line transmission through any transparent medium, we observe light rays to be reflected from some surfaces and to be refracted on traveling from one transparent medium to another. A stick appears bent at the interface between water and air in a glass; when removing the stick from the water proves it to be as straight as when it was put in, we must concede that it was the light that was bent. This bending of light due to its velocity being changed as it goes from one medium to another is the phenomenon of *refraction*.[1]

These phenomena, *reflection* and *refraction*, follow very simple laws and are the bases of our most common uses of optics, those involving mirrors and lenses. The Laws of Reflection may be stated as:

1. *A ray of light encountering a mirror at an angle i to the normal (perpendicular) to the surface will be reflected at the same angle r with respect to the normal* (Fig. 5.3).
2. *The incident ray, the reflected ray, and the normal all lie in the same plane.*

[1] An analogy to refraction of light may be made by considering the behavior of band members marching at an angle from the cinder track onto a muddy football field. They must at all costs stay in step; to do this, each person will slow down by taking smaller steps as he enters the thicker medium, mud, and consequently the line of march will turn slightly toward a line perpendicular (normal) to the cinder-mud boundary.

Application of these laws determines the formation of images by any reflecting surfaces. Consider a few.

1. When you look in an ordinary flat (plane) mirror to comb your hair, you know that, if you are right-handed, your mirror image is left-handed. It is right-side up, however, and life size (Fig. 5.4).

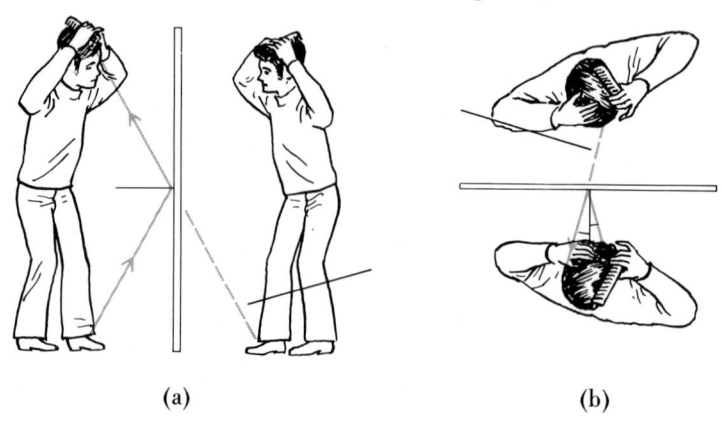

(a) (b)

Figure 5.4 Image formation by a plane mirror. The image is erect (a), but has right and left reversed (b) with respect to the object.

2. If you look into the inside of the bowl of a silver spoon you may see yourself as small and upside down, or if you are very close, you will get a right-side up, enlarged view of your eye. At some point in between, there will be no clear image at all—it will all go fuzzy.

3. If you look into the back side of the spoon, you will always see a very small, rightside up image of yourself no matter at what distance you hold the spoon.

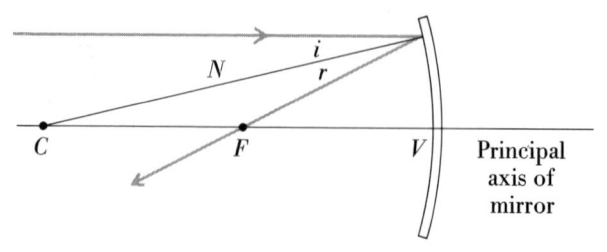

Figure 5.5 Concave mirror.

Geometric Optics

The center of curvature of the mirror is C in Figure 5.5. That is, it is the center of the sphere of which the mirror surface is a small part. So any line from C to the mirror is a radius, and therefore normal to the mirror. All rays parallel to the mirror's principal axis that fall on the mirror near the vertex, V, will cross the axis at the same point, called the *focus* of the mirror. We can use this observation plus the principle of reversibility, which points out that unless you see the source of light, you can't tell by looking at the middle of a light ray which way it is going, to construct geometrically, using standard rays, the image any mirror forms.

When the object is relatively far (farther than F in Figure 5.6a) from the mirror, the image is inverted. Whether it is larger or smaller than the object depends on how far the object is from the focus. In any case, the image formed is *real*. Light rays actually come to a focus and the image could be displayed on a screen placed at that distance from the mirror.

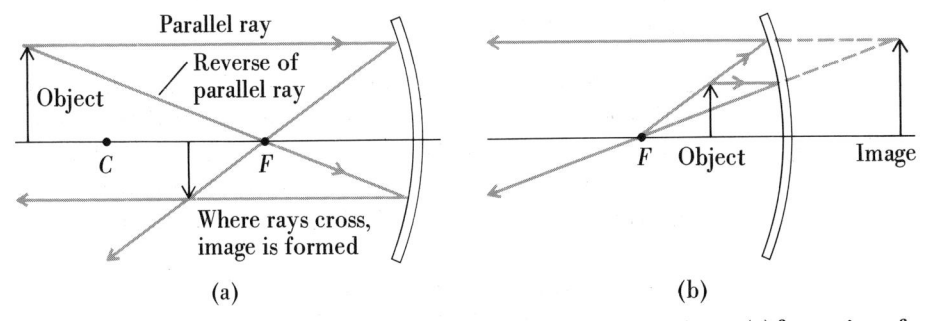

Figure 5.6 Image formation by a concave mirror: (a) formation of a real image; (b) formation of a virtual image.

If the object is between F and the mirror the reflected rays diverge (Fig. 5.6b); they never get together to form a real image. However, because we assume rectilinear propagation, we extrapolate the rays behind the mirror and see a *virtual*, or imaginary, image that is erect (right-side up) and enlarged with respect to the object.

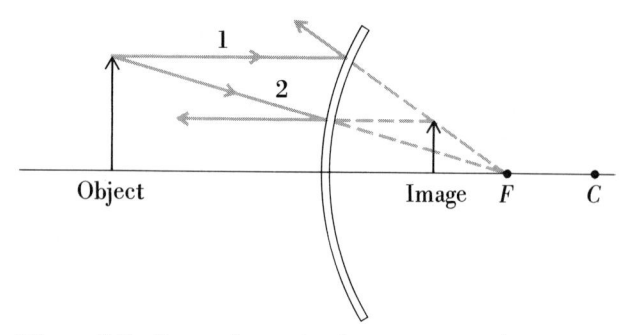

Figure 5.7 Image formation by a convex mirror.

Why does the image become fuzzy and disappear when the object is placed at the focal point F?

The center of curvature and the focus of a convex mirror (Fig. 5.7) are behind the mirror surface. Only one kind of image can be formed. It is virtual, erect, and reduced in size, as shown by the standard rays.

Law of Refraction: Because light is slowed down differently in different media, it is bent at different angles in different media.

A seventeenth-century Dutch physicist, Willebrod Snell, found that the angle of refraction and the angle of incidence for any medium have a simple relationship that defines a constant for the medium called the *refractive index*. This relationship is called *Snell's law*:

$$\text{refractive index of medium } M = n_M = \frac{\text{velocity of light in medium } M}{\text{velocity of light in vacuum}} = \frac{\sin i}{\sin r}$$

where i is the angle of incidence, r is the angle of refraction, and the sines of the angles are defined thus: the sine of an angle is the ratio of the side opposite the angle to the hypotenuse of the right triangle containing the angle. From Figure 5.8, it is seen that the sine of i (also written $\sin i$) is x/s, and the sine of r is x'/s'. You can see the connection between the ratio

Geometric Optics

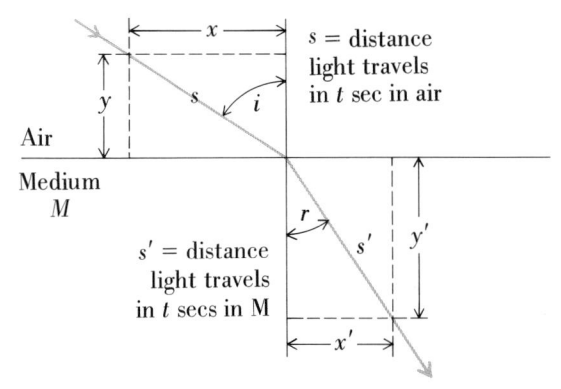

Figure 5.8 Refraction of light when ray goes from air to a denser medium. The diagram illustrates Snell's law.

of velocities and the ratio of sines by considering how far corresponding points on two parallel rays will move in a given time if one moves in, for example, air, and the other in water.

Light always goes faster in a vacuum than in anything else. The angle i is always greater than angle r for a ray going from a vacuum to something else; that is, light always bends toward the normal in the medium that slows it down.

The most widespread use of refraction is made with lenses (Figs. 5.9, 5.10). Light can be bent in any desired direction by appropriate shaping of glass surfaces.

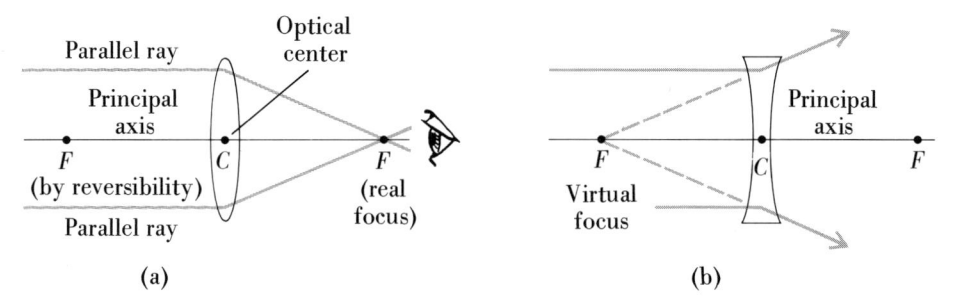

Figure 5.9 Lenses: (a) a convex, or converging lens brings light rays to a real focus; (b) a concave, or diverging lens has a virtual, or imaginary focus.

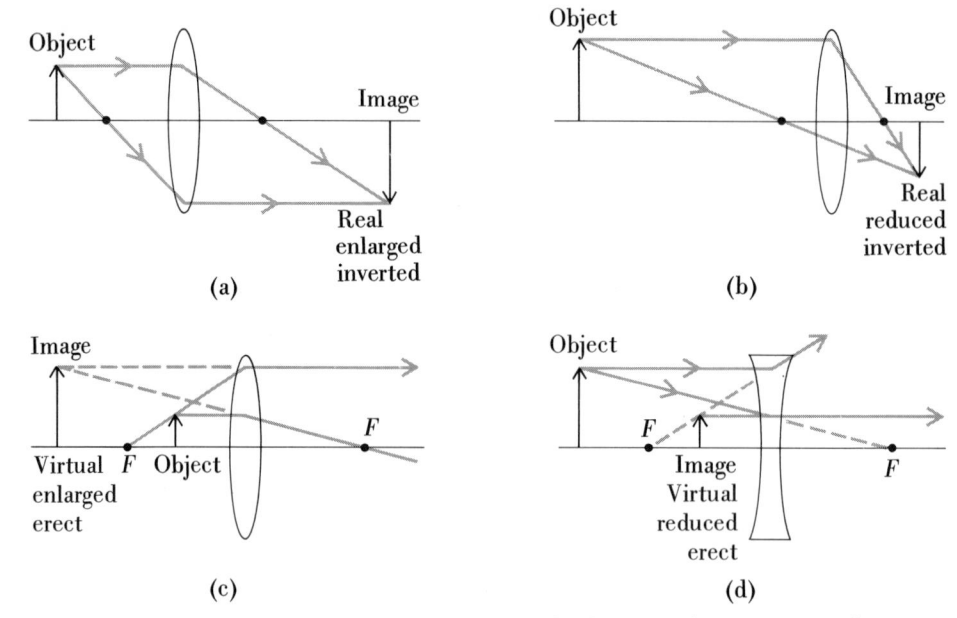

Figure 5.10 Image formation by lenses: (a) arrangement for projecting an enlarged image, as with a slide projector; (b) operation of lens correcting for farsightedness; (c) use of a magnifying glass; (d) operation of a lens correcting for nearsightedness. Notice that real images are always inverted, virtual images, always erect. In (b) and (d) the lenses are used to bend light rays so that the eye lens can focus them on the retina.

Of course, not every material is transparent. Light strikes many objects that neither transmit it nor reflect it. They absorb it. When this happens, the absorber gets warmer—convincing evidence that whatever else light is, it is energy, because it is converted into heat energy by absorption.

Color

Although a piece of black velvet will absorb all the light falling on it ("black" means the absence of light), most opaque things absorb only part of the light incident on them. They reflect the rest, and are seen to be the color of the light they reflect. Transparent colored things, like stained glass, are obviously only transparent to part of light, i.e., those colors they transmit. Isaac Newton was the first to demonstrate conclusively that *white light* is simply a mixture of all colors in equal proportions when he passed a narrow beam of white light through a prism into a darkened room and produced an array, or spectrum, of colors like the rainbow on the wall (Plate 7).

White light is dispersed into a spectrum of colors by a prism because the different colors are refracted through different angles by the glass. Red is least bent, then orange, yellow, green, and blue, and violet is bent through the greatest angle. If one considers light to be made of waves, this order corresponds to a decreasing order of wavelength. The shortest waves are bent most.

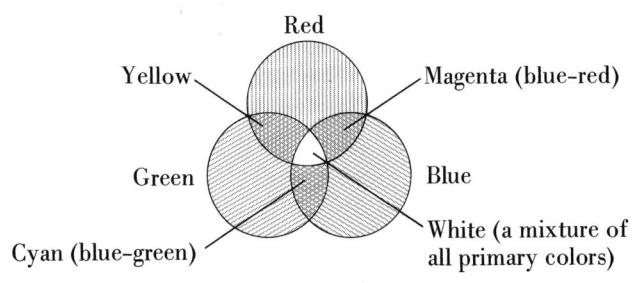

Figure 5.11 Primary colors of light, or additive primaries.

Two nineteenth-century physicists, Thomas Young, an Englishman, and the German von Helmholz, devised a theory of color vision that, although not wholly proven, is the basis for our treatment of color both in physics and in art. By this theory, the eye's cone cells are presumed to respond to blue, red, and green light, certain cells being specific for each color. The sensation of yellow is the result of roughly equal stimulation of "green" and "red" cones, etc. So we speak of red, blue, and green as the primary colors of light because all other colors can be produced by suitable mixtures of these three (Fig. 5.11). Conversely filters

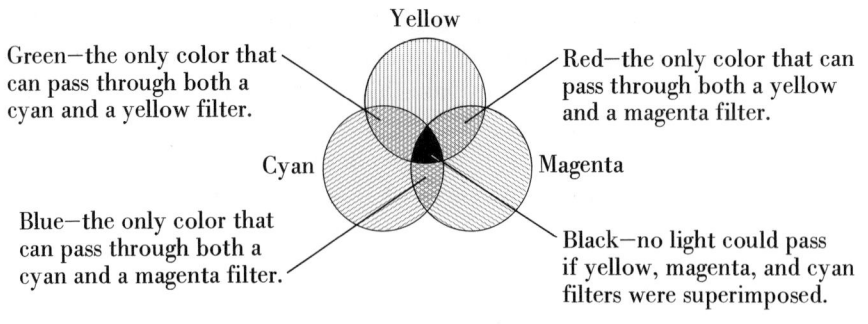

Green—the only color that can pass through both a cyan and a yellow filter.

Red—the only color that can pass through both a yellow and a magenta filter.

Yellow

Cyan

Magenta

Blue—the only color that can pass through both a cyan and a magenta filter.

Black—no light could pass if yellow, magenta, and cyan filters were superimposed.

Figure 5.12 Primary colors of filters, or subtractive primaries.

and pigments produce color by subtracting one or more of the primary colors from white light. Because yellow is a combination of green and red waves only, a yellow solution would not transmit blue light. It would be yellow to the eye because it absorbs the blue portion of white light falling on it. Those colors which represent the absorption of one of the primary colors of light are the *subtractive* primaries: yellow, magenta, and cyan (Fig. 5.12). Artists are likely to name red, yellow, and blue as their primary pigment colors. Physicists have no quarrel with this, but they might argue that their own designation is a shade more precise!

Colors opposite each other in the circular diagrams, which are seen to be the same in both, are called *complementary* colors (Fig. 5.13). Complementary

Additive combination

Red—Cyan

Blue—Yellow

Green—Magenta

Subtractive combination

Figure 5.13 Complementary colors. Combining complementary colors of light produces white light; a combination of complementary filters absorbs all light.

colors of light combine to produce white light, whereas complementary filters or pigments combined will absorb all light, producing blackness.

Colors and their charms and uses have doubtless been known to man as long as the species has existed. Just where color originates turns out to be a question answerable only by twentieth-century physics and chemistry, so we will defer our discussion of its origins until Chapter 8.

Physical Optics

The real question about the nature of light, whether it is a wave or a beam of particles, was apparently settled once and for all by an experiment done by Thomas Young about 1800 (Fig. 5.14). A single beam of light passing through a slit from a light source encounters a barrier that is opaque except for two narrow slits very close together, neither of which is aligned with the single slit. A pattern of alternating light and dark lines appears on the screen. The brightest line is the center one, which is aligned with neither slit, but is equidistant between them.

It is well nigh impossible to see how a stream of particles could produce this pattern of light and dark lines, called a *diffraction pattern*, but with waves, this is exactly the behavior expected. Two well-

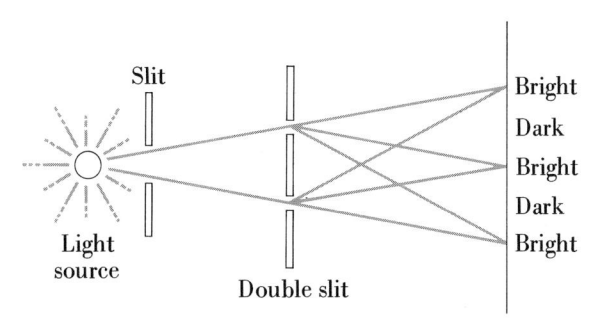

Figure 5.14 Young's double slit experiment.

established principles of wave motion, one going back to Newton's contemporary, Huygens, serve to account thoroughly for diffraction phenomena.

Huygens' Principle: Every point on a wave front acts as a new source of waves.

Superposition Principle: When two waves travel together in a medium, the resultant wave is their algebraic sum.

The behavior of the ripples formed by dropping a pebble in water illustrates these principles clearly (Fig. 5.15).

If we assume that light consists of waves, we can

Waves in phase interfere constructively.

Waves out of phase cancel each other out.

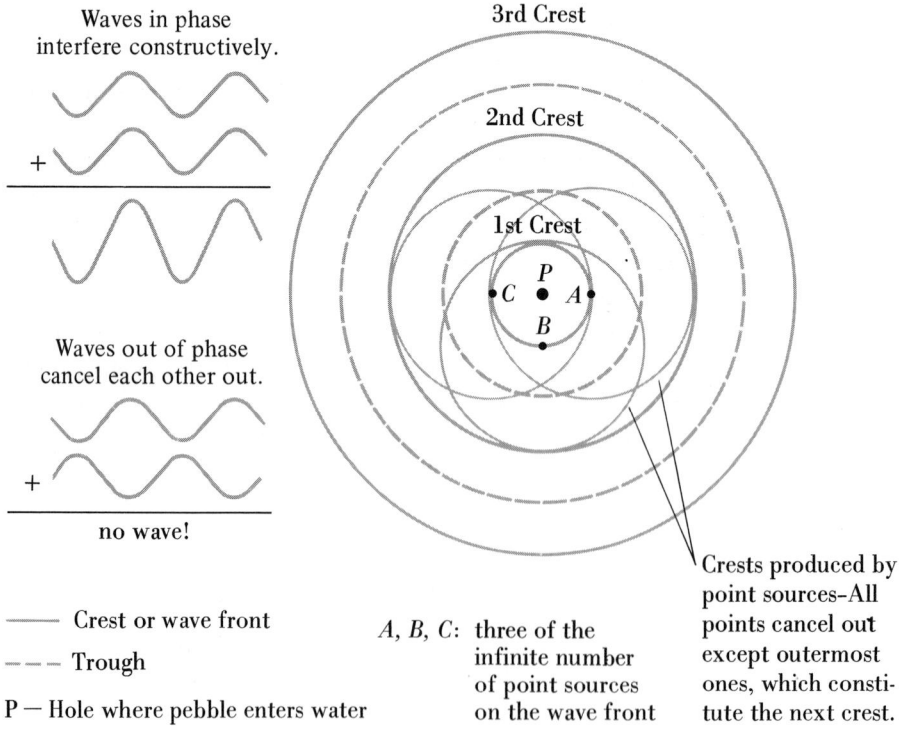

no wave!

——— Crest or wave front
--- Trough
P — Hole where pebble enters water

A, B, C: three of the infinite number of point sources on the wave front

Crests produced by point sources–All points cancel out except outermost ones, which constitute the next crest.

Figure 5.15 Interference of waves, illustrating Huygens' principle and superposition.

explain Young's experiment in this way: Wave fronts from the first slit spread out according to Huygens' principle. Eventually they fill the second set of slits, producing point sources there, from which waves spread out and reach the screen. If we use light of a single color (single wavelength) then waves that travel identical paths, or paths differing by a whole number of wavelengths, will arrive at the screen in phase, and when superimposed will reinforce each other to produce a bright spot. In between, waves will be out of phase when superimposed and will cancel each other. Of course, all degrees of phase mismatching are possible, so the pattern will shade from bright to dark over and over again.

Because each wavelength of light will produce its own diffraction pattern, white light can be dispersed by passing it through slits as well as through a prism. Actually, a diffraction pattern can be produced with a single slit, because the waves from different parts of the slit can interfere with each other, as anyone can see by looking at a light source through the narrow crack between two fingers. To disperse light for studying its component waves, however, the most effective slit system is the *diffraction grating:* a piece of glass ruled with a diamond point to contain as many as 13,000 or 14,000 lines (scratches) per inch. The clear glass areas between the scratches act as slits. The closer the spacing of the slits the greater the angle of diffraction, and so the greater the separation of the various wavelengths.

The Electromagnetic Spectrum

James Clerk Maxwell's synthesis of the ideas of electricity and magnetism into the laws of the electromagnetic field led to the prediction that whenever an electrically charged particle vibrates it produces a wave pattern in the electromagnetic field, which might be very short, like ultraviolet or visible light, or very long, or anywhere in between. This had two effects. It removed the last major difficulty in the way of complete acceptance of the wave picture of light by showing that no material medium was required to carry light waves, for it is the electromagnetic field that "waves." (A very peculiar medium, called *ether*, had been dreamed up to fill all of space for the purpose of carrying light waves. See Chapter 6.) The other effect was to start people looking for the predicted long waves. They were found by Heinrich Hertz, and are known as *radio waves* (Table 5.1).

Table 5.1   Electromagnetic Spectrum

Gamma rays, X rays	Ultra-violet	Infra-red	Micro-wave (radar)	Short wave	Radio Broadcast band
High frequency Very short waves ($\sim 10^{-8}$cm)		Visible light violet red 4×10^{-5}cm 7×10^{-5}cm			Low frequency Long waves (~ 200m)

Certain substances can alter the way in which the electric part of the field associated with a light wave vibrates. Reflection by nonconducting materials and passage through certain kinds of crystals produce light waves vibrating in a single plane rather than randomly (Fig. 5.16). Such light is said to be "plane polarized." Sunlight also becomes slightly polarized by scattering by particles in the air, and bees and ants have been shown to owe their remarkable direction-finding abilities to their ability to respond to the direction of polarization of sunlight. Polaroid sunglasses have lenses that selectively absorb polarized light. Because light reflected from nonmetallic surfaces is polarized to some extent, polaroid lenses cut down on glare due to reflection.

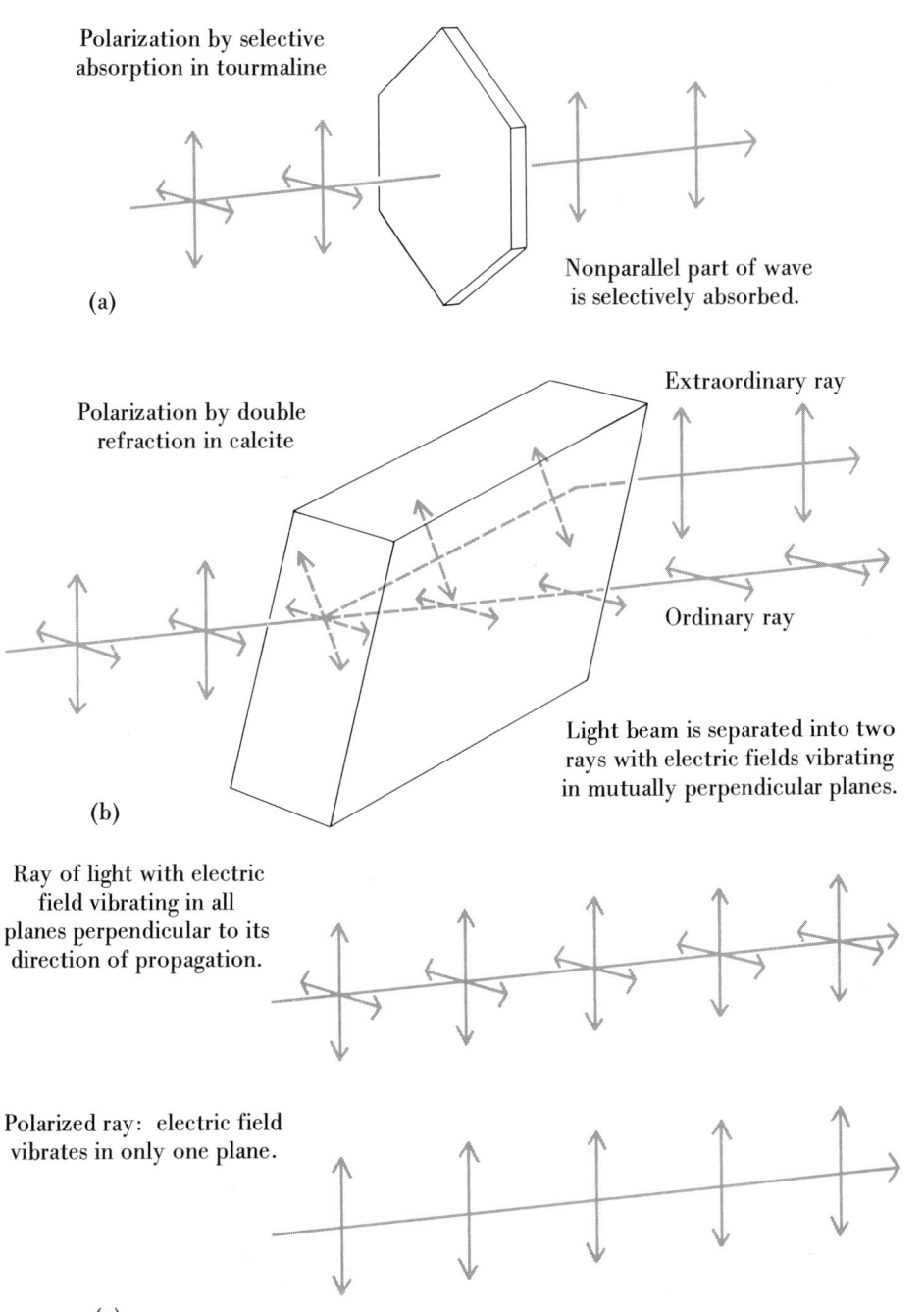

Polarization by selective absorption in tourmaline

(a)

Nonparallel part of wave is selectively absorbed.

Polarization by double refraction in calcite

Extraordinary ray

Ordinary ray

Light beam is separated into two rays with electric fields vibrating in mutually perpendicular planes.

(b)

Ray of light with electric field vibrating in all planes perpendicular to its direction of propagation.

Polarized ray: electric field vibrates in only one plane.

(c)

Figure 5.16 Methods of polarizing light: (a) selective absorption; (b) double refraction; (c) representation of unpolarized and polarized light.

Particle Nature of Light

Because waves can form diffraction patterns and particles cannot, the argument seemed to be over. Then came the photoelectric effect. J. J. Thomson, a British physicist, had found it possible to produce streams of electrically charged particles from a metal plate subjected to a high electrical potential difference. As convincing evidence that these negatively charged particles came from within atoms, heretofore deemed indivisible, he showed that these same particles could be produced by shining ultraviolet radiation (waves like light, but too short to affect our sense of sight) on certain metals. So the reality of electrons as subatomic particles was established, but a new question had to be asked: Why should light waves knock electrons out of atoms and give them energy to move through space?

This was not the only phenomenon that was puzzling the physicists of the 1880's. Two others, in particular, profoundly affected our understanding of light. One was the Michelson-Morley experiment that finally led to the unavoidable, if somewhat illogical (at the time) conclusion that the velocity of light in space always comes out to be the same, no matter how it is measured or how the measurer is moving. This experiment established the velocity of light as one of the fundamental constants of the universe (see Chap. 7). The other phenomenon, blackbody radiation, was as important as the photoelectric effect in disestablishing the wave picture of light, and in fact was the starting place for the formulation of a truly revolutionary view of radiant energy, the quantum theory, first proposed by Max Planck. This theory was so revolutionary, in fact, that its proposal may be said to have ended the era of classical physics and raised the curtain on the physics of the twentieth century. So profound a change was wrought that we had better stop a moment, as it were, on the threshold of the twentieth century to look both back and forward.

SUMMARY STATEMENTS

1. Light is that small portion of the electromagnetic spectrum to which our eyes are sensitive.

2. The electromagnetic spectrum consists of all that array of wavelengths produced by charged oscillators, ranging from gamma rays through X rays, ultraviolet rays, visible light, infrared rays, and microwaves, to radio waves. All travel at the velocity $c = 186,000$ mi/sec, and require no material medium for their transmission, since they are vibrations in the electromagnetic field.

3. When light is reflected (1) its angle of reflection equals its angle of incidence, and (2) both incident and reflected rays lie in the same plane. Image formation by mirrors is understood in terms of these principles.

4. When light passes from one medium to another of different optical density it is bent, or refracted; it is always bent toward the normal in the denser medium (in which its velocity is less). Image formation by lenses is understood in terms of this principle.

5. Young's double-slit experiment, in which light from a single source, separated into two beams, exhibited an interference pattern when recombined, established the wave nature of light.

6. Wave behavior, such as diffraction and interference, can be understood in terms of two principles:
   a. Huygens' Principle: Every point on a wave front acts as a new source of waves.
   b. Superposition Principle: When two waves travel simultaneously in a medium the resulting wave is their algebraic sum.

7. White light looks white because it contains an equal mixture of all colors. When we see a color it is because only certain wavelengths are transmitted to our eyes.

8. If an object looks colored when illuminated by white light, it is because some wavelengths are being absorbed and we are seeing only those the object reflects or transmits.

9. Certain substances, by selective absorption, double refraction, reflection, or scattering, line up light waves so

that all the electric field vibrations are in the same plane. Such light is polarized.

10. The photoelectric effect, discovered late in the nineteenth century, suggested that light also has a particle nature.

REVIEW QUESTIONS

1. List in order from short to long wavelengths the components of the electromagnetic spectrum.

2. By what methods did each of the following try to measure the velocity of light? Which were successful?

    Galileo   Fizeau
    Olaus Roemer Michelson

3. Define: reflection, refraction, absorption, diffraction, and interference with respect to light.

4. State the laws of reflection.

5. State the law of refraction.

6. State the two principles that govern wave behavior.

7. Describe Young's double slit experiment. Why was it important?

8. Show how a beam of parallel rays of white light behaves on encountering each of the following:

    concave mirror   concave lens   flat mirror
    convex mirror   convex lens   prism

9. What do we mean by: additive primaries; subtractive primaries; complementary colors?

10. Name four ways of polarizing a beam of light.

THOUGHT QUESTIONS

1. Defend with argument and experimental evidence the particle nature of light and the wave nature of light.

2. You can solve the following optical problems by applying the standard ray technique:

  a. You want to project a 2 cm x 2 cm slide so that it fills a 1 m x 1 m screen. About how far from the screen must the projector be?

  b. You are an illusionist's beautiful assistant. He will show the audience your head, life-size, on a platter—with a mirror—because you insist on keeping your head on your shoulders. What will be the relative arrangement of you, the mirror, and the image?

  c. What lens or mirror arrangement will allow you to

(1) read the fine print on your insurance policy; (2) watch for shoplifters in your crowded shop; (3) get a slightly enlarged view of your own face?

3. You have been hired to paint an apartment in a wild array of psychedelic colors, but not having been paid in advance you can afford only four buckets of paint. What colors will you buy if you must produce from them magenta, chartreuse, red, hot pink, purple, kelly green, yellow, orange, blue, and white? How will you achieve each color desired?

4. The human body can detect only a tiny fraction of the whole range of electromagnetic radiation. How would you detect the others, from gammas to radio waves?

READING LIST

Jaffe, Bernard, *Michelson and the Speed of Light*, Doubleday (Anchor), Garden City, N.Y., 1960. A biography that includes some historical background on theories of light and introduces the modern theories to which Michelson's work led.

Krogh, August, "The Language of the Bees," *Scientific American*, August 1948 (Reprint #21, W. H. Freeman, San Francisco). A fascinating study of how bees use the polarization of sunlight as a direction-finding aid and communicate information about food supplies to their hive mates in terms of it.

Land, Edwin H., "Experiments in Color Vision," *Scientific American*, May 1959 (Reprint #223, W. H. Freeman, San Francisco). This is a report on the continuing study of the mechanism of color perception, which is not yet very well understood.

6

Physical Science Circa 1900

During the three centuries that extended from the lifetime of Galileo to that of Maxwell an elegant system of natural laws was developed to explain the physical phenomena of the universe, at least on the macroscopic scale. The philosophic point of view that underlay this system of classical physics, and that seemed entirely justified by it, was *causality*. Every observed effect had its cause, immediate or remote, and every cause produced its predictable effect. From this scientific framework the seventeenth- and eighteenth-century deists and rationalists reduced the concept of God to that of First Cause or Prime Mover, and social philosophers viewed society and individual life as deterministic, set, authoritarian, unchanging—even in the midst of the profound changes we call the Industrial Revolution. That revolution itself was, of course, the result of the invention of machines run by newly available forms of energy: heat, electricity, magnetism, and mechanical energy. It produced not only a widespread machine technology, but also greater understanding of the forms of energy on which it depended. A great French mathematician of the early nineteenth century, Pierre Simon La Place, commented that if anyone could tell him the exact location and momentum of every material particle in the universe he could predict with certainty every future event. The discovery of the planet Neptune by Johann Gottfried Galle in 1846, using the calculations of Urbain Jean Joseph Leverrier[1] from the perturbation of the orbit of Uranus by an unknown planet according to the predictions of gravitation theory, seemed to justify La Place's statement. After he completed his theory of electromagnetism, putting that subject on as solid footing as Newton's mechanics, James Clerk Maxwell felt justified in saying that all the basic laws of physics were now known; it remained only to refine one's measurements within the known framework of natural law.

This is not to say that Maxwell or anyone else in the 1870's believed that every question had been answered. They certainly knew better than that, but they did believe that the remaining questions could be answered within the existing framework of classical physics. So they went to work to refine their measurements — and the whole system fell apart.

The Electromagnetic Theory

Maxwell's electromagnetic theory said that any oscillating charged body would produce electromagnetic waves, and that such waves were to be expected to occur with all wavelengths ranging from very short, through those sensed as visible light, to very long, so that light constitutes only a tiny central range out of the whole electromagnetic spectrum. Within the last three decades of the nineteenth century both very long (radio) waves and very short (X-ray) waves had indeed been produced. All these waves were found

---

[1] John Couch Adams in England had made the calculations earlier, but could not persuade the Astronomer Royal to look for the planet.

to travel through space at the same remarkably high velocity, which came to be designated by the letter c. Because a wave's speed may be measured by measuring the length of a wave and counting how many waves pass a given point in a unit time, we can write this relationship as

$$c = \nu\lambda$$

where c = velocity, ν = frequency (waves or vibrations per second), λ = length of each wave.

Michelson's previously mentioned determination of the value of c to a very high degree of accuracy set the stage for a bold attempt at measuring something that could never have been measured before: the absolute velocity of the earth through space. Michelson, along with Edward W. Morley, made the attempt, spending a major portion of the 1880's on the problem.

The Michelson-Morley Experiment

The Michelson-Morley experiment was based on a premise inherent in classical mechanics: the existence of an absolute frame of reference, a coordinate system built around a point in space that is absolutely still and fixed, with reference to nothing but itself. Time, as well as space, had an absolute quality, flowing always in one direction, evenly, so that one might paraphrase Gertrude Stein and say "a second is a second is a second." Whether or not the mysterious ether was needed to carry light waves in space, it remained in scientific thought in the role of the absolute frame of reference, in comparison to which the absolute motions of the bodies of the universe could be determined. Ether, sometimes called "luminiferous ether," was supposed to permeate all the universe, and to be at rest in the absolute sense except where disturbed by the motion of some body through it. Because ether was perfectly elastic, any deformation of it would be temporary, and it would spring immediately back to its original form when the disturbing body had passed. This temporary disturbance in the ether, producing an ether wind, was to be utilized in measuring the earth's absolute velocity in space.

To see how such a measurement might give the desired information, consider this situation: Suppose I am bicycling at a brisk rate through absolutely still air. My coat is blown back and the papers in the baskets flutter as in a brisk breeze. Clearly, though I may perceive the air as still when I am standing still, my motion through it will produce a wind that, relative to me, is blowing in the direction opposite to my motion. Analogously, the earth, moving rapidly along its orbital path through the ether, should create an ether wind whose velocity would in fact be the velocity of the earth through the ether in the other direction.

Because the earth's orbital velocity with respect to the sun is on the order of 60,000 mph, it is apparent that if the effects of an ether wind of similar magnitude were to be measured it would have to be done in terms of its effect on some other very high velocity indeed, which is where the speed of light came in.

The reasoning that led to the design of the experiment was based on an analogy to the different effects of a current on a boat depending on whether the boat were being paddled upstream, downstream, or across (Fig. 6.1). Let us see what effects would be expected. Suppose the oarsman can row v_r miles per hour through the water in any direction. If he goes upstream, that is, against the current of velocity v_c, his net rate of progress upstream as measured with respect to the bank of the river, will be only $v_r - v_c$. Coming back downstream, however, the current increases his speed to $v_r + v_c$.

If instead, our oarsman wants to go straight across from one bank to the opposite one, he must always

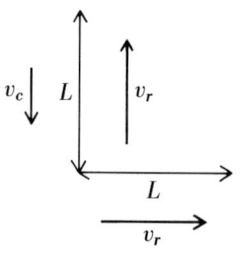

Figure 6.1 Velocity vectors and distances for oarsmen rowing parallel and perpendicular to current.

row upstream a little and let the current carry him back down, as Figure 6.2 shows. His actual velocity v with respect to the river bank can be shown to be $\sqrt{(v_r^2 - v_c^2)}$ by the Pythagorean theorem of plane geometry.

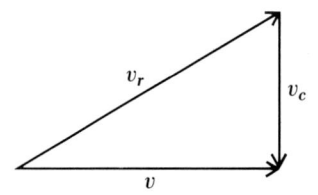

Figure 6.2 Addition of velocity vectors for cross-current trip.

If we mark out equal distances across the current and upstream-downstream for our oarsman, assuming v_r and v_c are always constant, a little messy but elementary algebra[2] will produce the result that the upstream-downstream round trip will always take longer than the cross-current round trip, no matter what the numerical values of v_r and v_c.

Michelson and Morley proposed to use light, traveling at the enormous velocity $c = 186{,}000$ miles per second, as their "oarsman," traveling in one case parallel to and in the other case perpendicular to the presumed ether wind.

We have seen that it is quite difficult to measure c,

---

[2] See Mathematical Appendix for details.

The Michelson-Morley Experiment

and we would expect it to be even more difficult to measure small differences in c due to the ether wind. But we can get around this difficulty by using the wave properties of light. Suppose we take a single beam of coherent light—waves all in phase with each other—and split it in two, so that half the beam goes in the direction parallel and the other half perpendicular to the ether wind. Now the beam that is slowed the most will arrive late and out of phase when it is reunited with the other beam, and the resulting interference pattern will display this fact, and indeed, allow the calculation of the difference, from which the ether wind's velocity can be calculated. The apparatus designed to produce this effect is the Michelson interferometer (Fig. 6.3).

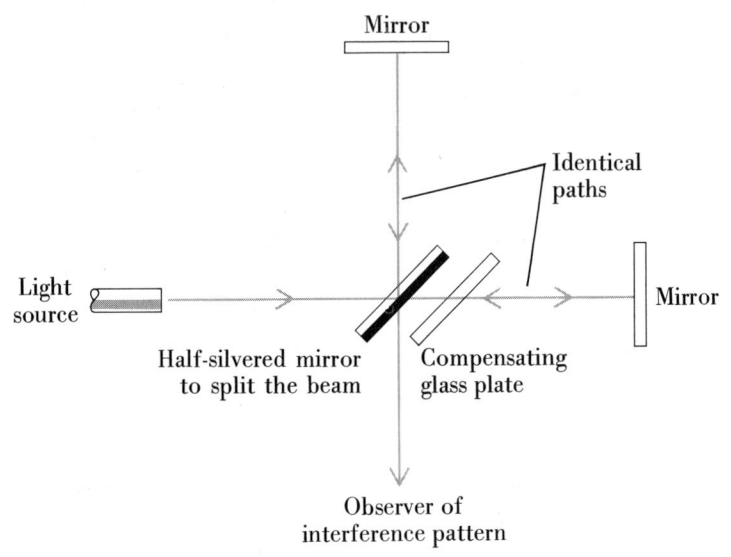

Figure 6.3 Michelson interferometer.

The experiment was beautifully simple. Shine the light along the two paths and observe the change in the interference pattern—the shifting of light and dark "fringes" as the arms of the interferometer are rotated with respect to the ether wind's direction—and calculate the velocity of the ether wind, which is really the absolute velocity of the earth with respect to the stationary ether. The only trouble was, it didn't work.

Michelson and Morley did their experiment repeatedly. Perchance the earth just happened to be going in a direction that compensated for the ether wind effect. Turn the apparatus around; do it at another season of the year; still failure to detect *any* shifting of fringes at all. Assume an ether drag, a layer of ether at rest with respect to the earth. That won't do: if that were so, the "fixed stars" would not appear fixed because their light would be refracted differently at the fixed ether—moving ether interface at different times of the year. Make a really wild assumption (this was from FitzGerald, an Irish physicist): assume the arm of the interferometer parallel to the ether wind is compressed just enough by the wind's pressure to compensate for the slowing down. Easy enough to check that out: make the interferometer arms unequal so the effects couldn't cancel each other out. Result: still nothing. To all appearance, if against all reason, the velocity of light is not affected by the ether wind. In fact, there is no evidence that the ether wind exists. So ended the hope of determining the earth's absolute velocity, or any absolute velocity. The absolute frame of reference can't be found.

The Michelson-Morley experiment is perhaps the scientific world's most famous failure. For two decades no one was able to say, reasonably, why it failed. The answer did not seem to lie within the framework of classical physics.

About the same time, after nearly a century of spectacular success in accounting for all kinds of optical phenomena, the wave picture of light was being called into question. First was the annoying problem of blackbody radiation. The properties of "blackbodies" had been studied thoroughly in the mid-nineteenth century. It had been established, experimentally by Stefan, theoretically by Boltzmann, that any body that absorbs energy will also radiate energy, in an amount proportional to the fourth power of its absolute temperature. This is summarized for ideal blackbodies—bodies that absorb and then radiate all the energy incident upon them—in the Stefan-Boltzmann Law:

$$R = \sigma T^4$$

where R is radiation in energy units per unit area per second; σ is the Stefan-Boltzmann constant; and T is absolute temperature. The puzzling thing about blackbodies was the distribution of wavelengths of the energy they radiated.

The best approximation to an ideal blackbody is a furnace with a pinhole opening and a rough, soot-coated cavity. When the furnace is heated—which is to say, when it absorbs energy—it begins to radiate energy, which may be visible light or longer or shorter waves, depending on the temperature of the furnace. This is the same phenomenon as the iron poker held in a hot fire becoming hot but not glowing, then red hot, yellow hot, and finally white hot. By using a prism or grating to disperse the radiation coming from a blackbody and allowing the separated colors to fall on a detector, such as a strip of film, the intensity of each color (wavelength) of radiation could be determined at each temperature. A graph of intensity vs. wavelength at several temperatures, $T_1 < T_2 < T_3 \ldots$, etc., looks like Figure 6.4.

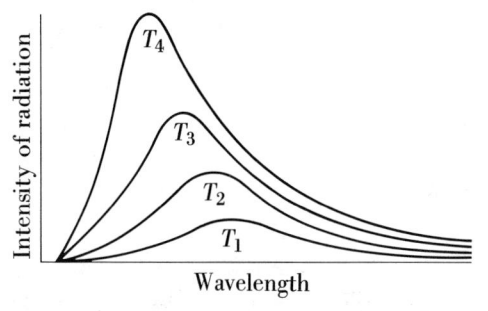

Figure 6.4 Variation of intensity with wavelength for a blackbody radiator at four temperatures, $T_1 < T_2 < T_3 < T_4$.

Intensity is clearly related to the amount of energy radiated. The area enclosed between an experimental curve and the *abscissa*, or horizontal axis, is a measure of the total energy being radiated at that temperature. As one would expect, the higher the temperature of the blackbody, the greater the total amount of energy radiated. It is more difficult to explain the shift of the maximum of the curve toward shorter wavelengths as the temperature increases. To answer this problem we must ask about the source of the energy radiated. What does heating the furnace do that subsequently causes it to radiate energy? The answer, as usual, goes back to our fundamental view of matter. Heat, the kinetic energy of molecules, when absorbed, results in the increased motions of the molecules in the furnace. They will vibrate more rapidly in their crystal lattices if they are molecules of a solid. In gases and liquids, if not always in solids, the atoms will vibrate, rotate, bend, twist with respect to each other, depending on their structure. Radiation occurs as a motion slows down, giving up its energy.

Physicists feel that they understand a process only when they can write a mathematical expression that will reproduce the experimental curve. That is, they like to have an equation for the lines on the graph that shows the relationship among the variables involved. In the blackbody curves, the variables are wavelength (λ), intensity (I), and absolute temperature (T). Several leading physicists of the late nineteenth century addressed themselves to the blackbody problem. The Austrian, Wien, came up with a fair approximation, but it deviated from the experimental curve noticeably at the long wavelength (infrared) end. In England, Lord Rayleigh and Sir James Jeans developed an equation that worked beautifully from the long wavelength end up to the maximum, but predicted that the intensity would continue to increase as wavelength decreased, infinitely. This glaring lack of agreement with observation at the short wavelength end was known as the "ultraviolet catastrophe." These efforts were based on the well-established principle that energy would be equally partitioned among all the vibrators present, and that the light absorbed to excite the oscillators would simply be that light which had previously been radiated, in a sort of eternal cycle at thermal equilibrium, independent of the nature of the vibrators (whose nature was, at that time, not very well known). It led to the untenable conclusion that the cavity would accumulate an infinite amount of energy, most of it at very short wavelengths. This simply did not happen. It didn't take a graph to say that blackbodies do not give off mainly X rays. At the time, X rays were in fact unknown.

Planck's Constant

The solution to the problem of the blackbody spectrum was found by Max Planck. It seemed so ridiculous even to him that he waited some five years to publish it, and it was laughed at, even then, but it turned out to be right. Planck's method was simply to write down a mathematical function that related the variables in a way that reproduced the curve, but to do this, he had to make a radically new assumption about energy.

Because heat is the kinetic energy of minute particles, that energy can usually be expressed as a simple function of temperature, and for a system in thermal equilibrium, i.e., at constant temperature, no other quantity is needed to specify energy. At least, that was the classical view. Implicit in this view was the assumption that any oscillator could absorb and emit energy in continuously varying amounts. However, the equations derived from these assumptions did not fit the experimental data.

The assumption that brought success in fitting the

curve was this: energy can be radiated only in certain discrete amounts by oscillators, rather than continuously. This is equivalent to saying that radiant energy, long described adequately as waves, is particulate. It is grainy, discontinuous, and, according to Planck, a function of the frequency of the associated wave.

So ended, although not all at once, the great wave-particle debate about the nature of light, and it ended in a draw. Radiant energy must be described both ways: it can be observed to travel as a wave, but its energy is determined by its frequency, or, as Planck put it,

$$E = h\nu.$$

The proportionality constant h is called Planck's constant. The packet of energy represented by $h\nu$ is called a quantum. Waves may have any frequency, but each frequency represents a discrete quantum of energy.

To make an analogy with a commonplace item, we may say that energy is like toothpaste. At your friendly neighborhood drug store you can buy toothpaste in the guest, personal, small, medium, large, family, or giant economy size, but not in any quantity in between. Toothpaste comes only in discrete packets; it is quantized. So is energy.

The Photoelectric Effect

About a decade later, in 1905, a young Swiss patent clerk published a paper that was to win him a Nobel prize in which he showed how the photoelectric effect could be explained by Planck's quantum theory of energy. However, the young man's name is known around the world today because of another pair of papers published the same year, on a new theory of his own—the Special Theory of Relativity, by Albert Einstein.

The photoelectric effect works like this: if a very

clean surface of an active metal is enclosed in an evacuated glass tube and light falls on the surface, electrons may or may not be emitted from the metal. If they are, they can be attracted to a positively charged electrode at the other end of the tube, and if the tube is built into an electric circuit a "photocurrent," electric current produced by light, will flow. The puzzling aspect of photoelectricity lies in the "may or may not" quality. Electrons may be shot out of potassium metal if violet light falls on it; electrons of low energy may be knocked out by blue light; whereas green, yellow, orange, and red light will be totally ineffective at producing a photocurrent, no matter how intense the light of these longer wavelengths might be. It is rather as if pounding surf composed of long waves could not budge pebbles that were easily dislodged by the tiniest ripple of short wavelength. Actually, it's a little hard to see why any wave should send electrons flying out of a metal. If you're playing marbles, you aim your shooter at the marble you want to knock out of the circle; you don't just wave at it. Furthermore, your shooter, which is glass like the other marbles, is more effective in knocking marbles out of the ring than the same size ball of styrofoam would be. On the other hand, a marble-sized lead sphere would really send the marbles scattering if it were used as the shooter.

This looked to Einstein like a quantized situation. Suppose light quanta hit the metal. If the quantum were energetic enough, it would knock out one electron at high speed. A lot of such energetic quanta would knock out a lot of electrons, one electron per quantum. On the other hand, a quantum of low energy (low frequency, long wavelength) would simply not be able to dislodge an electron, and no matter how many of these low energy quanta struck the metal, no photocurrent would flow. No electrons will be emitted by the metal until it receives energy in large enough packets to overcome the forces holding the electron in place. If larger quanta strike the metal, the energy left over from separating the electron from the metal will appear as the kinetic energy of the electron. Einstein wrote it simply in his photoelectric equation:

$$\tfrac{1}{2}mv^2 = h\nu - h\nu_0,$$

where $\tfrac{1}{2}mv^2$ = kinetic energy of emitted electron; $h\nu$ = size of quantum striking the metal; and $h\nu_0$ = minimum size of quantum required to dislodge electron. The photoelectric effect wasn't mysterious at all in terms of the absorption of radiant energy in quanta, just as assuming quantization of emission solved the blackbody mystery.

The principal mystery remaining was the quantum idea itself.

Radiation Theories

The last decade of the nineteenth century was as full of discovery in the laboratories of the western world as in the minds of its theoreticians. In 1895, Wilhelm Roentgen, in Germany, reported the discovery of the shortest electromagnetic rays yet observed. They were emitted by metals bombarded by electrons. Because their origin was unknown they were called X rays. Roentgen's report of his discovery, about Christmastime, 1895, spurred various other scientists to more radiation study. One, Henri Becquerel, pursued his study of fluorescence in certain rare minerals such as pitchblende, an ore of the heaviest known metal, a scarce and useless element called *uranium*. Fluorescence is the emission of visible radiation by substances that have previously absorbed radiation; for example, ultraviolet radiation from sunlight. Paris weather tends to rain in January, and Becquerel's researches were interrupted by a period of three or four sunless days. When the rain set in, he carefully wrapped the photographic plates he used to detect the fluorescent radiation in layers of black paper and put them in his desk drawer until the sun should shine again. He put the piece of pitchblende in the same drawer.

When the rain finally ended, Becquerel got out his plates and got ready to resume his fluorescence studies. Being a careful man, he developed one of his neatly wrapped plates just to make sure nothing had happened to them as they lay in the drawer—not that anything could! His surprise at finding the plate exposed—splotched where radiation had affected it— is understandable. It didn't take long, however, to establish that pitchblende was emitting radiation at a great rate, whether it had been exposed to sunlight or not, in fact, regardless of any external conditions or treatment. Becquerel called it *radioactive*.

Over the next few years the radioactive rays were analyzed and found to be three different types, which were designated α, β, and γ. The first two were shown to be very energetic particles, alphas (α) being nuclei of helium atoms, carrying a double positive charge, and betas (β) being simply those familiar unit negative charges, electrons. Gammas (γ) proved to be electromagnetic rays even shorter (and consequently more energetic) than the recently discovered X rays. Becquerel traced the origin of these rays to the uranium in the ore, although the slag left after uranium was extracted was still radioactive. The story of Marie Curie's extraction of minute amounts of still more strongly radioactive elements, which she named *polonium* and *radium*, from tons of pitchblende is familiar, marked by the remarkable care and perseverance of Madame Curie and her husband, and his tragic accidental death shortly after their great success.

The experimental facts of radioactivity called into question another area in which a simple picture had been taken for granted. It had been barely a century since Dalton had proposed the atomic theory of matter. The atom was considered ultimate, indivisible. If so, where were all these rays and particles coming from? By 1900 it was realized that virtually nothing was actually known about the ultimate nature of matter, and a number of physicists, led by J. J. Thomson, the discoverer of electrons, and Ernest Rutherford, newly arrived in England from New Zealand, set out to find out about it.

So dawned the twentieth century. In three decades scientists had gone from a state of confidence in a world of order and natural law, operating according to clear principles of cause and effect, to a state of uncertainty due to mysterious anomalies and dualisms, and questions no one yet quite knew how to ask. The first three decades of the new century found a brilliant array of scientific talent addressed to the problem of asking the right questions and coming up with even stranger and more unexpected answers. These decades are well-named by George Gamow as *The Thirty Years that Shook Physics*.[3]

SUMMARY STATEMENTS

1. Attempts to refine measurements of the physical world in the late nineteenth century led to anomalies that classical physics could not explain.

2. The Michelson-Morley experiment failed to determine the absolute velocity of the earth in space by failing to find an "ether wind" that affected the velocity of light.

3. After attempts to derive a theoretical equation that fitted the intensity-wavelength curves for blackbody radiation in classical terms failed, Max Planck proposed the new idea that radiant energy can be emitted only in quanta, or packets of definite size, such that their energy $E = h\nu$.

4. Other discoveries of the 1880's and 90's that did not fit with classical theories were radioactivity, the photoelectric effect, and X-ray emission from metals bombarded by electrons.

---

[3] This is the title of an excellent book on the period by George Gamow, published in paper by Doubleday (Science Study Series).

REVIEW QUESTIONS

1. Mention some great successes of classical physics.
2. Interpret this symbolic statement: $c = \nu\lambda$.
3. What was "luminiferous ether" supposed to be and why was it considered a theoretical necessity?
4. Discuss the Michelson-Morley experiment with respect to (1) purpose; (2) procedure; (3) results.
5. What is a blackbody? What is its importance in the development of modern physics?
6. What does the symbolic statement $E = h\nu$ mean?
7. What is the photoelectric effect? How did its discovery contribute to the acceptance of the quantum theory of energy?
8. What is radioactivity and how was it discovered?
9. Characterize each of these great names in modern physics in a few words, naming their principal discovery:

Max Planck	Albert Einstein
Marie Curie	J. J. Thomson
Wilhelm Roentgen	Henri Becquerel
A. A. Michelson	Edward Morley

THOUGHT QUESTIONS

1. Characterize the universe as science had delineated it in 1875. What were the predominant concepts in that description?
2. Why may the Michelson-Morley experiment be justifiably called "physics' most important failure"?
3. Why was Planck's suggestion that energy is discontinuous so shocking and so badly received by physicists? Do you think radically new ideas are more easily accepted in science than in other realms?
4. Write a brief summary of events and discoveries of the last two decades of the nineteenth century that portended drastic changes for physics.
5. The great names in physics at the turn of the century included Max Planck (German), A. A. Michelson (American), Henri Becquerel (French), Wilhelm Roentgen (German), Henrik Lorentz (Dutch), Marie Curie (Polish), Albert Einstein (German-Swiss), and J. J. Thomson (English). Each in some sense contributed to the work of many of the others; all were to some degree collaborators. Do you think science today offers an arena for international

cooperation that could promote peace? Can you cite any recent examples to support your opinion?

READING LIST

Gamow, George, *The Thirty Years That Shook Physics*, Doubleday (Anchor), Garden City, N.Y., 1966. An anecdotal account of the great ideas of quantum theory and the men and women who had them, by one of them.

Baker, Adolph, *Modern Physics and Antiphysics*, Addison-Wesley, Reading, Mass., 1970. A delightful book that should make even poets like physics, which is its intention.

Scientific American reprints (published by W. H. Freeman, San Francisco):

#205 Darrow, Karl K., "The Quantum Theory" (August 1948).

#321 Shankland, R. S., "The Michelson-Morley Experiment" (November 1964).

7

Relativity

The impact of science on society often produces contradictory effects. Solid citizens who would be horrified at the suggestion that they give up their antibiotics, automobiles, air conditioners, nylons, and permanent press suits long for the good old days when one could cling to eternal verities handed down from one's grandfather, when one could view the world in absolute terms and be sure one was right because no one ever questioned one's position. Of course, it's quite likely the "good old days" never existed. But responsibility for the modern practice of questioning everything, of refusing authoritarian answers, of insisting that merit might be found in a variety of points of view — which leads to the aforementioned nostalgia — can be laid at the door of science, just as fairly as can the credit for the technological advances that make modern life more comfortable for more people than ever before possible.

The scientific theory that has had perhaps the greatest impact on the thinking of society at large is the theory of relativity. It was popularly supposed to be incomprehensible to all but a handful of geniuses. It described the world in terms of a strange fourth dimension that no one could visualize. It also could be claimed as "scientific proof" that all things are relative, whether in the realm of physics or in the realm of moral values. The public might not comprehend this fourth dimension, but repercussions from a doctrine of moral relativism could be felt by everyone, and it did indeed seem that the old safeguards of Western society were breaking down.

It perhaps illustrates the lack of communication between scientists and others that the theory of relativity has very little in common with these suppositions. Let us then look into its origins, its postulates, and its predictions in an attempt to see it as it is.

Transformation

The principle of relativity dates back to Galileo and Newton, and was enunciated by Newton like this: "The motions of bodies included in a given space are the same among themselves, whether that space is at rest or moves uniformly forward in a straight line." It means simply that the laws of motion are the same whether one is at rest or in uniform motion, so that no one could tell, for instance, from internal evidence whether the railway car in which he was performing the experiment was parked at the station or moving uniformly along the track. Of course if an observer on a uniformly moving train and another observer standing on the platform were each to specify the location of some landmark, each in terms of distance from himself, their specifications would differ, except at the moment the train was passing the platform, but there would be no problem in comparing the two distances if the train's velocity and the elapsed time since the two observers passed each other were known (Fig. 7.1). Because the train will travel distance vt in the time t, the difference in the distance, x', from the moving observer to the landmark and the distance

Transformation

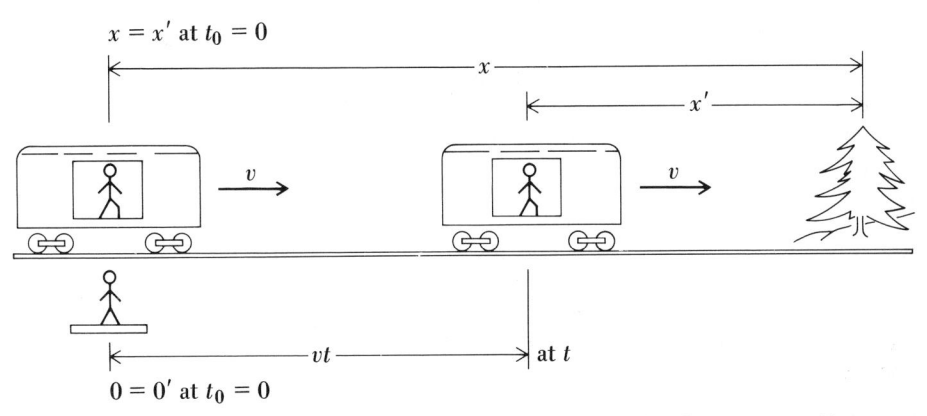

[figure: two reference frames, one stationary on a platform and one on a moving train, with labels $x = x'$ at $t_0 = 0$, x, x', v, vt at t, and $0 = 0'$ at $t_0 = 0$]

Figure 7.1 Comparison of two frames of reference, one "stationary" and the other "moving."

of the stationary observer to the same landmark, x, is vt. The two can compare their observations by using the relation

$$x = x' + vt \quad \text{or} \quad x' = x - vt.$$

Such an equation, which permits data taken in one frame of reference or coordinate system, i.e., the platform, to be translated into another coordinate system, i.e., the moving train, is called a *transformation equation*. The equation we have just used is called the *Galilean-Newtonian transformation*. It rests on an assumption that classical physics took for granted: the passage of time is the same for everyone. Our two observers may have different x coordinates, but their t will be the same. All they need to compare x's by the Galilean-Newtonian transformation is a reasonably accurate clock apiece that can be set together so that both can determine their x coordinate simultaneously.

Under the Galilean-Newtonian transformation, observers moving relative to each other would observe different x coordinates for objects (always assuming for convenience a Cartesian coordinate system, x and y axes perpendicular to each other and so oriented in

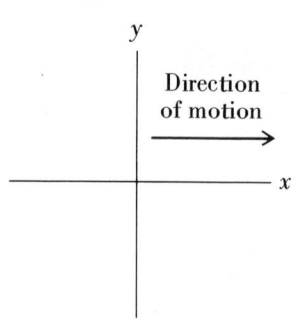

Figure 7.2 Cartesian coordinate system, showing the most convenient orientation of the velocity vector.

space that the moving observer is moving parallel to the x axis; that way the y coordinate is the same for both observers—see Fig. 7.2), and if they measured the velocity of something moving relative to both of them, say an airplane flying overhead, they would get different results, which they could compare by saying

$$v' = v + u$$

where v' is the velocity observed by the moving observer, v is the velocity observed by the stationary observer, and u is the velocity of the moving observer relative to the stationary one. Both observers would get the same results, however, if they measured the length of the airplane. And certainly there would be no difference in their measurement of its mass. So we say that although coordinates and velocity vary, the three fundamental measurable quantities, mass, length, and time, are *invariant* under the Galilean-Newtonian transformation. Two hundred years of classical mechanics bore this out.

Mechanics and Electromagnetic Problems

Maxwell's laws of electromagnetism quite clearly do not follow the Galilean-Newtonian principle of relativity. This is apparent from considering the fields around charges. If a charge is stationary it is surrounded by an electrostatic field, but if the charge is moving it has both an electric and a magnetic field. In Figure 7.3, suppose a test charge Q is located some distance R from a charged wire. An observer at rest with respect to this system will observe an electric force on the charge Q due to the field of the wire. However, an observer moving at some velocity v parallel to the wire will observe a magnetic force also acting on the charge Q. Not only would the two observers disagree on the magnitude of the force on Q, they would write their force laws (equations) in different forms, in violation of the relativity principle.

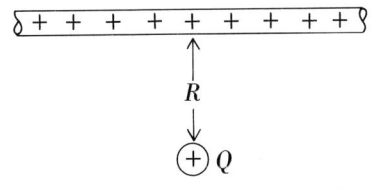

Figure 7.3 Test charge in the field of a charged wire.

Something was clearly wrong here. There were three alternatives as to where the error lay:

1. The relativity principle holds for mechanics but not for electrodynamics. Therefore a preferred reference system exists in which the laws of electromagnetism hold, and only in this system does light travel with velocity c.
2. The principle of relativity holds for all laws of physics, but the Maxwell equations of electromagnetism are in error.
3. The principle of relativity holds for all laws of physics, but Newton's laws of mechanics are in error.

The first alternative was unacceptable because it violates the symmetry commonly observed in nature, and (more practically) because the Michelson-Morley experiment effectively proved the velocity of light to be c in every frame of reference.

The second alternative was the preferred choice of physicists at the end of the nineteenth century. Maxwell's equations were still "new." They had been tested only twenty years or so, whereas Newton's laws had stood up under two centuries of testing. Nevertheless, the solution to the dilemma lay in making the third assumption: the laws of motion need correcting.

Albert Einstein deserves all the credit he gets for the theory of relativity. Nevertheless, it would be a mistake to think the idea came to him full-blown without any help from anyone else. One source was the French theoretical physicist Jules Henri Poincaré, who suggested that the third alternative was correct; another was the Dutch mathematician H. A. Lorentz, who had developed a new set of transformation equations in which the Maxwell equations could be made to follow the relativity principle. Einstein performed the synthesis by stating two postulates and proceeding to show how they could be universally applied if the Lorentz rather than the Galilean-Newtonian transformations were applied to mechanics. The two postulates of Einstein's special theory of relativity are:

1. *All the laws of physics hold equally well for all observers regardless of their relative uniform motion.*
2. *The velocity of light in free space is constant, regardless of the motion of the observer or the source.*

The first postulate is very restrictive. It is simply the relativity principle applied to unaccelerated systems. (Such systems are called *inertial frames of reference*.) Notice also the role of the observer. The relativity principle does not say that bodies will be described identically by observers moving relative to each other but that they will obey the same laws, and each observer's description of a given system will be equally correct—for his own frame of reference. The second postulate simply confirms the Michelson-Morley results and rules out the unsymmetric solution according to the first alternative.

The strange and wonderful description of the world that is associated with relativity appeared when the Lorentz transformations were applied to the laws of mechanics. Table 7.1 is a comparison of the relativistic with the classical results.

Table 7.1 Comparative Transformation Equations

Classical	Relativistic
Transformation equations	Transformation equations
$x' = x - vt$	$x' = x/\sqrt{1 - v^2/c^2}$
$t' = t$	$t' = \dfrac{t + \dfrac{vx}{c^2}}{\sqrt{1 - v^2/c^2}}$
Results[a]	
$v' = v + u$	$v' = \dfrac{v + u}{1 + \dfrac{uv}{c^2}}$
$l = l_0$	$l = l_0 \sqrt{1 - v^2/c^2}$
$m = m_0$	$m = \dfrac{m_0}{\sqrt{1 - v^2/c^2}}$
$t = t_0$	$\tau = \dfrac{\tau_0}{\sqrt{1 - v^2/c^2}}$
Length, mass, and time are invariant	

[a] The subscript $_0$ refers to the value observed by the observer at rest with respect to the system. τ represents a time interval, and t represents the time coordinate.

Classically, we have seen that length, mass, and duration of time do not vary from one coordinate system to another. In relativity, all these quantities change as their velocities with respect to the observer change. One's first reaction to this idea is that it is ridiculous! A thing *has* some length, some mass, as intrinsic, objective properties, we feel. But who is to specify what values these quantities have? The observer at rest will find a meter stick l_0 long, but another observer passing (or being passed by?) the meter stick will observe its length as only l. Who is right? Both are right, each in his own frame of reference. It is of very little use to speak of the objectively true length, for who is to measure it? Let length be hereafter operationally defined to be what any observer measures it to be, all measurements being equally valid if done with comparable equipment and care. FitzGerald's proposal about the shortening of the interferometer arm wasn't so ridiculous, after all. The relativistic change in length of an object moving relative to the observer is now called the Lorentz-FitzGerald contraction.

This conclusion may not seem so strange if we consider a case in which we accept this kind of thing as normal. Suppose a stewardess is serving coffee to the passengers in an aircraft. The passengers will tell you that her velocity along the aisle of the cabin is one meter per second (1 m/sec). An observer in a balloon sees the plane pass him at 300 m/sec, and so observes the stewardess to be passing at 301 m/sec. However, the man in the control tower on the ground knows a 10 m/sec wind is blowing both plane and balloon along, so that the stewardess is actually doing 311 m/sec past the tower observer. An interplanetary observer hovering somewhere beyond the earth's orbit in his saucer sees earth tearing along in its orbit at 30,000 m/sec and consequently computes the stewardess's velocity to be 30,311 m/sec. But the interplanetary observer is whirling along with the galaxy, as anyone in Andromeda could tell you. . . .

Grant, then, the possibility that the only valid definition of a quantity is in terms of its measurement, and that there is no one frame of reference in which the measurement is *right*. Mass, length, and duration of time vary with the velocity of the observer relative to the thing being measured in such a way that length contracts in the direction parallel to the motion (that's what FitzGerald said; when Lorentz's mathematics confirmed it the phenomenon came to be known as the Lorentz-FitzGerald contraction), and its mass increases, and moving clocks run slow.

Proving Mass Increase

The phenomenon of mass increase is the most easily proved of the strange predictions of relativity. Once big machines to accelerate subatomic particles were built, the relativistic mass effect showed up and had to be taken into account. In a cyclotron, charged particles are accelerated by giving them a "kick" with a strong electric field each time they reach the gap between the hollow "dees" in the magnetic field in which their spiral path lies (Fig. 7.4). When light particles, such as electrons, were accelerated, after a few turns of their spiral path they began falling behind schedule. When the regularly pulsed electric field came on at the gap, no particles had arrived to be ac-

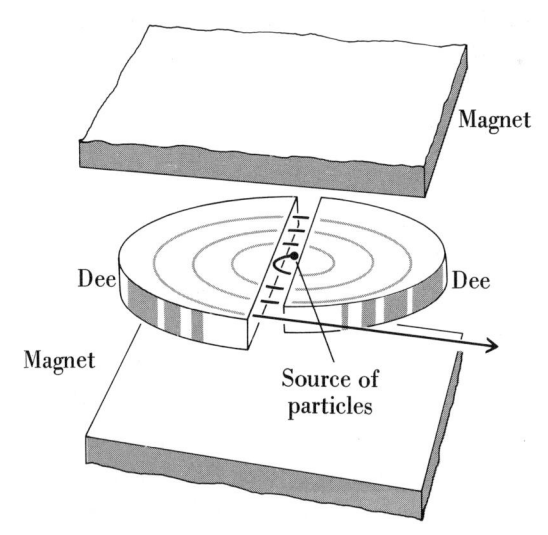

Figure 7.4 Cyclotron for accelerating charged particles to very high velocities.

celerated. They were gaining mass as well as velocity as energy was added to them, and as their speed approached c, their mass increased faster than their velocity. Where did the increase in mass come from? The only possible source was the energy added, according to the most famous result of Einstein's work, the equivalency of mass and energy as stated in the famous little equation $E = mc^2$. This idea of equivalency came about as the result of another somewhat arbitrary choice. It was clear from the transformation equations that mass and energy could not be assumed to be conserved independently. Einstein decided that momentum, which includes both mass and velocity (and so is related to kinetic energy), was the essentially conserved quantity. If you rigorously conserve momentum, you must define mass as a function of velocity, and this will lead to the idea of mass as energy in a highly concentrated form, so concentrated, in fact, that one gram of mass converted to energy would produce more than 2×10^{10} of those Calories dieters are always counting. (That's about 50,000,000 hamburgers with all the trimmings.) After all, the conversion factor c^2, the square of the velocity of light is a very large number ($9 \times 10^6 m^2/sec^2$).

Time Dilation

The strangest of all relativistic phenomena to comprehend is time dilation, but it, too, finds experimental support in the behavior of high energy subatomic particles. One type of particle, the μ meson, or *muon*, is unstable and tends to fly apart into three smaller particles. No one can say when any particular muon will decay, but for a large collection of them, one can calculate with a probability that is almost certainty how long it will take them to disintegrate on the average. This time period is called the muon's *mean life*. Muons can be produced in high-energy accelerators in the laboratory; they also occur in the upper atmosphere as a result of bombardment of atoms by high-energy cosmic rays. Atmospheric muons have velocities up to $0.9c$, and they are observed to have longer mean-lives than their slower laboratory counterparts—an observation best explained by time dilation.

One ramification of time dilation still being argued is the so-called twin paradox. It goes like this. Suppose one member of a set of twins departs at the age of 20 on a space flight to Arcturus, a distance of 40 light years, or some 240 trillion miles. (A light year is the distance light can travel in one year, about 6 trillion miles. Because 6,000,000,000,000 is awkward to write, scientists usually use the notation 6×10^{12}.) The rocket, once launched, will travel at a constant speed of $0.9c$, so that it will take him 40.4 years to get there. Then he will turn around and come back. The question is, how old will the brothers be when they are reunited? The stay-at-home brother, who has measured off 80.8 years on his calendar since the traveler's departure, is now 100.8 years old. To his aged amazement, the traveler springs spryly from his spacecraft, a young fellow of only 31.4 years! The time dilation factor $\sqrt{(1 - v^2/c^2)}$ induces the stay-at-home to conclude that the traveler's clock was running only $\sqrt{(1 - (0.9c)^2/c^2)} = 0.141$ times as fast as his own, accounting for the traveler having aged only 11.4 years during his trip. Because this trip has not yet been made, we have no "Exhibit A: Returned Spaceman" to introduce as concrete evidence. So let us consider the arguments pro and con.

The first objection is based on the symmetry argument. Surely the spaceman can draw exactly the same conclusion about his earthbound brother as the earthman drew about him. In the spaceman's frame of reference, the earth was hurtling away at $0.9c$, and its clocks were all running slow. It is the earthman who appears young to the aging traveler. Now obviously, once they are reunited in the same frame of reference, you can't have it both ways. Each can't be

69.4 years younger than the other. By symmetry, it would appear that the traveler must age all the faster on the homeward trip to make up for the slowing down on the outward bound half of the journey, and that he must be as old as his earthbound brother when he gets back.

The answer usually given to this objection is that the traveler's situation is inherently asymmetric because he changes reference frames when he turns around at Arcturus, while the earthling does not. The argument is that because the traveler must slow down, stop, and then speed up in the opposite direction, he is not in an inertial frame of reference during this time, thus breaking the symmetry.

A more general defense of the predicted difference in aging is based on the argument that not only moving mechanical clocks run slow, but so do moving "atomic clocks." That is the fundamental atomic, molecular, and electronic processes within an organism are slowed down, with a subsequent retardation of the aging process.

At the present time, the time-dilation effect is hardly the key to perpetual youth. We are a long way from producing velocities near c for anything larger than subatomic particles. As to the reality of the twin paradox, we shall have to say that the experimental evidence is not yet in.

In examining the mass, length, and time relationships in relativity we find that the velocity of light has a role to play in defining another of those natural limits the universe seems to have set. Note the anomalies that would occur if a material body acquired a velocity $v = c$:

$$m = \frac{m_0}{\sqrt{(1 - c^2/c^2)}} = \infty$$

$$l = l_0 \sqrt{(1 - c^2/c^2)} = 0$$

$$\tau = \frac{\tau_0}{\sqrt{(1 - c^2/c^2)}} = \infty$$

That is, mass becomes infinitely large, the dimension of length in the direction of motion vanishes, and time stands still. These conditions are absurd, if for no other reason than that it would take an infinite amount of energy to reach them. So we accept c as an upper limit on the velocities of material bodies, a limit that may be approached arbitrarily closely, but never reached. On the other hand, if we are to consider light quanta as particles (usually called *photons*) we must specify that their rest mass is zero, because they do travel at velocity c.

The Fourth Dimension

There is nothing really mysterious about the fourth dimension. It is simply that because time is not the same for all observers, it must be specified as a coordinate in each observer's frame of reference. It is customary in relativity to describe events in terms of four coordinates, three in space: x, y, and z, and one, t, in time. In classical physics a space interval (length) was invariant to the Galilean-Newtonian transformation; in relativity, the interval between two sets of coordinates x_1, y_1, z_1, t_1, and x_2, y_2, z_2, t_2 is invariant to the Lorentz transformation. That t must be treated as a coordinate that changes from one reference frame to another means that we cannot assume simultaneity in the classical sense. Two observers may set their clocks together, but when they separate, acquiring some velocity relative to each other, their clocks will run at different rates, so there will be no way for them to be sure that they are ever making simultaneous observations.

The Special Theory of Relativity provided the needed corrections to classical mechanics to make it work rigorously in inertial (unaccelerated) frames of reference regardless of velocity, and it is easily seen that, if v is very much smaller than c in the relativity equations, the "fudge factor" $1/\sqrt{(1 - v^2/c^2)}$ reduces to unity, and we are back to the equations of classical mechanics. But the universe is made up of masses and electric charges all exerting forces on each other, and forces produce accelerations, and there are no actual inertial frames of reference. In 1915 Einstein published a sequel to special relativity, called the *General Theory of Relativity*, which dealt with accelerated systems.

General Relativity

Einstein himself once wrote that the simpler and more comprehensive an idea in science, the more complex the mathematics required to describe it. The mathematics of General Relativity is quite difficult, but the ideas are rather simply stated. We begin with the relativity principle, but we remove the restriction that the relative motion must be uniform. Then we consider the most prevalent unbalanced force in the universe, gravitation, and say that the acceleration it causes should not be intrinsically different from any other acceleration. Classically, it had been mildly surprising that *inertial mass* (mass defined in terms of the laws of motion) is identical with *gravitational mass* (mass defined in terms of the law of gravitation). In General Relativity this result is an equivalence principle, which states that no experiments may be done entirely within a frame of reference that will indicate whether that frame of reference is in a gravitational field or is being accelerated by some non-gravitational force.

Einstein illustrated the Principle of Equivalence with thought experiments (Fig. 7.5). Suppose an observer is enclosed in an elevator-like room. He can see nothing of his external environment. Now suppose he releases a ball from his hand one meter above the floor of the elevator. If the ball accelerates down-

General Relativity

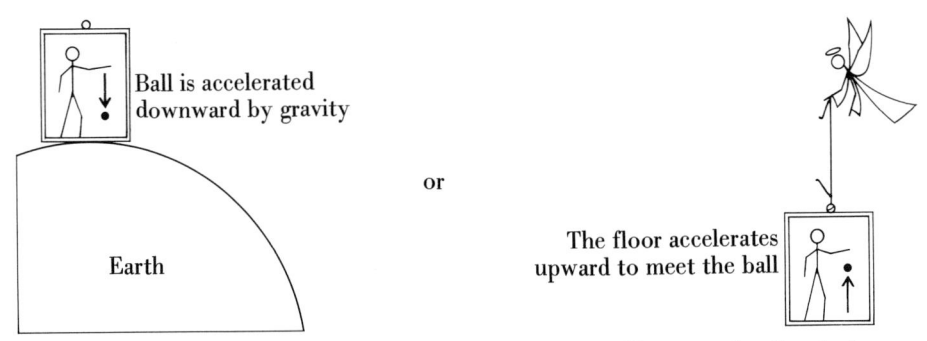

Ball is accelerated downward by gravity

Earth

or

The floor accelerates upward to meet the ball

Figure 7.5 A thought experiment to illustrate the Equivalence Principle.

ward, he is quite likely to conclude that his elevator is a gravitation field, such as on the earth. However, an external observer may be able to see that his elevator is not on earth or in any gravitational field at all, but is being pulled upward through empty space with an acceleration equal to gravitational acceleration on earth. (You can have an angel yanking on the rope if you like; in thought experiments no pains are spared and all arrangements that are thinkable are possible.) To the external observer, the ball doesn't fall at all; the floor is pulled up to meet it.

However, the man inside the elevator will know whether or not he is in an accelerated system. All he has to do is drop something and watch it fall. A third principle that is sometimes incorporated into General Relativity was first enunciated by Ernst Mach. It suggests that although acceleration can be defined absolutely, in the sense that one can tell from internal evidence that a system is accelerated, acceleration nevertheless can only occur because of the motions of other (external) bodies. Thus Mach's principle relates the motion of any body to that of all others in the universe in an ultimate sense of interdependence and interaction.

The world of General Relativity is one of space and time defined only by their content. Space becomes curved in the vicinity of masses, so gravitation is simply a geometrical effect in curved space. This can be illustrated in this way: suppose we have a rubber membrane stretched on a frame (Fig. 7.6). If we drop a handful of BBs on the membrane, these lightweight spheres will scatter in straight lines from the point of landing and presently come to rest on the essentially flat surface. However, if we place a bowling ball in the middle of the membrane and then drop the BBs, the small spheres roll down the curved membrane into the depression around the bowling ball. It does not occur to us to say the shots were attracted to the bowling ball. We simply say the BBs rolled downhill, following the curvature of the membrane surface, although we are aware that the curvature is caused by the presence of the ball.

BB's on flat membrane

BB's rolling down curved membrane

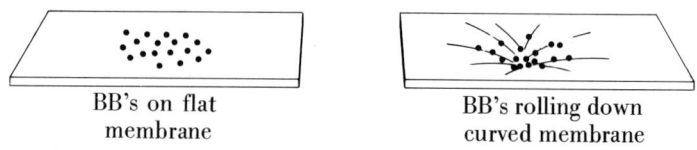

Figure 7.6 How the shape of space affects the state of motion of bodies.

The relativistic universe is one in which no frame of reference is preferred, but in which the same natural law applies everywhere. Every particle influences every other particle, just by being there. Some of the old absolutes are gone—but they were never really there. In their place are myriad approaches, but the same unifying principles for all.

SUMMARY STATEMENTS

1. To resolve the discrepancies between the laws of mechanics and the laws of electromagnetism, Albert Einstein chose to correct the laws of mechanics.

2. The theoretical basis for the corrections consisted of two postulates, known as the postulates of the Special Theory of Relativity.

    a. Relativity Principle: The laws of nature hold

equally for all observers, regardless of their relative uniform motion.

   b. The velocity of light in space is invariant (unchanging) regardless of the motion of either the light source or the observer.

3. To make the laws of mechanics as well as the laws of electromagnetism conform to the postulates, it was necessary to use the Lorentz transformation of coordinates rather than the Galilean-Newtonian transformation.

4. Under the Lorentz transformation, mass, length, and time all depend on velocity, whereas all three are invariant under the classical transformation.

5. The velocity of light is the upper limit that can be approached by the velocity of material particles. As the velocity of a body increases, its mass increases, its length parallel to its motion diminishes, and time slows down for the moving body.

6. If momentum is to be conserved, mass must vary with velocity. A consequence of this is that mass and energy can be shown to be equivalent, and interconvertible according to the equation $E = mc^2$.

7. The General Theory of Relativity extends the relativity principle to accelerated systems.

8. General Relativity postulates a principle of equivalence that says that acceleration due to gravity cannot be distinguished from any other acceleration by evidence internal to the accelerated system.

9. Internal evidence, such as the behavior of objects released from the observer's hand, can demonstrate that a system is accelerated.

10. According to Mach's principle, acceleration of a body is always due to motions of other bodies, making everything in the universe dependent on its interaction with everything else.

11. A corollary to Mach's principle is that the shape and properties of spacetime itself depend on its contents and their configuration at any given moment.

REVIEW QUESTIONS

1. What do the following terms mean?

   invariant
   coordinate system
   coordinates (of a point)
   transformation equation
   simultaneity

2. State: (1) the classical relativity principle (Galilean-Newtonian); (2) the principle of relativity according to Einstein. What is the real difference between them?

3. Interpret the symbolic statement $E = mc^2$.

4. What does the General Theory of Relativity do that the Special Theory (STR) does not?

5. What quantities are (1) invariant classically, but vary according to the observer, according to STR? (2) vary with the observer, both in classical physics and STR? (3) are invariant even in STR?

6. How did the Special Theory of Relativity change the Conservation Laws?

THOUGHT QUESTIONS

1. How do the philosophical bases of classical and relativistic physics differ?

2. Why do you think relativity seemed, and still seems, so strange to many people?

3. What does it mean to say that time is a "fourth dimension"?

4. How has relativity theory influenced thinking and attitudes beyond the realm of physics?

5. Does the physical theory of relativity justify moral relativism?

6. What does "frame of reference" mean in science? in common life?

7. Cite evidence for such predictions of relativity as time dilation, mass increase, and mass-energy equivalence.

READING LIST

Barnett, Lincoln, *The Universe and Dr. Einstein,* Bantam Books, New York, 1968. One of the best popular accounts of the theory of relativity.

Bondi, Hermann, *Relativity and Common Sense,* Doubleday (Anchor), Garden City, N.Y., 1962. An excellent introduction to relativity theory by a noted scientist who removes the mystery and replaces it with a logical development of the way things are.

Coleman, James A., *Relativity for the Layman,* New American Library (Signet), New York, 1969. Another good popular account by a scientist, with emphasis on experimental evidence.

Gamow, George, *One, Two, Three . . . Infinity,* Bantam Books, New York, 1967. A fascinating and thought-provoking book about the universe as seen by modern physics.

———, *Mr. Tompkins in Paperback*, Cambridge University Press, Cambridge, 1967. A combination of two earlier fantasies, *Mr. Tompkins in Wonderland* and *Mr. Tompkins Meets the Atom*, and a delightful way to get acquainted with relativity and quantum theory.

Scientific American reprints (W. H. Freeman, San Francisco):

#209 Einstein, Albert, "On the Generalized Theory" (April 1950).

#273 Gamow, George, "Gravity" (March 1961).

#291 Bronowski, J., "The Clock Paradox" (February 1963).

#309 Gardner, Martin, "Can Time Go Backward?" (January 1967).

8

Quantum Theory: The Fundamental Structure of Energy and Matter

Once physicists got over the shock of thinking of radiant energy as having a particle aspect they found the concept tremendously useful. Planck's assumption of the quantization of energy solved the mysteries of the blackbody spectrum and the photoelectric effect. While these questions were being answered, a host of new problems about the fundamental structure of matter were clamoring for attention.

By 1900, it was fairly clear that the atom was by no means indivisible. It had to be the source of electrons, which were discovered in the 1880's by J. J. Thomson and others and shown to be the ultimate electric charge by Robert A. Millikan in 1909. It had also to be the source of α, β, and γ rays from radioactive substances. The question was whether these particles had any permanent identity within atoms, and if so, how they were arranged.

J. J. Thomson proposed a model of the atom that promptly became known as the *watermelon model* (Fig. 8.1). He pictured negative electrons scattered through a positively charged pulp—because atoms have no net charge—like seeds in a watermelon.

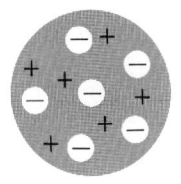

Figure 8.1   J. J. Thomson's atomic model.

The Structure of Matter

In 1911, Ernest Rutherford and his coworkers performed a classic experiment that established what we still believe to be the fundamentally correct picture of the atom (Fig. 8.2). He directed a beam of alpha particles from radium onto a sheet of very thin gold foil and observed the angle through which the alphas were scattered by the gold atoms by noting where scintillations were produced on a zinc sulfide screen by the alpha particles striking it. Three things happened to the alphas when they were fired at the foil: Most alphas went right on through, striking the zinc sulfide screen as if the foil weren't there. Some alphas came straight back from the foil. (Rutherford expressed his surprise at this by saying it was rather like firing a cannon ball at a sheet of tissue paper and having it come back and hit you.) Some alphas were deflected at various angles, as though repelled by the foil.

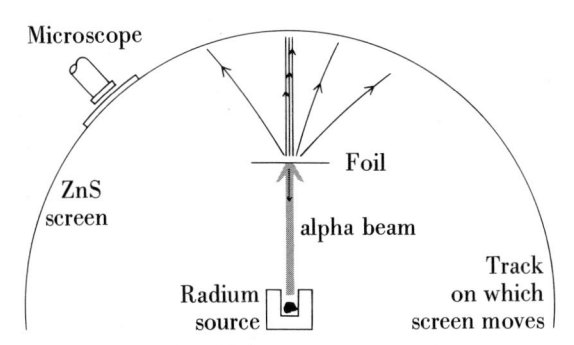

Figure 8.2   The alpha-scattering experiment.

From these observations, Rutherford concluded that (1) atoms are mostly empty space; (2) they have a dense nucleus which accounts for only a tiny fraction of the atomic volume; and (3) the nucleus is positively charged, and the electrons surround it like a cloud (Fig. 8.3).

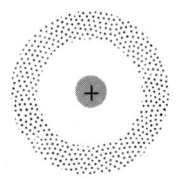

Figure 8.3 Rutherford's atomic model.

Such a picture of the atom immediately raised a difficulty. According to Maxwell's electromagnetic theory, a moving charge radiates energy and subsequently slows down. Electrons revolving around the nucleus should spiral in toward the center, radiating energy all the while, until they fall into the nucleus. And it shouldn't take very long for them to get there. According to classical electrodynamics, all electrons should have fallen into their respective nuclei within a very few minutes of the creation of atoms in the beginning of the universe. Because Rutherford still found electrons outside the nucleus some few billion years after the "beginning" the classical prediction must be incorrect.

At this point Niels Bohr, a student of Rutherford's, used the quantum theory to solve the dilemma. Because electrons do not in fact spiral into the nucleus, there must exist certain "radiationless states," orbits in which electrons could circle the nucleus ad infinitum without radiating any energy. There could be any number of these orbits, but for an electron to move from one orbit to another, energy would have to be either absorbed or radiated (Fig. 8.4). At about this point his friend H. M. Hansen suggested to Bohr that he had the solution to a long-standing mystery: line spectra.

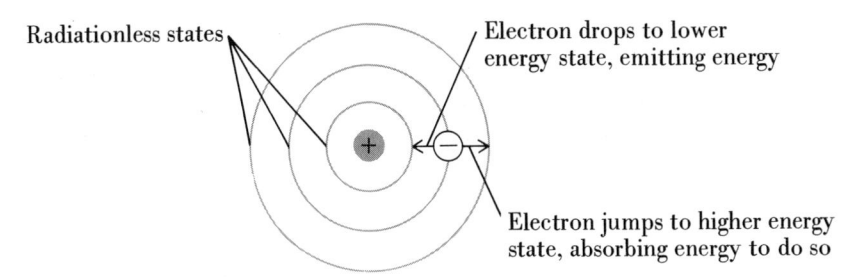

Figure 8.4 Bohr's atomic model.

Line Spectra

When certain elements or their compounds are heated in a gas flame, they impart brilliant colors to the flame, each element with its own characteristic color — red for calcium, blue-green for copper, yellow for sodium, etc. If the light from the flame is dispersed by a prism or grating, the resulting spectrum will consist of discrete, sometimes widely separated colored lines, the strongest being those corresponding to the flame color, and each of these line spectra is at least as distinctive for its element as a fingerprint is for a person. Line spectra had been known for a long time — since Bunsen and Kirchhoff observed them in the 1850's — and had been used routinely for analyzing and identifying substances. New elements were found by their spectra, including helium, whose spectrum was observed in sunlight some years before the rare gas was ever found on earth. But the origin of line spectra was completely unknown. Classical radiation theory certainly had no accounting for it.

The element whose spectrum was best known was hydrogen, the lightest weight, and presumably simplest atom of all, consisting of a single electron orbiting a nucleus carrying one positive charge. Two series of spectral lines were known, one in the ultraviolet range (Lyman series) and the other in the visible range (Balmer series). Equations could be written expressing the regular variation of the frequencies of lines in a series, but they were strictly empirical, that is, based on observation, not on theory. The empirical formula for the line frequencies looked like this:

$$\nu = cR\left(\frac{1}{n_1^2} - \frac{1}{n_2^2}\right)$$

where c is the velocity of light and R is called the Rydberg constant. For the Lyman lines, $n_1 = 1$ and n_2 could have any integral value greater than 1. For the Balmer series, $n_1 = 2$, and n_2 could be any integer greater than 2.

Suppose, suggested Bohr, the n's are the numbers counting successive radiationless states. Lyman lines result from electrons falling back to the lowest state ($n = 1$) from some higher state, emitting energy of the calculated frequency in the process. Balmer lines result from electrons falling into level 2 from higher levels. Series for which $n_1 = 3$, $n_1 = 4$, etc., should exist, and indeed, the next three were found, by Paschen, Brackett, and Pfund, respectively. In each case, the emitted frequency corresponded to a quantum of energy $E = h\nu$, according to Planck's quantum equation.

Bohr proceeded to calculate the specifications of orbits to which the observed spectra would correspond. Noticing that Planck's constant, h, has the same dimensions as angular momentum (mass × velocity × radius of rotation) he guessed that the angular momentum of the electron in orbit might be quantized, that is, the electron's angular momentum could have only certain discrete values, so that it could occupy only the corresponding orbits. On this basis, he calculated the radii of the radiationless states, or "allowed" orbits, and the energy differences between them. The energy differences exactly coincided with the observed frequencies of the hydrogen spectrum. The Bohr Atom became the model for understanding the structure of all the atoms of the known chemical elements.

For all its powerful mathematical tools, physics to this day remains stumped on what appears to be a rather simple mechanical problem, the "three body problem." It is impossible to solve rigorously (without any approximations) the problem of the positions and energies of three mutually interacting bodies, for example, the nucleus and two electrons that comprise a helium atom. "Many body" problems are completely insoluble except by approximation. This means that we cannot make exact calculations of orbits and energies for any atoms except hydrogen, but we can make satisfactory approximations that show that the Bohr model represents the structure of all atoms in general outline, if not in detail.

By 1912, when Bohr published his calculations on the hydrogen atom, virtually all ninety naturally occurring chemical elements, those simplest building blocks of all matter, were known. They could be arranged in order of their atomic weights, from hydrogen up to uranium, and it was assumed that all were made up of the same fundamental units, as Prout had suggested in 1815 when he proposed that all atoms were merely multiples of hydrogen. Both the chemical and physical properties of elements were known to vary periodically with atomic weight, and Mendeleev's construction of a periodic table of the elements in 1869 had led to the discovery of new elements to fill the slots that were clearly vacant when the periodic arrangement was first set up (see Chapter 9). Atomic weights, however, did not vary regularly. Mendeleev switched the position of several pairs of elements when their periodic properties were clearly out of line with their weights. Some yet unknown quantity or quality of atoms appeared to be more fundamental in describing them than their atomic weights.

Atomic Numbers

The simplest kind of spectra of elements is their X-ray emission when they are bombarded by high speed electrons. Each element emits several series of lines (discrete frequencies). In 1913, a young English physicist, H. G. J. Moseley, arranged the frequencies of the first lines of the highest energy series, the so-called K_α lines, in ascending order and found that he had arranged the elements in exactly the same order as they fell in the periodic table. He even was obliged to leave the same gaps for elements not yet discovered. Here was, apparently, an indication of that fundamental quantity that identified an element. The number of an element on Moseley's list was called the *atomic number*, and came to be identified with the number of positive charges in the nucleus of the atom. We now identify elements by their atomic numbers, for atoms of the same element may differ in weight, for reasons that came to be known somewhat later on. The concept of atomic number is a fitting memorial to a promising career cut short by war. Moseley was killed at Gallipoli in 1915.

By the 1920's a great many lines of evidence were converging to form a coherent picture of the fundamental structure of matter. From X-ray spectra came the concept of atomic number, which must equal the number of electrons, as well as positive charges in the nucleus, since atoms have no net charge. The appearance of several series of X-ray lines was taken to indicate the arrangement of electrons in several shells or orbits around the nucleus. The existence of a group of chemically inert gaseous elements, one at the end of each period of the periodic chart, coupled with the known chemical combining power, or *valence*, of the other elements, suggested (1) that electrons were involved in chemical bonding, and (2) that electron orbits had just so many spaces and that chemical stability was related to the orbits being filled. Further study of optical spectra—the sets of colored lines by which elements had long been recognized—led to characterizing electrons and their orbits by a set of four numbers, called *quantum numbers,* because they described the positions and energies of electrons in their allowed orbits according to rules (called *selection rules*) determined by the requirements of quantization of energy and angular momentum.

In 1925, Wolfgang Pauli, one of the great theoretical physicists of the era, proposed a rule by which the electron structure of any atom could be built up. The Pauli Exclusion Principle states that:

No two electrons in any atom may be described by identical sets of quantum numbers.

These quantum numbers, designated n, l, m, and s, respectively specify the orbital level (Bohr orbit), the shape of the orbital within that level, its orientation in space, and finally the direction of the electron's spin (clockwise or counterclockwise). Figure 8.5 is a rather crude pictorial representation of the results of this principle (see also Plates 8 and 9).

These pictures should not be taken too seriously. It is not to be supposed that electrons are little hard spheres following exactly delineated paths. Several theoretical physicists proposed mathematical models that have proved satisfactory in giving theoretical physicists an idea of what an atom is like, but they are not much help to anyone else. Nevertheless, they are tremendously important to fundamental understanding of matter, especially, in a practical way, to chemistry (see Chapter 9).

Wave Nature of Matter

The mystery of why electron orbits should be quantized at all, and why they should have the particular radii they are calculated to have, still remained. A solution was proposed in 1925 by Louis de Broglie, in

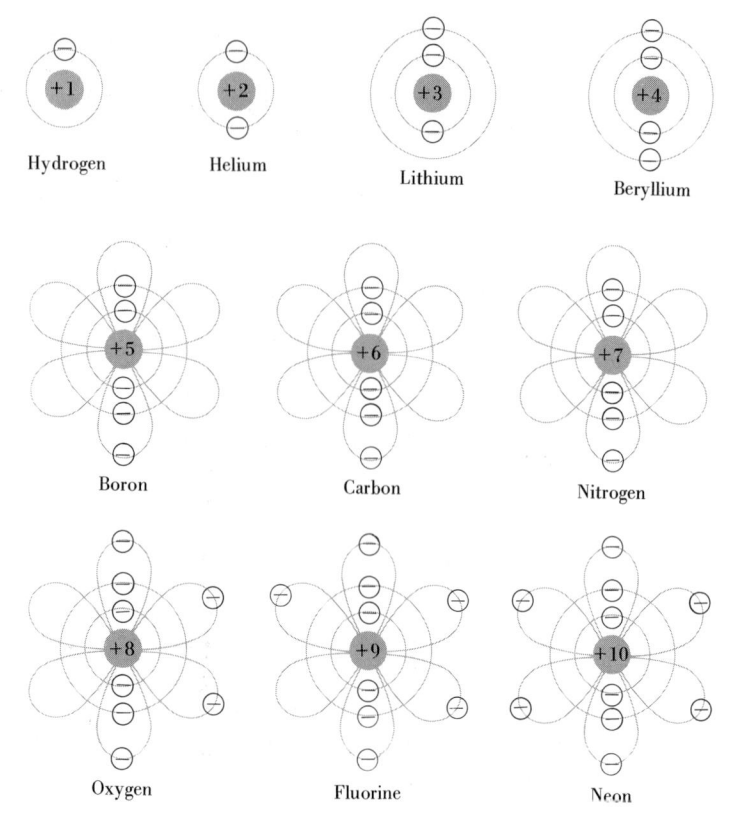

Figure 8.5 Atomic models based on Bohr's Theory and the Pauli Principle.

the only doctoral dissertation so far to win its author a Nobel Prize in physics. Simply put, de Broglie's reasoning went something like this: if energy has a quantum or particle nature, may not matter have a wave nature? (See Fig. 8.6) He proceeded to consider orbits as spaces into which standing waves must fit, and found that the circumference of a Bohr orbit would exactly contain a wave of such length that $\lambda = h/mv$, where mv is the linear momentum of the electron in the orbit, and h is the ever-recurring Planck's constant.

It seemed a little strange to think of an electron as a wave, but there is a simple way to find out if a thing is a wave or not, and within the year the evidence was

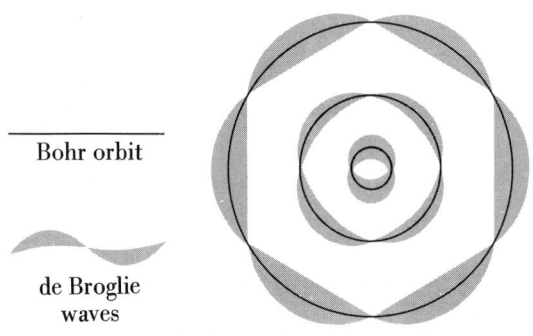

Figure 8.6 De Broglie waves in Bohr orbits.

in. Two physicists at the Bell Telephone Laboratories, Davisson and Germer, shot a beam of electrons at a single crystal of nickel that acted as a diffraction grating with very narrow slits (i.e., the spaces between the atoms in the crystal lattice) and observed a diffraction pattern. Waves are diffracted; particles are not. If electrons are diffracted, they are behaving like waves.

It began to appear that it is as hard to say what matter is as to say what light is. Both display the dual characteristics of wave and particle behavior. The mathematical models of the atom were built upon this dualism. The so-called matrix mechanics of Heisenberg (published in 1925) provided a way of calculating the behavior of atomic systems, using such observable quantities as the frequencies and intensities of spectral lines, and did not concern itself with visualizable atomic models. Although mathematically elegant, it has had less appeal even to physicists than the wave mechanics approach of Erwin Schrödinger, who set out to describe mathematically the mechanics of particles such as electrons in terms of waves.

These mathematical models were successful in permitting the calculation of orbits and energies quite precisely for the hydrogen atom and, in the hands of theoretical chemists who were willing to use approximation methods quite horrifying to the theoretical physicists, they were successful in predicting the structures of quite complicated molecules. Still, there was the uncomfortable feeling that particles and waves were not quite the same thing, and the question remained: what is it that waves? Max Born was the first to propose a physical interpretation of Schrödinger's wave equation that was adopted by Bohr, Heisenberg, Pauli, and many other leading physicists. In his interpretation of *quantum mechanics,* as this whole approach to matter and energy is now called, Born suggested that the waves in question represent probabilities, and that one can specify only where an electron is most likely to be, not where it is, in any absolute sense. If, after the theory of relativity had been at work for twenty years, any belief in a deterministic universe remained, quantum mechanics finished it off.

The Uncertainty Principle

The idea of indeterminism in atomic systems was further developed by Werner Heisenberg, and summarized in the Uncertainty Principle:

It is impossible to determine simultaneously with any arbitrary degree of exactness both the velocity and the position of a particle.

Thus an electron cannot exhibit both wave and particle properties simultaneously; it will always be observed as one or the other. The basic reasoning behind this principle is that one cannot measure a system without changing it. This always is true, but our measuring devices make so little difference to the operation of macroscopic systems that we are unaware of any effect. When our system is one electron or one atom our measuring technique will make a great deal of difference. Heisenberg illustrated this with a thought experiment.

Suppose we can control a beam of electrons so

accurately that we can admit one electron at a time into the viewing chamber, which is equipped with a microscope so sensitive that it can respond to one photon. Now in order for the one photon to provide us with any information about the electron, the two must collide. Because photons are subject to diffraction, there will be some smearing out of the photon which will introduce some uncertainty in the location of the electron, but we can minimize the diffraction effect by using photons of shorter wavelength. However, if we decrease the wavelength, we increase the frequency, and consequently the energy, and so the effect of the collision on the electron is increased, with the result that we now know less than before about the momentum of the electron after the collision (Fig. 8.7).

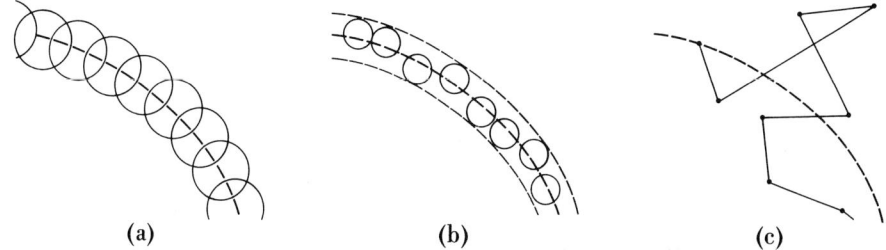

(a) (b) (c)

Figure 8.7 Thought experiment illustrating the impossibility of making simultaneous exact measurements of position and momentum. (a) A long wave does not disturb the electron much, but it does not pinpoint its location, either. (b) A wave of intermediate energy locates the electron within narrower limits, but changes its momentum noticeably. (c) A very short wavelength pinpoints the electron's location rather definitely, but changes its momentum by a large factor. Regardless of the wavelength of light used, the product of the uncertainty in the electron's position, Δx, and the uncertainty in its momentum, Δp, will never be less than $\Delta x \cdot \Delta p = h/2\pi$.

Diffraction Patterns

Another thought experiment may give further insight about quantum mechanics. Suppose that this time we aim our beam of electrons at a barrier containing a

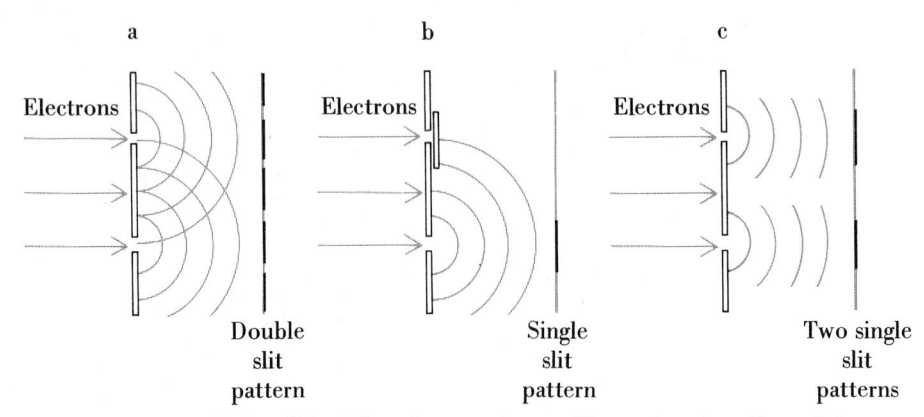

Figure 8.8 Thought experiment illustrating the effect of observation on the results of the experiment. (a) Electrons passing "single file" through either of two slits produce a double slit diffraction pattern. (b) Electrons passing "single file" through one slit produce a single slit diffraction pattern. (c) Electrons "monitored" to see which slit they go through produce two single slit patterns.

double slit (Fig. 8.8). Again, we will allow only one electron to reach the slit at a time, and we may safely suppose that the one electron must go through one slit and not the other. We shall record where it goes after passing through a slit by having it strike a film, where it will leave a spot. After a sufficiently long time to allow a great many electrons to reach the film, we shall develop it and observe a two-slit diffraction pattern. Even though each electron went through only one slit, the presence of the other slit somehow influenced it. If next we cover one slit, but otherwise duplicate our previous procedure, we will observe a single slit diffraction pattern. Our observation technique has influenced the behavior of the electrons. Suppose on the third run we leave both slits open but observe them alternately. The result? Two single slit patterns. The electrons seem to "know" where we are going to look for them! How we make our observation influences what we observe. We appear to have reached another of those limits set by nature.

The world view based on the Uncertainty Principle admits that we cannot get at any "objective reality" because all our attempts at observation change what we observe. It is pointless to say "light is a wave" or "light is a particle" or "an electron is a wave" or "an electron is a particle." These statements have no meaning. The only meaningful statements we can make are such as this: "an electron exhibits wave behavior on passing through a system of slits." We are reduced to operational definitions entirely. This philosophic view of the results of quantum mechanics is known as the *Copenhagen Interpretation,* and is held by most theoretical physicists today, following Born, Bohr, and Heisenberg. However, Einstein and Schrödinger were never reconciled to the idea of indeterminacy, and in 1969, de Broglie, then past 80, chided physicists for making the Uncertainty Principle a dogma that limited both their experimental and theoretical efforts.

SUMMARY STATEMENTS

1. Planck's quantum theory proved to be widely applicable in explaining the behavior and structure of atoms and molecules.

2. Rutherford's alpha-scattering experiment demonstrated that atoms have positive nuclei surrounded by electrons.

3. Bohr's application of quantization to electron orbits explained why electrons do not spiral into the nucleus, but can exist forever in certain "radiationless states."

4. The distinctive line spectra of elements corroborated the idea that energy must be supplied in discrete amounts to promote an electron to an orbit of higher energy, and that the same size quantum will be emitted when the electron drops back to its original orbit.

5. Corrections and refinements of the Bohr theory of the atom took several mathematical forms, developed by Heisenberg, Schrödinger, and Dirac.

6. Schrödinger's wave mechanics became the most popular representation of orbits and energy levels, and Max Born gave the now widely accepted interpretation that

the wave functions represent probabilities of finding the electron in certain configurations.

7. The original idea of attributing wave properties to electrons came from Louis de Broglie, who suggested that Bohr's allowed orbits were simply those corresponding to wavelengths such that $\lambda = h/mv$.

8. Pauli formulated the exclusion principle that formed the basis for understanding the structure of atoms more complex than hydrogen. It said that no two electrons in an atom may have identical sets of the four quantum numbers required to describe them.

9. Not only did the quantum mechanical picture of atoms become hazier, but also Heisenberg enunciated a principle that suggested that an absolute limit on the accuracy of measurement exists. This Uncertainty Principle points out that any attempt to measure a system changes it, and that the changes are very large compared to the measurements in atomic systems. The product of the uncertainties in measurement of two related variables like momentum and position will never be less than $h/2\pi$.

REVIEW QUESTIONS

1. What did each of these men contribute to our understanding of the nature of matter?

    Democritus (see Chap. 1)   Niels Bohr
    John Dalton (see Chap. 3)   Wolfgang Pauli
    J. J. Thomson   H. G. J. Moseley
    Ernest Rutherford

2. Describe the procedure and results of Rutherford's alpha-scattering experiment and the interpretation of the results.

3. What, in the view of classical physics, made Rutherford's picture of the atom unlikely if not impossible? What way out of the dilemma did Bohr suggest?

4. How did line spectra, observed for sixty years by Bohr's time, substantiate his theory?

5. What is a quantum number?

6. State the Exclusion Principle; describe how it relates to the electronic structure of atoms.

7. What does the term "quantum mechanics" mean?

8. How did de Broglie attempt to explain Bohr's quan-

tized orbits? Is there evidence for the correctness of his idea? If so, what is it?

9. The following names are closely associated with the development of quantum mechanics. What was the contribution of each?

>Heisenberg
>Schrödinger
>Born

10. State the Uncertainty Principle. Why can you not observe an electron acting simultaneously as a particle and as a wave?

THOUGHT QUESTIONS

1. How has the wave-particle argument been resolved — or has it?
2. What is the experimental basis for the quantum theory?
3. Do you see any general effect on thinking and culture of quantum theory comparable to the widespread effect of relativity?
4. Are essential limitations set on scientific endeavor by the principle of uncertainty? Louis de Broglie recently cautioned physicists against making uncertainty into a dogma. What was his point?
5. Why wasn't the idea of observation interfering with the system, the basis of the uncertainty principle, discovered any sooner?
6. How do the philosophical bases of twentieth-century science differ from those of nineteenth-century science?
7. What is the overall effect on today's society and culture of the twentieth-century "revolution in physics"?

READING LIST

Hoffmann, Banesh, *The Strange Story of the Quantum*, Dover, New York, 1959. One of the best accounts of the development of quantum theory that requires almost no mathematics.

Andrade, E. N. daC., *Rutherford and the Nature of the Atom*, Doubleday (Anchor), Garden City, N.Y., 1964. History of the exciting developments of the first third of the twentieth century by a student of Rutherford.

Moore, Ruth E., *Niels Bohr*, Alfred A. Knopf, New York, 1961. A personal as well as scientific biography of one of the most winning personalities of modern physics (as well as one of the greatest thinkers).

Romer, Alfred, *The Restless Atom,* Doubleday (Anchor), Garden City, N.Y., 1960. A very readable history of the experimental and theoretical work leading to our present understanding of atomic structure.

Thomson, George P., *J. J. Thomson,* Doubleday (Anchor), Garden City, N.Y., 1966. The story of the discoverer of the electron and long-time head of the Cavendish Laboratory, by his son, also a Nobel laureate in physics.

Scientific American reprints (W. H. Freeman, San Francisco):

#212 Gamow, George, "The Principle of Uncertainty" (January 1958).

#241 Schrödinger, Erwin, "What Is Matter?" (September 1953).

#264 Gamow, George, "The Exclusion Principle" (July 1959).

See also the reading list for Chapter 6.

9

Chemistry: The Periodic System

In the Middle Ages they called it *alchemy,* and its practitioners had one goal: the transmutation of base metals into gold. They never succeeded, but they learned a great deal about many other substances, and as the Age of Reason superseded the ages of superstition, alchemy transmuted itself into the useful and practical science of chemistry. It was suggested previously that physicists and chemists have a somewhat different basic approach to the world: the physicist simplifies and abstracts until he has an idealized model he can describe with very precise mathematics, while the chemist must cope with the world as it is, and though his theories may use wild approximations that make the physicist tear his hair, in practice the chemist comes up with quite concrete accomplishments in the way of new forms of matter.

By the middle of the twentieth century, some half million different chemical compounds were known. In the next twenty years that number increased by a factor of six, the chemical literature having indexed about three million different "pure substances" by 1970. All are made from only ninety naturally occurring elements, although a disproportionate number are formed primarily of carbon, which is wonderfully versatile in its ability to form chemical combinations. It is clear that the first requirement for a chemist would be to have a prodigious memory, if he were to begin to keep the factual knowledge of his science straight, unless some system of organization exists to help him. Fortunately, there is a system.

The French aristocrat Antoine Lavoisier, who literally lost his head in the French Revolution, is often called *the father of chemistry* because of his role in stripping it of its alchemical trappings and putting it on a quantitative basis. Burning, which we take for granted as a simple chemical reaction, was not understood in the eighteenth century. Combustible substances were supposed to contain a mysterious fluid (they loved mysterious fluids in those days—remember caloric and electrical fluids?) called *phlogiston,* which was given off in the process of burning, leaving the "dephlogisticated" ash behind. When it was demonstrated that the ash of burned metals weighed more than the original metal, phlogiston had to assume the extremely peculiar property of negative mass. Lavoisier, in careful experiments involving the burning of metals in sealed vessels, showed that indeed, although the ash or *calx* weighed more than the original metal, the whole system remained at constant weight, and when the vessel was opened, air rushed into it to fill the partial vacuum resulting from the removal of some component of air by the metal in forming the calx. Further studies showed the removed component to be Joseph Priestley's recently discovered "dephlogisticated air," which Lavoisier renamed *oxygen.*

A large number of elements not previously recognized as such were identified in the last decades of the eighteenth century. Many of these were gases: oxygen, hydrogen, nitrogen, and chlorine. Other gases were also identified and shown not to be elements, but *combinations* of two or more other elements. A great many new metals were also discovered. In

Sweden Karl Wilhelm Scheele classified a group of oxygen-containing substances that had sour tastes called *acids*. (Scheele had an unfortunate habit of tasting all his discoveries, and because they included such highly toxic substances as hydrogen cyanide, hydrogen fluoride, and arsenious acid, he died at 44.)

Classifying Substances

By the middle of the nineteenth century a great many chemical substances (not mixtures) were known, and they were classified in various ways. We might list them as in Table 9.1. These are still useful categories, but they have little predictive value. A more useful organization would arrange the elements in such a way as to recognize their common properties and would allow newly discovered elements to be fitted in and their properties predicted. As more and more chemical facts accumulated, some elements were found to fit into "families." In 1829, the German chemist Döbereiner found several "triads," groups of three elements with similar or regularly varying properties, such as the "halogens" or salt-formers (chlorine, bromine, and iodine) and the "alkaline earths" (calcium, strontium, and barium). The atomic weights of the elements in the triads also showed a regular variation,

Table 9.1 Pure Substances

Elements	Compounds
Metals (mostly solids, a few liquids) all shiny, malleable, conductors of heat and electricity	Calxes (oxides) Acids (sour) Bases (bitter)
Nonmetals (may be solids, liquids, or gases) poor conductors, few physical properties in common	Salts (made by neutralization of acid by base) Organic compounds (supposed until 1848 to come only from living organisms)

which seemed significant. John A. R. Newlands, in 1864, arranged the known elements in order of increasing atomic weights and found the properties repeating after every eighth element. Newlands' "octaves" were the first suggestion of a periodicity of atomic properties.

Since the practices of weighing and measuring and doing quantitative experiments were introduced by Lavoisier, many quantitative laws of chemistry have been established. Two particularly important ones are the Law of Definite Proportions and the Law of Multiple Proportions:[1]

Law of Definite Proportions: When two or more elements combine to form a compound they do so in definite proportions by weight.

Law of Multiple Proportions: When two elements combine in two or more ways to form different compounds, the weights of the one that combine with a fixed weight of the other are in the ratio of small whole numbers.

These laws may be illustrated by considering the two common oxides of sulfur. One always contains 32 grams of sulfur to 32 grams of oxygen (definite proportions), and the other always has 32 grams of sulfur to 48 grams of oxygen (definite proportions). It is seen that the amounts of oxygen combining with 32 grams of sulfur, 48 and 32 grams, respectively, have a ratio of 3:2 (multiple proportions).

A further relationship among volumes was observed:

Law of Combining Volumes: when gaseous elements combine, definite volumes are involved, and these combining volumes have small whole number ratios.

[1] When substances are combined in any arbitrary proportions, without chemical reaction between them, the result is a *mixture*. Often a mixture may be separated by mechanical means: if iron filings and salt were mixed, the iron could easily be pulled out with a magnet. Sugar and salt would be harder to separate, but it could be done by taking advantage of the differences in their solubilities in water.

If the product formed was also gaseous, the ratio of its volume to that of the reactants was also expressible in small whole numbers. For example, one volume of hydrogen combines with one volume of chlorine to produce two volumes of hydrogen chloride, whereas two volumes of hydrogen combine with one of oxygen to produce two volumes of water vapor. This result strongly suggested the generalization known as Avogadro's Hypothesis:

Equal volumes of gases contain equal numbers of molecules.

That is, if one molecule of hydrogen combines with one molecule of chlorine to produce two molecules of hydrogen chloride, and one liter (~ 1000 cm^3) of hydrogen combines with one liter of chlorine to produce two liters of hydrogen chloride, Avogadro's hypothesis must be inferred. The volumes must of course be measured at the same pressure and temperature.

It was then apparent that elements, the simplest chemical substances, must, as Dalton had suggested, consist of some basic units, or *atoms*, and that compounds resulting from combination of elements must consist of ultimate particles (*molecules*) made of definite numbers and kinds of atoms, and that all molecules of a pure compound are essentially identical. On the basis of their combining weights and ratios, each element could be assigned a relative weight that could be designated as its *atomic weight*. Hydrogen, having the smallest combining weight, was assigned an atomic weight of unity, and all the others followed in order. Remember Prout's Hypothesis, that all elements were built of hydrogen units.

The existence of the Law of Multiple Proportions indicates that many elements have more than one combining weight. Therefore, although a given kind of atom must have one definite atomic weight, it is evident that it may have more than one "combining power." For example one unit (atomic weight) of sulfur could combine with either two or three atomic weights of oxygen. The combining power of an atom came to be called its *valence*. Hydrogen had only one combining weight; its combining power was taken to be one. Oxygen would then have a valence of two while chlorine has a valence of one, on the basis of the observation of combining volumes mentioned previously.

Periodic System of the Elements

By 1869, the stage was set for the introduction of a comprehensive scheme of classification, and the Russian chemist Dmitri Mendeleev put it all together. He combined the ideas of ordering by atomic weight, variation in valence, and family resemblances and presented the relationships in an easily readable form, the Periodic Table of the Elements (Fig. 9.1).

The Periodic Table met with instant acceptance. Perhaps its most spectacular success was predicting

Series	I	II	III	IV	V	VI	VII	VIII
1	H(1)							
2	Li(7)	Be(9.4)	B(11)	C(12)	N(14)	O(16)	F(19)	
3	Na(23)	Mg(24)	Al(27.3)	Si(28)	P(31)	S(32)	Cl(35.5)	
4	K(39)	Ca(40)	—(44)	Ti(48)	V(51)	Cr(52)	Mn(55)	Fe(56),Co(59), Ni(59),Cu(63)
5	[Cu(63)]	Zn(65)	—(68)	—(72)	As(75)	Se(78)	Br(80)	
6	Rb(85)	Sr(87)	?Yt(88)	Zr(90)	Nb(94)	Mo(96)	—(100)	Ru(104),Rh(104), Pd(106),Ag(108)
7	[Ag(108)]	Cd(112)	In(113)	Sn(118)	Sb(122)	Te(125)	I(127)	
8	Cs(133)	Ba(137)	?Di(138)	?Ce(140)				
9								
10			?Er(178)	?La(180)	Ta(182)	W(184)		Os(195),Ir(197), Pt(198),Au(199)
11	[Au(199)]	Hg(200)	Tl(204)	Pb(207)	Bi(208)			
12				Th(231)		U(240)		

Figure 9.1 The Periodic Table of the Elements according to Mendeleev, 1872.

the properties of the missing element between gallium and arsenic. Because it lay below silicon, which is a solid nonmetal with a valence of four and a density of 2.33 grams/cm^3 (density is the mass per unit volume of a substance), Mendeleev made the predictions in Table 9.2 about the missing element, which he called *ekasilicon*. A few years later, germanium was discovered.

Table 9.2   Mendeleev's Predictions

	Predicted Eka-silicon (Es)	Found Germanium (Ge)
Atomic weight	~ 72	72.6
Source and method of isolation	EsO$_2$ or K$_2$EsF$_4$ by sodium reduction	K$_2$GeF$_4$ by sodium reduction
Physical description	dark grey metal high melting point specific gravity 5.5	dark grey metal melting point 958°C specific gravity 5.35
Oxide	EsO$_2$, high melting, sp gr 4.7	GeO$_2$, mp 1100°C sp gr 4.7
Sulfide	EsS$_2$, insoluble in water; soluble in ammonium sulfide	GeS$_2$, insoluble in water; readily soluble in ammonium sulfide

By mid-twentieth century, no gaps remained, atomic number rather than atomic weight had been established as the ordering principle for arranging elements in the table, and the reason for periodicity of properties was understood in terms of electron structure of atoms. More elements and many more compounds are known today than in Mendeleev's time. The periodic chart has become more useful as the years pass, for the working chemist need only remember trends, rather than thousands of specific facts of chemistry. Why bother to memorize that francium fluoride is the most stable compound when a mere glance at the table tells you so—if you know the system.

Let us look at the chart first from the point of view of the electron structure of the atoms in Figure 9.2. We see first that the number of elements in any row (Period) is determined by the number of electrons that can fit into the orbit whose n value equals the number of the period. Thus, Period 1 has only 2 elements; Period 2 has 8, and Period 3 has 8.[2] Each period ends with a "noble gas," one of the rather rare gaseous elements that forms no ordinary chemical compounds at all. (Several compounds have been made of noble gases with fluorine and with water but they are rather "extraordinary"!) Each of these elements has its outer orbit filled by 2 or 8 electrons. This stable or inert gas structure seems to be the one all other atoms tend toward in their chemical behavior. Thus, elements in the first column (Group I) have one electron in an outer level, beyond a completed octet (or doublet, in the case of lithium), so that to say these elements have a valence of 1 is to say they would like to be rid of the outer electron. Because elements in Group VII have space for one more electron to complete their octets, it should not be surprising if these elements readily absorbed the electrons the Group I elements have to give away, forming such stable combinations as NaCl (sodium chloride) or LiBr (lithium bromide) or that most stable combination FrF (francium fluoride).

The Guiding Principles of Chemical Reactions

Chemistry, then, is to be understood, and chemical behavior is to be predicted, in terms of the outer shell of electrons of atoms. Atoms with complete octets

---

[2] In quantum theory, Period 3 elements could have up to 18 electrons in the third, or $n = 3$, orbit. As we will see shortly, 10 of these electrons would have more energy than those in a 4s orbit. That is why Period 3 ends with argon.

The Guiding Principles of Chemical Reactions

1 H 1.008																	2 He 4.00
3 Li 6.94	4 Be 9.01											5 B 10.8	6 C 12.0	7 N 14.0	8 O 16.0	9 F 19.0	10 Ne 20.2
11 Na 23.0	12 Mg 24.3											13 Al 27.0	14 Si 28.1	15 P 31.0	16 S 32.1	17 Cl 35.5	18 Ar 39.9
19 K 39.1	20 Ca 4.01	21 Sc 45.0	22 Ti 47.9	23 V 50.9	24 Cr 52.0	25 Mn 54.9	26 Fe 55.6	27 Co 58.9	28 Ni 58.7	29 Cu 63.5	30 Zn 65.4	31 Ga 69.7	32 Ge 72.6	33 As 74.9	34 Se 79.0	35 Br 79.9	36 Kr 83.8
37 Rb 85.5	38 Sr 87.6	39 Y 88.9	40 Zr 91.2	41 Nb 92.9	42 Mo 95.9	43 Tc (99)	44 Ru 101.1	45 Rh 102.9	46 Pd 106.4	47 Ag 107.9	48 Cd 112.4	49 In 114.8	50 Sn 118.7	51 Sb 121.8	52 Te 127.6	53 I 126.9	54 Xe 131.3
55 Cs 132.9	56 Ba 137.3	see below 57-71	72 Hf 178.5	73 Ta 180.9	74 W 183.9	75 Re 186.2	76 Os 190.2	77 Ir 192.2	78 Pt 195.1	79 Au 197.0	80 Hg 200.6	81 Tl 204.4	82 Pb 207.2	83 Bi 209.0	84 Po 210	85 At (210)	86 Rn (222)
87 Fr (223)	88 Ra (226)	see below 89-103	104 Rf (261)	105 Ha (260)													

57 La 138.9	58 Ce 140.1	59 Pr 140.9	60 Nd 144.2	61 Pm (147)	62 Sm 150.4	63 Eu 152.0	64 Gd 157.3	65 Tb 158.9	66 Dy 162.5	67 Ho 164.9	68 Er 167.3	69 Tm 168.9	70 Yb 173.0	71 Lu 175.0
89 Ac (227)	90 Th 232.0	91 Pa (231)	92 U 238.0	93 Np (237)	94 Pu (242)	95 Am (243)	96 Cm (247)	97 Bk (245)	98 Cf (251)	99 Es (254)	100 Fm (253)	101 Md (256)	102 No (254)	103 Lw (257)

Figure 9.2 Periodic table of the elements.

(doublets for Period 1) in their outer orbits are stable and chemically inert, and do not enter into chemical reaction under ordinary circumstances. Other atoms tend to achieve such a stable configuration of electrons in their outer orbits by loss or gain of a few electrons, or in some cases, by sharing. Hydrogen gas, for example, proves to have a molecular weight of 2 rather than 1. Two atoms must join together to form the molecule by sharing their two electrons between them. This may be very crudely represented as shown in Figure 9.3.

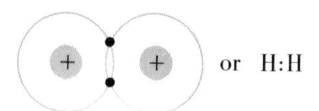

Figure 9.3 Bohr type of model of H_2.

The sharing of electrons, usually in pairs, produces a "covalent bond," which may be represented in several common substances as in the case of hydrogen (Fig. 9.4). More than one pair of electrons can be shared, and the shape of molecules is determined by the directions in which the bonds are formed. As will be noted in the case of water, electrons and nuclei are not always symmetrically arranged about the same point. When this is true, the molecule will be an electric dipole, or *polar molecule*, because the center of negative electricity does not coincide with the center of positive electricity. Similarly, a covalent bond may be polar because the shared electrons are held more closely to one nucleus than to the other. The elements lying in the middle groups (III, IV, and V), because they would have to lose or gain quite a few electrons to achieve complete octets by that method, most often form covalent bonds. Carbon (Group IV, Period 2) forms covalent bonds almost exclusively. Elements in Groups III and V will form covalent bonds that are more or less polar, and in some cases they will lose

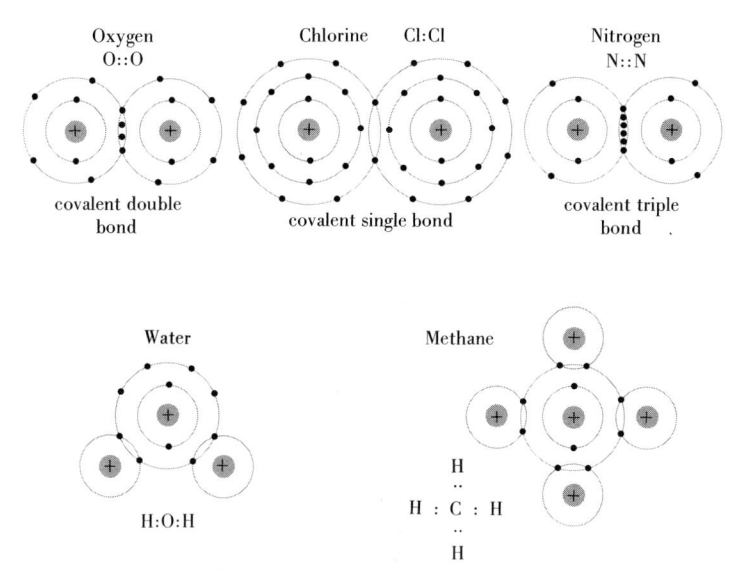

The Guiding Principles of Chemical Reactions   181

Figure 9.4 Bohr type of models of covalent molecules.

(Group III) or gain (Group V) three electrons to form ionic compounds, especially if they are metals.

In chemical reactions when one element loses electrons another must gain them, for molecules, like atoms, have no net charge. However, when a Group I metal loses an electron to a Group VII element, the result is not a "molecule" in the sense of a single particle, but rather a pair of oppositely charged ions held together by strong electrostatic attraction called an *ionic* or *electrovalent bond*. Using our former scheme of representation we have the picture of the formation of sodium chloride in Figure 9.5. The Na⁺ particle is called a sodium ion; Cl⁻ is a chloride ion. Such ionic compounds as sodium chloride are generally crystalline solids with very high melting points. Their crystals are built up of separate ions as units.

As is usual in nature, there are relatively few clear-cut cases and many shades of variation in between. But in general, the farther separated laterally two elements are in the periodic table the more ionic the bond that forms when they combine chemically.

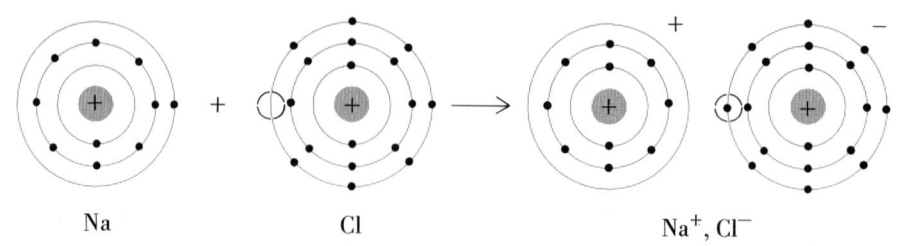

Figure 9.5 Bohr type of representation of the way in which an ionic compound is formed.

There are variations from top to bottom of the table as well as from right to left. Clearly the elements in a group are successively heavier as you go down the column, and their atoms have more shells of electrons. As more electron shells intervene between the positive nucleus and the "valence" electrons, these outer electrons are held less and less tightly. Thus francium, at the bottom of Group I, is the most likely of all elements to lose its outer electron, and is the most electropositive element, whereas fluorine, at the top of Group VII, is most likely to acquire an additional electron, and is the most electronegative. This is the basis for our prediction that FrF is the most stable compound possible. Its bonding would be expected to be entirely ionic. (Francium is a rare, radioactive element not readily available for chemical studies.)

The elements we commonly call metals are generally electropositive; nonmetals are electronegative or form covalent bonds. Many more or less common metallic elements lie in the middle of the Periodic Table in columns labeled, from left to right, III B, IV B, V B, VI B, VII B, VIII B, I B, and II B. These are the so-called transition elements, and their electronic structure can be understood in terms of the Pauli Principle and quantum mechanical descriptions of orbits. We recall that each principal electronic level in an atom contains one spherical sublevel (s) with space for two electrons with opposite spins. Each level after $n = 2$ has three additional p subshells, each of which can contain 2 electrons with opposite spins. By element number 20, calcium, shells or orbitals, as they are called, are filled as follows:

Orbital	Electrons (e^-)
1s	2
2s	2
2p	6
3s	2
3p	6
4s	2

Where does the next electron go? Furthermore, how do you decide where it will go? The answer to the second question is easy: the spontaneous process is always the one that results in the lowest energy state of the system; the electron will go where it can exist in the lowest state of energy. For electron number 21, this is not in a 4p orbital, but lower down in a d orbital of level 3. It turns out that the relative energies of orbitals line up as in Figure 9.6. Thus the transition series of elements in Period 4 (elements 21–30) and Period 5 (elements 39–48) are characterized by two electrons in their outer (4s or 5s) shells and varying numbers, from 1 to 10, in the next lowest d orbitals. Consequently, transition elements have much more in common with each other than with the A family elements of their group. On the other hand

Diagonal lines connect orbitals of nearly equal energies; energy increases from upper right to lower left along a diagonal

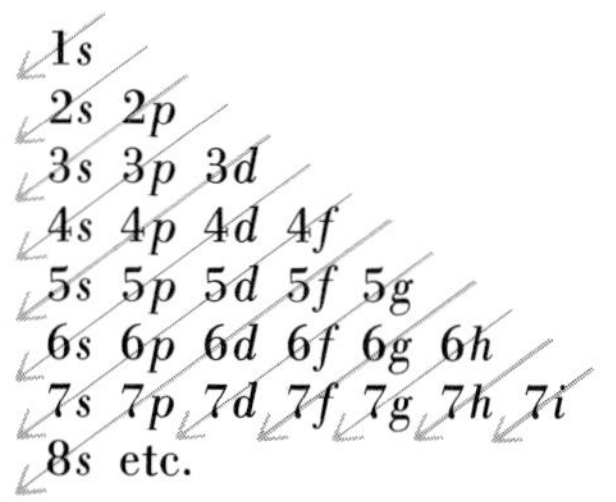

1s
2s 2p
3s 3p 3d
4s 4p 4d 4f
5s 5p 5d 5f 5g
6s 6p 6d 6f 6g 6h
7s 7p 7d 7f 7g 7h 7i
8s etc.

Allowed number of electrons

s : 2
p : 6
d : 10
f : 14

Figure 9.6 Comparative energies of electron subshells, showing order in which subshells are filled.

B family elements will show the same kind of gradation of properties going down a group as do the A families. For example, in Group IB, copper, silver, and gold have "family characteristics" just as do the Group IA elements sodium, potassium, and rubidium, although in the B family, the similarities are less pronounced.

Another characteristic of transition elements is their variable combining power. Nearly all commonly exhibit more than one valence. This is because the d electrons lying very close to the outer shell can also act as valence electrons. Iron, for example, has stable compounds in which its valence is 2, and equally stable compounds in which its valence is 3.

Two additional rows of elements are usually printed at the bottom of the Periodic Table as numbers 58–71 and 90–103. These constitute second transition series of Periods 6 and 7, respectively, and in these elements the orbitals two levels down are being filled with electrons. That is, their electron arrangements are:

Ce (no. 58) $1s^2$, $2s^2$ $2p^6$, $3s^2$ $3p^6$ $3d^{10}$, $4s^2$ $4p^6$ $4d^{10}$ $4f^1$, $5s^2$ $5p^6$, $6s^2$ $6p^1$

Pr (no. 59) $1s^2$, $2s^2$ $2p^6$, $3s^2$ $3p^6$ $3d^{10}$, $4s^2$ $4p^6$ $4d^{10}$ $4f^2$, $5s^2$ $5p^6$, $6s^3$ $6p^1$ etc.

The transition series of Period 7 is somewhat less regular, suggesting that the $5f$ and $6d$ orbitals are very close in energy.

Several times we have deferred explanations of physical properties of materials until we had a better understanding of their structure. In Chapter 4, we discussed the phenomena of electricity and magnetism without delving into their basic nature, and in Chapter 5 we put off the "why" of color phenomena. Now we are ready to tie up these loose ends.

When we think of good conductors of electricity, and of heat, as well, we think of metals. Because electric current is simply the net shift of electrons in a given direction, we can now see that metals, having only a few loosely held and thus easily shifted electrons in their outer orbits, are naturally the best conductors, while those substances whose electrons are normally involved in covalent bonding or tightly held in completed shells would be poor conductors. Freely vibrating electrons in a metal would and do add to its heat conductivity, also.

Magnetism is a little more complex. We commonly associate it with iron, and iron's near relatives, cobalt and nickel. Actually, the most strongly magnetic element is the rare earth *gadolinium*, and many rare earth and transition metals and their compounds have *paramagnetic properties* (that is, they align themselves parallel to a magnetic field). Moreover, oxygen at low temperatures where it is a liquid is strongly paramagnetic. What do these diverse substances have in common that makes them magnetic?[3] We find the answer in their electron structure, usually in the arrangement of the electrons in those partially filled subshells characteristic of both first and second transition series elements. The rules of the game of placing electrons in d or f subshells specify that each subshell shall get at least one electron before any of them get two. (When there are two electrons in a *d orbital*, as chemists call these subshells, they differ only in the direction of their spin.) This means that an atom may have several electrons spinning one way and none or fewer spinning the other way. Take iron, for example. With twenty-six electrons in all, iron has this arrangement: $1s_2$, $2s_2$, $2p_6$, $3s_2$, $3p_6$, $3d_6$, $4s_2$. The $3d$ subshell, which can hold ten electrons (five pairs) has only six, so it will have one pair of electrons with paired spins and four single "unpaired" electrons. An electric current always produces a magnetic field. Four electrons spinning the same way constitute an atom-sized electric current. In a piece of iron bar,

[3] "Magnetic" in common usage means paramagnetic or *ferromagnetic*, a term applied to strongly magnetic substances like iron. Some materials line up crosswise to a magnetic field; they are called *diamagnetic*.

little neighborhoods, or *domains*, of atoms will have their electrons all spinning one way, creating small magnetic fields, but neighboring domains with different orientations will cancel each other's magnetic effects. When we magnetize an iron bar by placing it in a strong magnetic field we are letting the field reorient the magnetic domains until most of them are aligned with the field. Because it takes energy also to "unalign" these domains, the bar stays magnetized permanently, or at least until heating or jarring it severely disrupts the alignment.

The impossibility of severing a "north" from a "south" pole is explained by this picture of magnetism, too (Fig. 9.7). A pole is not a "thing" in the sense that an electric charge is a thing, or at least a property of a material particle. "Pole" is simply a designation for a direction, and "north pole" has no more meaning apart from "south pole" in magnetism than "east" has without "west" in geography.

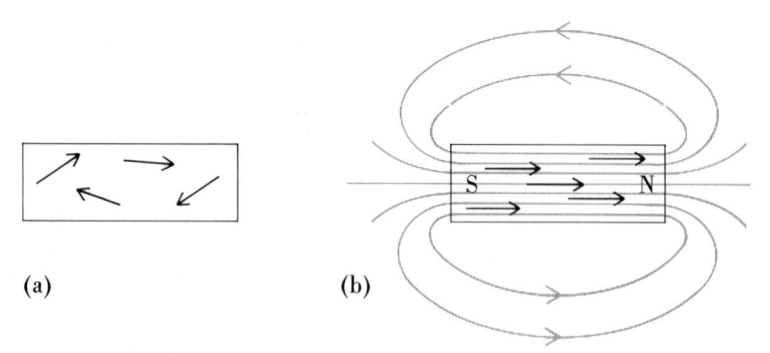

(a)   (b)

Figure 9.7 Model for the domain theory of ferromagnetism. (a) Unmagnetized iron bar has domains aligned at random. (b) Magnetized iron bar has domains lined up in the same direction.

We have seen how the emission of light of specific colors is explained by the quantum theory. The explanation of the colors of solutions, dyes, and pigments is very similar. We observe that all solutions containing Cu^{+2} ions are cyan in color. Therefore, some electrons in Cu^{+2} must be able to change orbits in such a way as to require the energy of "red" photons, leaving only blue and green photons to be transmitted by the Cu^{+2} solution. Of course, the excited electrons fall back down and red light is emitted, but we do not see a red glow emanating from the solution because the red light is too diffuse to see.

Most dyes are quite complicated molecules, but their color is still due to the selective absorption of certain frequencies of light corresponding to the energies of their possible electron transitions, leaving the other frequencies to be transmitted or reflected, and therefore seen.

As atomic weights came to be very accurately determined, especially by T. W. Richards at Harvard, it became quite obvious that chemical atomic weights were not whole numbers in many cases. In 1913 Richards determined that the lead resulting from the radioactive decay of uranium did not have the same atomic weight as "ordinary" lead. It was realized that the atoms of an element could vary in weight while retaining their elemental identities. Because these atoms of different weights would have the same atomic number and so occupy the same place in the Periodic Table, they were called *isotopes*. The chemical atomic weight is simply a weighted average of all the commonly occurring isotopes in their naturally occurring proportions. The chemical atomic weight of hydrogen is 1.008, indicating that the isotopes heavier than 1 are very rare, whereas the value of 35.46 for chlorine indicates that the isotope of mass 37 is somewhat less common than the isotope of mass 35.

Because hydrogen, with a usual "atomic mass number" of 1, contained one proton, whereas helium, mass number 4, contained 2 protons, apparently something else was in the nucleus besides protons— something not electrically charged. These neutral particles were not positively identified until 1932, when James Chadwick discovered a particle about equal in mass to a proton, but uncharged, which fit the specifications. This particle is called a *neutron*. Isotopes can be defined as atoms having the same number of protons but different numbers of neutrons.

It is somewhat surprising that stable nuclei containing a large number of protons can exist at all, because positive charges repel each other strongly. Within the tiny nucleus — diameter of the order of 10^{-13} cm — very strong short-range attractive forces seem to exist which are not yet very well understood, but which apparently are associated with the ability of neutrons and protons to turn into each other by transferring the charge from one particle to another. On a very simple level we can consider the neutrons in the role of "insulation," and if we compare atomic number (protons) to atomic weight (protons + neutrons) we note that the elements of higher atomic number have one to one and one-half times as many neutrons as protons. Nevertheless, it seems impossible for the scheme of increasing the proportion of neutrons to work indefinitely to hold nuclei together, and we find that for elements of atomic number greater than 83, no stable isotopes exist. All these larger nuclei fall apart by one or more methods of radioactive disintegration.

Although much evidence shows that nuclear structure may also be understood in terms of a system of shells for *nucléons* (protons and neutrons) somewhat analogous to electron shells, nuclear properties and chemical properties are quite independent of each other. Chemical behavior, diverse as it is, is still simply behavior involving the electrons orbiting the nucleus. The nucleus itself, is, as it were, another world.

The more one knows of chemistry, the more useful the Periodic Table becomes. The trends we have noted and the predictions they make possible are only a small beginning to the possibilities of the system.

The Language of Chemistry

Students frequently find chemistry difficult because they must learn a new language along with the new facts and ideas of the science. Once the language barrier is breached, however, the chemical symbolism proves to be an extremely convenient and versatile means of communication.

The chemical symbol is the abbreviation for the name of an element. For the longest-known elements, the symbols are often taken from their Latin names. But the symbol means much more. For example, Fe

Table 9.3

Water H_2O
1. The compound contains two elements, hydrogen and oxygen.
2. The more electropositive element is written first.
3. The subscript $_2$ indicates that the molecule has two atoms of hydrogen.
4. The absence of a subscript by oxygen is interpreted as "1" (i.e., the molecule has one atom of oxygen).
5. The molecular weight of water is $2 \times 1 + 1 \times 16 = 18$ on the atomic weight scale.
6. The formula represents 18g or 1 gram molecular weight of water, usually called *1 mole.*

Sugar $C_{12}H_{22}O_{11}$ (sucrose)
1. There are three elements, carbon, hydrogen, and oxygen, in the compound.
2. Carbon, though not the most electropositive, is the dominant element in "organic" compounds like sugar, and is customarily written first.
3. According to the subscripts, a sugar molecule contains 12 carbon atoms, 22 hydrogen atoms, and 11 oxygen atoms.
4. One mole of sugar consists of $12 \times 12 + 22 \times 1 + 11 \times 16 = 342$ grams.

Salt NaCl (sodium chloride)
1. The compound contains two elements, sodium and chlorine.
2. The "molecule" actually consists of two separate ions, Na^+ and Cl^-, but they will always be paired up one-to-one, and so the formula represents the proportions of ions that are associated correctly.
3. One mole of salt consists of $1 \times 23 + 1 \times 35.5 = 58.5$ grams.

---

stands for: (1) iron (Latin *ferrum*); or (2) one atom of iron; or (3) one atomic weight of iron in grams (sometimes called a "gram atom"). Symbols consist of one or two letters; the first is always capitalized.

Chemical formulas represent compounds in the same sense that symbols represent elements, and are made up of symbols arranged in a conventional way to show the chemical composition of the compound. For example, Table 9.3 gives the formulas for some familiar compounds and their "translations."

Chemists describe what happens in chemical reactions by putting formulas together in equations. Let us take as an example a familiar household reaction: combining baking soda and buttermilk to make biscuits.

$$NaHCO_3 + HC_3H_3O_2 \rightarrow NaC_3H_3O_2 + CO_2 \uparrow + H_2O$$
(baking soda)   lactic acid   sodium   carbon   water
   sodium   (from   lactate   dioxide
bicarbonate   buttermilk)   (a salt)

When lactic acid reacts with sodium bicarbonate, the products are sodium lactate, a compound of the type known as a *salt*; carbon dioxide, a gas (the arrow tells you that), which was the whole reason for the reaction, because it is what makes the biscuits "light"; and water. Any acid will react with baking soda to produce carbon dioxide, so you can use baking powder, which contains an acid, instead of buttermilk to make biscuits, unless you are a traditional southern cook.

A chemical equation, then, provides the following information:

1. The reactants and products are represented by their correct formulas.

2. The relative numbers of molecules involved in the reaction are shown when the equation is "balanced."

3. A balanced equation demonstrates the *Law of Conservation of Matter: Matter is neither created nor destroyed in any chemical reaction.*

4. An equation is balanced when equal numbers of each kind of atom are on both sides of the arrow. (In the example above, if there are 5 oxygen atoms on the left side of the equation, there must also be 5 atoms of oxygen on the right, though they are now differently combined.)

5. The arrow means "yields" and indicates the direction in which the reaction proceeds. The reactants are written on the left and the products to the right of the arrow.

If an electric spark is introduced into a mixture of hydrogen gas and oxygen gas, they will combine to form water. To write this in a chemical equation, we use this procedure:

1. Write the reactants and products correctly:

$$H_2 + O_2 \rightarrow H_2O$$

Hydrogen and oxygen exist as diatomic molecules in the free state.

2. Balance the equation so that the Law of Conservation of Matter is observed:

$$2H_2 + O_2 \rightarrow 2H_2O$$

If we start with two O's on the left (or reactant) side of the equation, we must have two O's in the product on the right. Because we can't change the formula for water, H_2O, we must multiply the whole thing, giving $2H_2O$ on the right. But if we have four H's in the product, we are required to start with four H's in the form $2H_2$.

Carbon, the element in the middle of the Periodic Table, has the remarkable property of combining with itself to form long chain or complicated ring structures that also contain hydrogen, and perhaps also oxygen, nitrogen, sulfur, or other elements. There are between two and three million of these so-called organic compounds. We will return to them in the next chapter.

The remaining few hundred thousand compounds and their formation and behavior comprises the field of inorganic chemistry. Inorganic compounds are mainly of mineral origin and are not primarily associated with living things, although a few are essential to life. The "inorganic-organic" classification of chemical compounds is an old and useful one, but not very precise. That is, there are a lot of borderline cases. Most inorganic chemical reactions fall into four main types: direct combination, decomposition, displacement, and ionic.

1. Direct Combination
   a. Magnesium burning in oxygen, as in a flash bulb.

   $$2\ Mg + O_2 \rightarrow 2\ MgO$$

   b. Iron rusting (at low temperatures, it only works if water is present, too).

   $$4Fe + 3O_2 \rightarrow 2Fe_2O_3$$

   c. Sodium hydroxide formation when sodium oxide dissolves in water.

   $$Na_2O + H_2O \rightarrow 2NaOH$$

2. Decomposition (the opposite of Combination)
   a. Separating mercury metal from its oxide by heat.

   $$2HgO \rightarrow 2Hg + O_2$$

   b. Removing water from a hydrated salt, copper sulfate.

   $$CuSO_4 \cdot 5H_2O \rightarrow CuSO_4 + 5H_2O$$

3. (Single) Displacement
   a. Producing hydrogen from an acid and an active metal.

   $$Zn + 2HCl \rightarrow H_2 + ZnCl_2$$

b. Producing a pure metal by displacement by a more active metal.

$$Cu + 2AgNO_3 \rightarrow Cu(NO_3)_2 + 2Ag$$
copper   silver nitrate   copper nitrate   silver

4. Ionic or Double Displacement Reactions
   a. Neutralizing an acid by a base to form a salt and water.

$$HCl + NaOH \rightarrow NaCl + H_2O$$
hydrochloric acid   sodium hydroxide (base)   sodium chloride (salt)   water

   b. Forming an insoluble product (precipitate)

$$BaCl_2 + Na_2SO_4 \rightarrow BaSO_4 \downarrow + 2NaCl$$
barium chloride   sodium sulfate   barium sulfate (precipitate)   sodium chloride

These reactions call attention to several facts and principles of chemistry. As we have seen before, elements combine when one can supply the electrons needed by the other. An oxide such as MgO is ionic, as the locations of Mg and O in the Periodic Table would lead one to expect. It is also sometimes possible for two compounds to combine. Most acids are formed by dissolving the oxides of nonmetals or transition metals in water, as the formation of the most common strong mineral acids illustrates:

Nitric acid $N_2O_5 + H_2O \rightarrow 2HNO_3$
Sulfuric acid $SO_3 + H_2O \rightarrow H_2SO_4$
Phosphoric acid $P_2O_5 + 3H_2O \rightarrow 2H_3PO_4$

These nonmetal oxides are covalent, but the acids that they form break into ions in a characteristic way when they are dissolved in water. In fact, acids are commonly defined as compounds that ionize in water to give hydrogen ions:

$$HNO_3 \rightarrow H^+ + NO_3^- \text{ (nitrate ion)}$$
$$H_2SO_4 \rightarrow 2H^+ + SO_4^= \text{ (sulfate ion)}$$
$$H_3PO_4 \rightarrow 3H^+ + PO_4^\equiv \text{ (phosphate ion)}$$

In the latter two cases, the hydrogens ionize one by one, so that a solution of H_2SO_4 also contains HSO_4^- ions, and a solution of H_3PO_4 will have substantial amounts of $H_2PO_4^-$ and $HPO_4^=$. Because H^+ is only a bare proton, its independent existence is quite unlikely. It is much more probable that it sticks to a water molecule, which is polar (Fig. 9.8). Groups of atoms like $SO_4^=$, the sulfate ion, are joined together by strong covalent bonds. They act as a single entity in many chemical reactions. A general name for such groups of atoms is *radical*. Most common inorganic radicals are the negative ions of acids.

(a) Water molecule

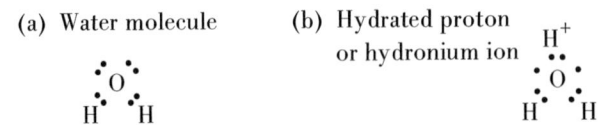

(b) Hydrated proton or hydronium ion

Figure 9.8 The role of water in the ionization of acids.

When we speak of a *polar molecule*, such as water, we mean that because when oxygen forms covalent bonds with hydrogen the angle H—O—H is about 105°, the molecule has a relatively negative end where the electrons not involved in bonds are located and a relatively positive end, where the two hydrogen nuclei are. "Polar" means that the center of negative charge and the center of positive charge do not coincide.

We also speak of molecules having polar bonds, which means that the electrons are held more closely by one nucleus than by the other. The bond between hydrogen and chlorine is a very polar bond, and when this gas is dissolved in water it also ionizes to give an acid: $HCl \rightarrow H^+ + Cl^-$. In general the compounds of hydrogen with nonmetals in Groups VI and VII behave in this way—"hydrogen oxide" (water) being a notable exception.

Oxides of metals also combine with water, and these compounds form ions in water solution, but the result is quite opposite to that observed with nonmetal oxides. For example, sodium oxide reacts with water thus:

$$Na_2O + H_2O \rightarrow 2 \overset{+}{Na}\overset{-}{OH}.$$

We often write the + and − signs to indicate that, like Na^+Cl^-, this compound is never molecular, but in its crystal lattice as well as in solution it is made up of ions. The OH^- ion is characteristic of bases. Simply put, a base is the chemical opposite of an acid. Many metal hydroxides do not dissolve to any great extent in water, but many of those that do act as bases. The exceptions here are some transition metal oxides, which ionize like acids. The oxides of Group IA metals dissolve to form strong bases, or alkalies. Soaps are salts formed from sodium hydroxide, or lye, and fatty acids, organic compounds we will discuss in Chapter 11.

The only common inorganic radical that is a positive ion is the ammonium ion, NH_4^+. It forms when ammonia, the covalent compound of nitrogen and hydrogen, dissolves in water:

$$NH_3 + H_2O \rightleftarrows NH_4^+ + OH^-.$$

Since one product of this reaction is OH^-, we might expect that the compound NH_4OH is a *base*. A solution of ammonia in water does act as a weak base, that is, one having only a few OH^- ions, as opposed to a strong base like $NaOH$, which separates virtually completely into Na^+ and OH^-. In fact, the molecule NH_4OH probably does not exist, and the double arrow ($\rightleftarrows$) in the equation above indicates that, while NH_3 is dissolving in water to form NH_4^+ and OH^-, the reverse reaction is also going on. Such a system will reach a point of equilibrium at which the forward and backward reactions are proceeding at equal rates, resulting in no net change in the amounts of the four species (NH_3, H_2O, NH_4^+, and OH^-) in the solution.

We will return to this idea of chemical equilibrium in the next chapter.

It would be strictly correct to say that every chemical reaction has an equilibrium point, but in those that we say "go to completion," such as the neutralization of HCl by NaOH, the extent of the back reaction is negligibly small.

The ionic reaction of neutralization is really just the formation of the covalent molecule H—O—H from the characteristic H^+ of an acid and OH^- of a base, thus "neutralizing" the characteristics of both. Indeed, a mixture of ions can hardly be said to have "reacted" unless one pair of ions has somehow been removed from the action, by forming a covalent molecule, by forming a gas that bubbles out of the solution, or by forming an insoluble compound that precipitates or falls down to the bottom of the reaction vessel. The other product of neutralization is a salt, composed of the + ion of the base and the − ion of the acid, but many salts are soluble in water and may be recovered only by evaporating the water. Salts are solids with almost completely ionic bonding in their crystal lattices.

These are only the broadest outlines of inorganic chemistry, but they may make the reader conversant with a few of the important kinds of compounds and reactions in this field, and most important, introduce him to the great organizing scheme of the Periodic Table.

SUMMARY STATEMENTS

1. Chemical compound formation follows two laws:
   a. Law of Definite Proportions: When two or more elements react to form a compound, they always combine in definite proportions by weight.
   b. Law of Multiple Proportions: When two elements combine to form two or more compounds, the weights of the one combining with a given weight of the other are always in a ratio of small whole numbers.

2. The great organizing principle of chemistry is the Periodic Law of the Elements: the physical and chemical properties of elements are periodic functions of their atomic numbers.

3. A chemical reaction involves electrons in the outermost orbits of atoms. Chemical bonds, which hold atoms together in molecules, range from electrostatic attraction between two ions formed when electrons from one atom are transferred to the orbital of another atom to the equal sharing of pairs of electrons between two atoms.

4. An ionic bond consists of electrostatic attraction between ions of unlike charge formed by transfer of electrons from one atom to another.

5. A covalent bond consists of a pair of electrons shared between two atoms. If they are shared equally, the bond is nonpolar; if they are held more tightly by one atom than by the other, the bond is polar.

6. The most stable state of atoms appears to be that in which the outer shell of electrons contains eight electrons (two in Period 1). Elements on the left of the Periodic Table reach this state by losing electrons in chemical reactions; those on the right by gaining; those in the middle by sharing.

7. Electrical conduction, magnetism, and colors of materials depend on the electronic structure of atoms and molecules.
   a. Electric conduction depends on the mobility of outer electrons.
   b. Magnetism depends on the possibility of aligning atoms with a net electron spin in one direction.
   c. Color depends on the energy absorbed by electron transitions between orbits.

8. The masses of atoms are approximately integral multiples of the mass of hydrogen, because all nuclei are made up of protons and neutrons, units of roughly equal mass. Atomic weights are often fractional because atoms of the same element may contain different numbers of neutrons, and the chemical atomic weight is an average of the weights of the several isotopes according to their abundance. (The weight of an electron is not quite 1/1800 that of a proton.)

9. Among the most important classes of inorganic com-

pounds are acids, bases, and salts. Salts and water are the products of the neutralization of acids with bases. A base is characterized by OH$^-$ in its water solution; an acid is characterized by H$^+$ (or H$_3^+$O) ions in solution.

10. A chemical symbol represents all the information contained in the idea of one atom of an element; a formula represents all the information contained in the idea of one molecule of a compound. An equation uses symbols and formulas to state what happens in a chemical reaction.

REVIEW QUESTIONS

1. What is the difference between alchemy and chemistry?

2. About how many chemical compounds are now known?

3. About when did chemistry begin to develop as a science? Name three or four early chemists and their achievements.

4. Outline the attempts at classifying chemical elements before 1860.

5. What two laws govern the formation of compounds from elements?

6. Chemists characterize all material substances as being either elements, compounds, or mixtures. What does each of these terms mean?

7. What was Mendeleev's contribution to chemistry? What was the basis of his organizational scheme? How effective was it? Have corrections been necessary?

8. Explain the structure and predictive value of the Periodic Table in terms of the atomic theories of Bohr and Pauli.

9. Define: chemical bond; covalent bond; nonpolar bond; polar bond; electrovalent (or ionic) bond; ion; molecule (in the chemical sense).

10. What do transition elements have in common? Why do they occur where and as they do in the Periodic Table?

11. List all the things a chemical symbol stands for. List all the information given by a chemical formula.

12. Using the Periodic Table write formulas for the following:

        silver iodide   aluminum chloride
        copper (II) oxide   carbon dioxide
        hydrogen sulfide   magnesium nitride

Chemistry: The Periodic System

13. We have seen that acids ionize to give ions made up of several atoms, such as the sulfate ion $SO_4^=$. Such groups are also called *radicals*. They are very stable, and act as a unit in chemical reactions. Combined with metals, they form salts. Write the formulas for the sulfate ($SO_4^=$), nitrate (NO_3^-), and phosphate ($PO_4^\equiv$) salts of the following elements:

sodium	titanium
magnesium	iron (II)
aluminum	nickel (IV)

14. Write chemical equations for the following and label each as to the type of reaction it exemplifies.
    a. Iron + sulfur yields iron (II) sulfide.
    b. Carbonic acid (hydrogen carbonate) yields water + carbon dioxide.
    c. Zinc + copper (II) chloride yields copper + zinc chloride.
    d. Phosphoric acid + calcium hydroxide yields calcium phosphate and water.

THOUGHT QUESTIONS

1. A well-known chemical company has long used the slogan "Better things for better living through chemistry." Comment on whether or not this slogan gives the whole picture of the effect of chemistry on our lives.

2. The "language" of chemistry is usually forbidding to people with no particular bent toward science. What purpose does this special language serve? To what extent is it useful for the "average citizen" to know or understand this language?

3. What is the use of the Periodic Table?

4. Name some household processes that are chemical in nature.

5. What is the relation between physics and chemistry? between biology and chemistry? Do physicists and biologists need to know some chemistry? Why or why not?

READING LIST

Asimov, Isaac, *A Short History of Chemistry*, Doubleday (Anchor), Garden City, N.Y., 1965. A brief, clear account of the history and principles of chemistry by one of the very best writers on science.

Horrigan, Philip A., *The Challenge of Chemistry*, McGraw-

Hill, New York, 1970. A topical treatment of chemical principles and their present-day applications in medicine, industry, and agriculture, and of ecology.

Young, Louise (ed.), *The Mystery of Matter*, Oxford University Press, New York, 1965. A collection of readings around the theme of the structure of matter. Part II is of special interest here.

10

Two and One-Half Million Compounds of Carbon

The earliest chemists recognized that there was something special about the compounds of carbon, but they were partly wrong about what it was. All the carbon compounds they knew about were the result of some natural function of living organisms. Oxygen-breathing organisms exhaled CO_2, carbon dioxide. Decaying plants produced swamp gas, CH_4. Carbon in the form of coal turned out to be remains of ancient forests, and even limestone, calcium carbonate, which can be metamorphosed into marble, originally came from the shells of minute sea organisms. It was absolutely correct to trace the origins of carbon compounds to living organisms, but the early chemists were wrong in assuming that such "organic" compounds could not be made synthetically. (If you want to argue on the other side, of course, you can point out that when organic compounds are synthesized by humans, they are the product of "organisms," but it isn't exactly the same thing.)

Synthesizing Compounds

The proof that man could synthesize compounds identical to those occurring as natural organic products came in 1828, when Friedrich Wöhler, a former pupil of the great Swedish chemist Berzelius (to whom we owe the use of chemical symbols), found that heating ammonium cyanate, an inorganic salt, converted it into urea, a nitrogen-containing compound found in urine:

$$NH_4CNO \xrightarrow{\Delta} H_2N-C\overset{\overset{\displaystyle O}{\parallel}}{\underset{NH_2}{}}$$

(Formulas of organic compounds are generally written so as to indicate their structures.)

It might be argued that NH_4CNO, because it contains carbon, is itself organic, and that the "synthesis" was merely a rearrangement, but Wöhler's work is nonetheless important because it started chemists to considering the possibility of actually synthesizing carbon compounds, and broke the stranglehold of vitalism, the idea that a life-force is necessary to produce organic substances. In 1845, Wöhler's pupil, A. H. H. Kolbe, made acetic acid (the acidic compound in vinegar) from its elements, carbon, hydrogen, and oxygen, and this achievement could not be refuted.

Since the 1840's, and even more, since the 1940's, the stream of new synthetic organic compounds threatens to inundate the chemical literature, if not the planet.

Coal tar dyes were among the first commercially important groups of synthetic organic chemicals. The first of these was discovered by an English student, William Henry Perkin, who was assistant to the noted German chemist, August Wilhelm von Hofmann, then living in England. So popular was his "aniline purple" (or *mauve*, as the French called it) that the

1860's were called the "Mauve Decade." Drugs, explosives, fibers, plastics, detergents, pesticides followed — all the vast array of synthetic materials that make our lives easier and contaminate our environment. Unfortunately, sometimes the same materials do both — an example of the moral neutrality of science and the need for moral responsibility on the part of the exploiters of science.

Table 10.1 Landmarks in Chemotherapy

Discoverer	Event	Use
1909 Paul Ehrlich	synthesis of arsphenamine	syphilis therapy
1932 Gerhard Domagk	discovery of bacteriocidal properties of sulfanilamide	bacterial infections
1928 Alexander Fleming	discovery of the first antibiotic penicillin[a]	bacterial infections

[a] The discovery was not exploited until about 1940.

Unique Property of Carbon

What unique property of carbon enables it to be the backbone ingredient of so many compounds? Its place in the Periodic Table (p. 179) gives the first clue. In Period 2, Group IV, it has two electron shells, the outer of which contains four electrons, enough to fill exactly half the spaces in that orbit. Rather than losing or gaining four electrons, the carbon atom tends to share, forming four identical electron pair bonds that are usually not very polar.

These bonds form very readily between carbon atoms, leading to the formation of chains of carbon atoms of any arbitrary length, and of rings of (mainly) five or six carbon atoms. If all the bonds of a carbon are identical they must be formed symmetrically about the nucleus, and this symmetry would take the form of a tetrahedron (Fig. 10.1). The angle between any two bonds is about 109.5°. Carbon atoms will form rings of such a size that the bond angles are not

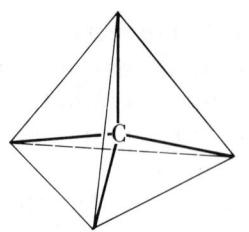

Figure 10.1 Tetrahedral bond angles in carbon.

too different from the tetrahedral angle, i.e., 108° in a 5-carbon ring and 120° in a planar 6-membered ring. Smaller or larger rings are less stable because they are more strained. The fact that the four bonds are identical can be explained only in quantum-mechanical terms. We change our picture of chemical bonding from a simple shared pair of electrons to one of molecular orbitals formed by combination and hybridization of the orbitals of the atoms entering into combination, which shouldn't be unexpected. After all, we recognize a chemical change by the formation of one or more totally new substances whose properties are not merely the additive properties of the components but different altogether.

Types of Organic Compounds

Perhaps the most effective way to take a brief look at the vast field of organic chemistry is to consider the kinds of organic compounds that can exist.

Hydrocarbons

The simplest and among the stablest of all carbon compounds are the saturated hydrocarbons, or alkanes. In these compounds carbon atoms form all four bonds either with hydrogen or with each other to make long straight or branched chains. "Saturated"

Types of Organic Compounds

means that all the bonds are single electron pairs. A series of compounds differing each from the next by one carbon atom and its two attached hydrogen atoms is called a *homologous series*. The homologous series of alkanes starts like this (using structural formulas in two dimensions to represent carbon atoms that actually form bonds at tetrahedral angles, and representing each electron pair bond by —).

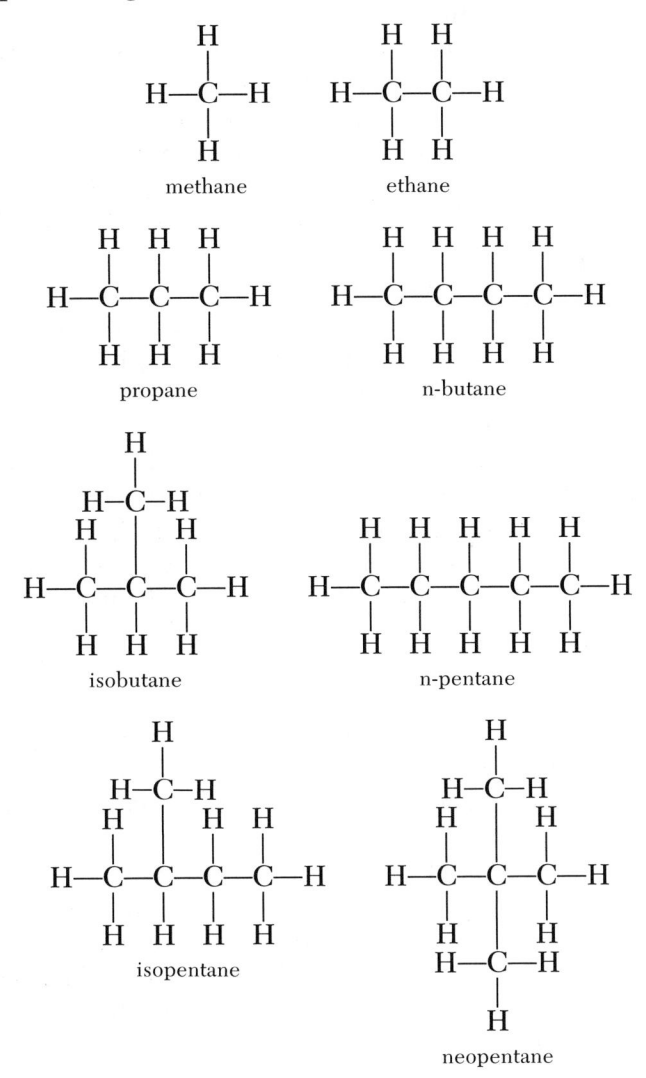

```
        H              H   H
        |              |   |
    H—C—H         H—C—C—H
        |              |   |
        H              H   H
     methane           ethane


   H  H  H          H  H  H  H
   |  |  |          |  |  |  |
 H—C—C—C—H      H—C—C—C—C—H
   |  |  |          |  |  |  |
   H  H  H          H  H  H  H
    propane            n-butane


        H
        |
      H—C—H
    H   |   H        H  H  H  H  H
    |   |   |        |  |  |  |  |
  H—C—C—C—H       H—C—C—C—C—C—H
    |   |   |        |  |  |  |  |
    H   H   H        H  H  H  H  H
     isobutane          n-pentane


        H                  H
        |                  |
      H—C—H              H—C—H
    H   |   H   H      H   |   H
    |   |   |   |      |   |   |
  H—C—C—C—C—H      H—C—C—C—H
    |   |   |   |      |   |   |
    H   H   H   H      H   |   H
       isopentane        H—C—H
                           |
                           H
                        neopentane
```

As these structural formulas show, as soon as you have four carbon atoms in a compound, an interesting thing happens: they can be put together in more than one way. The more carbon atoms, the more possibilities. Compounds identical in composition (having therefore the same empirical formula) but differing in structure are *isomers*. We will encounter several varieties of isomerism in our introduction to organic compounds.

The naming of organic compounds is something of an art in itself. The names *methane, ethane, propane,* and *butane* antedate any need for a system of nomenclature, and have simply been incorporated into the system that developed later. From 5 carbons on, the root name is the Greek number word root: *pent-* = 5, *hex-* = 6, *hept-* = 7, *oct-* = 8, etc. The suffix *-ane* means a saturated hydrocarbon. For isomers, *n-* is normal; *iso-* means the simplest branched chain; *neo-*, the next simplest. This system soon gets unwieldy, however, so the systematic way of naming a compound is to name it by its longest straight chain and to designate saturated hydrocarbon side chains as groups named for their number of carbons, the name ending in *-yl*. Thus -CH_3 is the methyl group, and "neopentane" is more properly called 2, 2-dimethyl propane. "Dimethyl" indicates the presence of two methyl groups; the 2, 2- indicates their location on the second carbon atom from the end of the chain. (You start counting so that the numbers are as small as possible.) Such hydrocarbon groups are called alkyl groups (alkane − ane + yl) and are represented by *R* in writing general formulas for homologous series of compounds.

Alkanes burn, that is, they combine with oxygen to produce light and heat. Methane is the main constituent of natural gas, propane and butane are the familiar "bottled gas" used for heating and cooking in rural areas, and branched chain octanes are the best

Types of Organic Compounds

fuel for internal combustion engines. The end products are carbon dioxide and water:

$$CH_4 + 2O_2 \rightarrow CO_2 + 2H_2O.$$

In general, however, alkanes are not very reactive chemically. Nevertheless, other organic compounds are named and classified as if they were derived from the alkanes.

The two other series of hydrocarbons are said to be "unsaturated" because one (or more) pairs of carbon atoms are joined by two (alkene series) or three (alkyne series) electron pair bonds. For example:

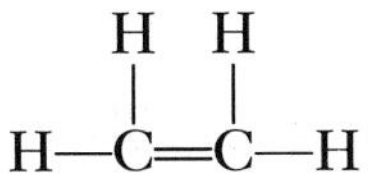

$$H_2C=CH_2$$
ethene
(common name: ethylene)

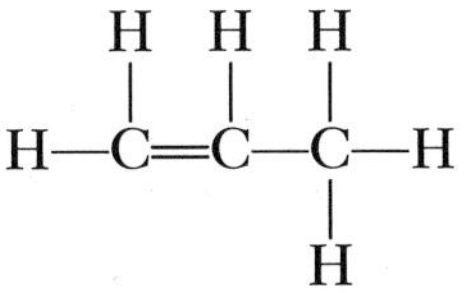

$$H_2C=CH-CH_3$$
propene
(propylene)

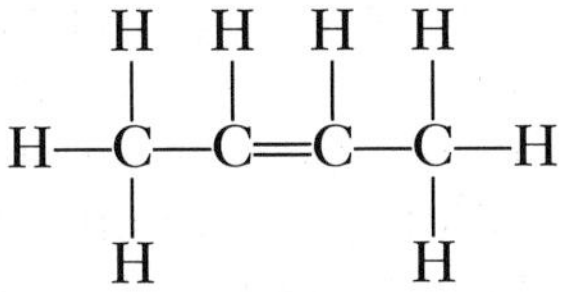

$$CH_3-CH=CH-CH_3$$
2-butene

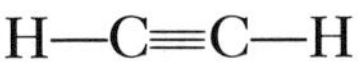

$$H-C\equiv C-H$$
ethyne
(acetylene)

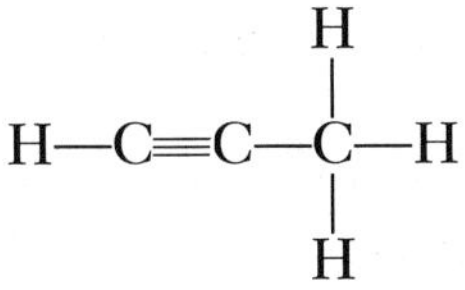

$$H-C\equiv C-CH_3$$
propyne

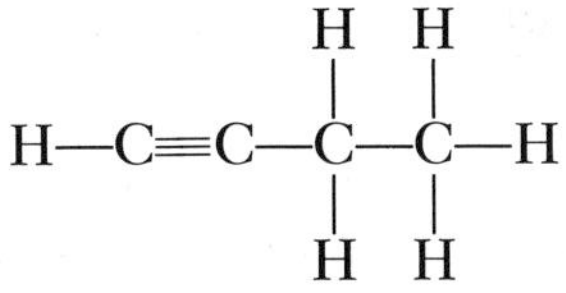

$$H-C\equiv C-CH_2-CH_3$$
1-butyne

These compounds also burn (you've heard of oxyacetylene torches for welding, perhaps), but because double bonds and triple bonds represent a "strained" situation — bonds bent far from their normal tetrahedral angles — they are reactive in many ways. Halogens such as F_2, Cl_2, and Br_2 add readily across a double or triple bond. One of the most important reactions is the adding of double bonded compounds to each other in long chains. The process is called *polymerization,* the products, *polymers.* When ethylene polymerizes, the product is the plastic, polyethylene.

$$H-\underset{\underset{H}{|}}{\overset{\overset{H}{|}}{C}}=\underset{\underset{H}{|}}{\overset{\overset{H}{|}}{C}}-H + H-\underset{\underset{H}{|}}{\overset{\overset{H}{|}}{C}}=\underset{\underset{H}{|}}{\overset{\overset{H}{|}}{C}}-H + \text{etc.}$$

$$\rightarrow H-\underset{\underset{H}{|}}{\overset{\overset{H}{|}}{C}}-\underset{\underset{H}{|}}{\overset{\overset{H}{|}}{C}}-\underset{\underset{H}{|}}{\overset{\overset{H}{|}}{C}}-\underset{\underset{H}{|}}{\overset{\overset{H}{|}}{C}}-\text{etc.}$$

All synthetic plastics, fibers, and rubbers, as well as natural fibers and rubbers, are polymers, although not all are hydrocarbon polymers (see Plate 12). Wool, hair, skin, and muscle are complicated nitrogen-containing polymers or *proteins.*

We can consider most other organic compounds to be formed from alkanes by replacing a hydrogen atom by some other atom or group of atoms. Such groups are called *functional* groups because their presence in the molecule is the principal determinant of the properties and functions of the compound.

Alkyl Halides and Other Halogenated Hydrocarbons

One or more hydrogen atoms can be replaced by a halogen atom. Properly speaking, an alkyl halide is an alkyl group attached to a halogen. Some familiar halogenated hydrocarbons are (see Plate 11):

methyl chloride

trichloromethane (chloroform, an anaesthetic)

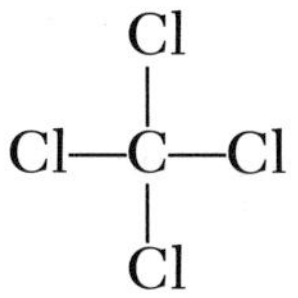

Types of Organic Compounds 209

Cl—C(Cl)(Cl)—Cl
carbon tetrachloride
(cleaning fluid)

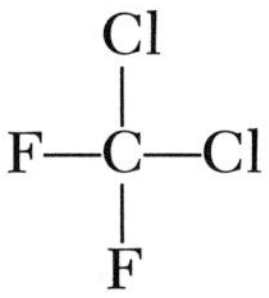

F—C(Cl)(F)—Cl
"freon"
(refrigerant)

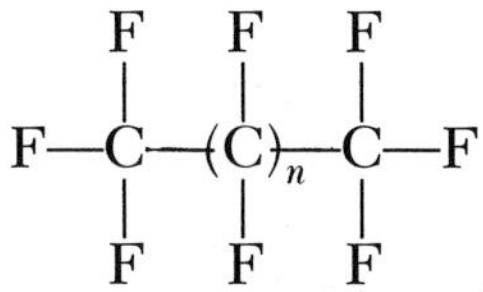

F—C(F)(F)—(C)$_n$—C(F)(F)—F
teflon
(n is a large number)

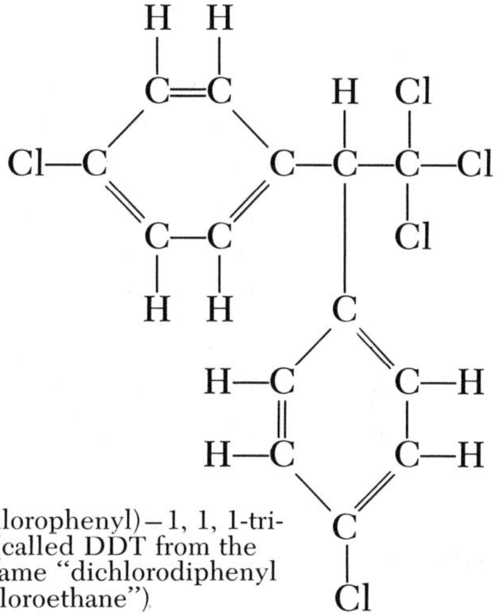

2,2-bis-(paradichlorophenyl)–1, 1, 1-tri-chloroethane (called DDT from the unsystematic name "dichlorodiphenyl trichloroethane")

Alcohols

The functional group characteristic of alcohols is the —OH, or hydroxyl group, and its presence in a molecule is denoted by a name ending in *-ol*. For example:

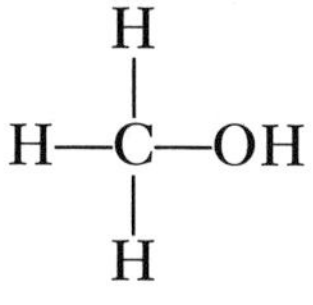

H—C(H)(H)—OH
methanol
(methyl alcohol)

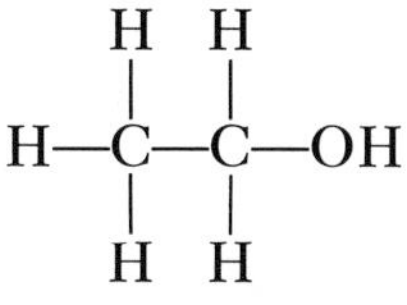

H—C(H)(H)—C(H)(H)—OH
ethanol
(ethyl alcohol)

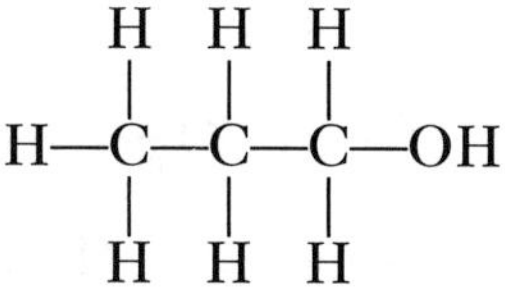

H—C(H)(H)—C(H)(H)—C(H)(H)—OH
n-propanol
(propyl alcohol)

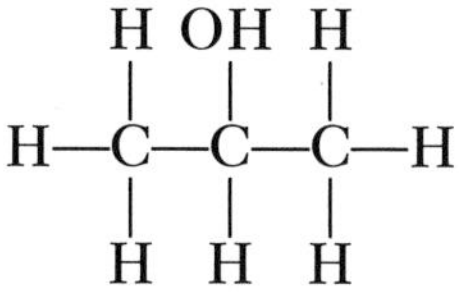

H—C(H)(H)—C(OH)(H)—C(H)(H)—H
isopropanol
(isopropyl alcohol)

There are two isomeric propanols because the -OH group can be attached to the end or to the middle carbon. The two ends are equivalent, so only two isomers are possible.

Alcohols are poisonous compounds useful as solvents and as intermediates in the synthesis of other organic compounds. Those with small alkyl groups, such as the examples, are miscible ("mixable") with water, as their structure, similar to water, suggests. Alcohols with six or more carbon atoms are not miscible with water, although they are liquids up to rather long carbon chains, because the carbon chain, which is similar in structure to the alkanes, dominates the molecule rather than the water-like -OH group. It may also be noted here that the boiling points and melting points of compounds increase as the length of the carbon chain increases, in any homologous series.

Ethers

Isomeric with alcohols but very different in properties are the ethers. The functional group characteristic of this homologous series is an oxygen linking two carbon atoms. The first few members of the series include:

$$H_3C-O-CH_3 \qquad H_3C-O-CH_2-CH_3$$

dimethyl ether   methylethyl ether

$$H_3C-CH_2-O-CH_2-CH_3$$

diethyl ether

Ethers are solvents for nonpolar materials. They are volatile and burn readily. Diethylether is the one commonly used as an anaesthetic.

Aldehydes

As the name suggests, if you take a couple of hydrogen atoms away from an alcohol molecule you have left an aldehyde, and this is in fact a way in which these compounds are formed in biological systems. The aldehyde group is

$$-\overset{\overset{\textstyle H}{|}}{C}=O.$$

The systematic names of these compounds end in -*al*. The first two members of the series are the most familiar.

$$H-\overset{\overset{\textstyle H}{|}}{C}=O \qquad H-\overset{\overset{\textstyle H}{|}}{\underset{\underset{\textstyle H}{|}}{C}}-\overset{\overset{\textstyle H}{|}}{C}=O$$

formaldehyde (methanal)   acetaldehyde (ethanal)

The aldehyde group also occurs in many sugars.

Ketones

Ketones, which are useful solvents, have

$$\diagdown_{\diagup}C=O$$

as their characteristic functional group. They may be named in two ways systematically, but the first member of the series goes by a common, nonsystematic name.

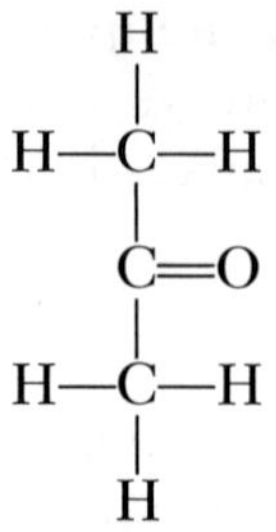

```
      H
      |
   H—C—H                    H     H H
      |                     |     | |
      C=O                H—C—C—C—C—H
      |                     |     | | |
   H—C—H                    H O H H
      |                   methylethyl ketone
      H                        butanone

      acetone
   dimethyl ketone
      propanone


   H     H H H              H H     H H
   |     | | |              | |     | |
H—C—C—C—C—C—H         H—C—C—C—C—C—H
   |     | | |              | |     | |
   H O H H H              H H O H H
   methylpropyl ketone      diethyl ketone
      2-pentanone             3-pentanone
```

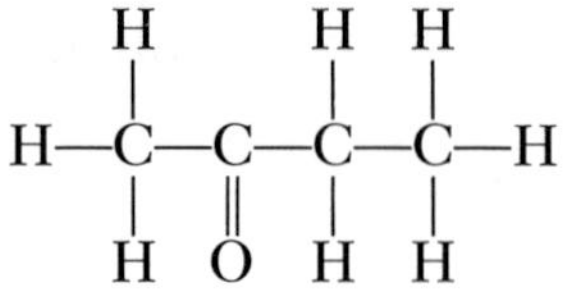

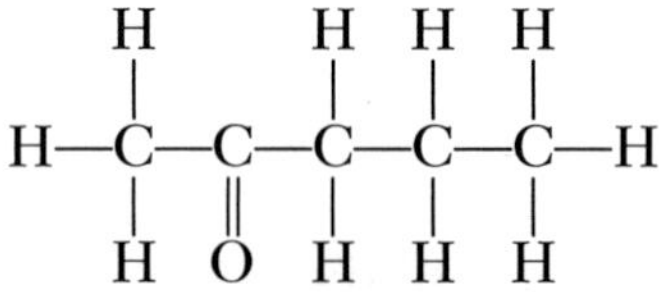

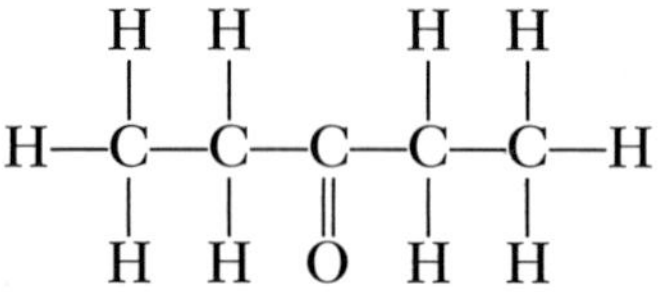

Carboxylic Acids

All organic compounds mentioned up to here have been covalent, although those containing oxygen are more or less polar because the oxygen atom forms its two single bonds at roughly 100° to each other. Pure carboxylic acids are polar, but in water solution they also ionize, like any other acids, to give hydrogen ions. The functional group, called the *carboxyl group*, is

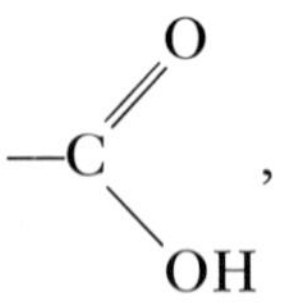

$$-C\begin{matrix}\nearrow O \\ \searrow OH\end{matrix},$$

and the H of the carboxyl group comes off in solution. Several familiar organic acids have two carboxyl groups.

Types of Organic Compounds

$$\underset{\text{formic acid}}{\text{H}-\overset{\displaystyle\text{O}}{\underset{\text{OH}}{\text{C}}}}\qquad\underset{\text{acetic acid}}{\text{H}-\overset{\text{H}}{\underset{\text{H}}{\text{C}}}-\overset{\displaystyle\text{O}}{\underset{\text{OH}}{\text{C}}}}\qquad\underset{\text{oxalic acid}}{\overset{\displaystyle\text{O}}{\underset{\text{HO}}{\text{C}}}-\overset{\displaystyle\text{O}}{\underset{\text{OH}}{\text{C}}}}$$

citric acid

tartaric acid

These are not systematic names. If you wanted to be systematic, you would have *methanoic acid* for formic acid, and acetic acid would be *ethanoic acid*. Tartaric acid would be 2, 3-dihydroxy butanedioic acid. Figure the others out for yourself!

Many carboxylic acids are components of fruit flavors. Vinegar is a 5% water solution of acetic acid. Vitamin C is ascorbic acid. Formic acid is the pain-producing substance injected by bee stings and ant bites. Oxalic acid removes rust stains.

Organic acids are relatively weak compared to such "strong" mineral acids as sulfuric or hydrochloric acid. "Weak" means that their solutions contain relatively few hydrogen ions. We may express this in a chemical equation:

$$\text{H}-\overset{\text{H}}{\underset{\text{H}}{\text{C}}}-\overset{\displaystyle\text{O}}{\underset{\text{OH}}{\text{C}}} \rightleftarrows \text{H}-\overset{\text{H}}{\underset{\text{H}}{\text{C}}}-\overset{\displaystyle\text{O}}{\underset{\text{O}^-}{\text{C}}} + \text{H}^+.$$

This represents the ionization of acetic acid. The double arrow indicates that the reaction is reversible, and so an equilibrium is established. In mechanics, an equilibrium situation arises from the action of balanced forces. In a chemical system, an equilibrium situation is characterized by two opposing reactions proceeding at the same rate. To understand how this comes about, let us start with pure acetic acid and pure water and consider what happens as equilibrium is reached.

1. Time t_0 Pure (covalent) acetic acid is added to pure water to make a 1M (molar) solution. (That means one gram molecular weight of $HC_2H_3O_2$, 60 grams, is dissolved in one liter (1000 cm³) of water.)

2. Time $t_0 + \Delta t$ (Δt = a few nanoseconds[1]) The hydrogen-oxygen bonds, bombarded by water molecules, break in a large number of molecules. The hydrogen ions stick to some of the water molecules to produce H_3^+O (hydronium ions), but an occasional H^+ reattaches itself to an acetate ion:

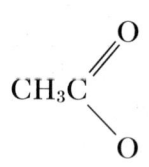

$$CH_3C\begin{matrix}\nearrow O \\ \searrow O\end{matrix}$$

3. Time t_1 (a few nanoseconds later) As more and more acetic acid molecules ionize, the rate at which hydrogen ions reattach themselves to the acetate ions increases. The rate at which ionization proceeds decreases as the number of ions increases, and so on until the two rates are the same.

4. Time (any time later) The rate of ionization of acetic acid molecules equals the rate of recombination of hydrogen and acetate ions. Both reactions keep going on, but because their rates are the same, there

---

[1] The prefix *nano-* means one billionth.

is no net change in the amount (concentration) of any substance—the condition of chemical equilibrium.

Chemical equilibrium is a dynamic state. Neither reaction stops. There is no *net* change in concentrations only because the forward and backward reactions continue to occur at the same rate.

Thioalcohols or Mercaptans

According to the Periodic Chart, sulfur should behave very much like oxygen, chemically. It does, and some kinds of oxygen-containing organic compounds have sulfur analogs. Mercaptans, so named because they "capture" or combine with mercury, are the sulfur (thio) analogs of alcohol. If you have ever met a disturbed skunk you know about mercaptans.

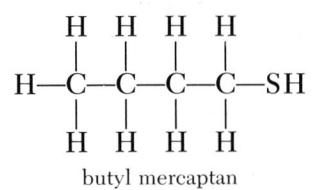

$$\begin{array}{c} \text{H} \ \ \text{H} \ \ \text{H} \ \ \text{H} \\ | \ \ \ | \ \ \ | \ \ \ | \\ \text{H—C—C—C—C—SH} \\ | \ \ \ | \ \ \ | \ \ \ | \\ \text{H} \ \ \text{H} \ \ \text{H} \ \ \text{H} \end{array}$$

butyl mercaptan

Some biologically important compounds contain -SH as one of their functional groups, and some protein chains are connected in parallel to each other by sulfide (-S-) linkages. Permanent waving solution breaks these sulfide linkages in hair so that they can be reformed in another place after the hair is set in a curled form. The new link holds the curl "permanently."

Amines

Nitrogen also forms covalent bonds with carbon. Such compounds can be considered as derivatives of ammonia, by replacing a hydrogen atom with an organic radical (alkyl group), so they are called *amines*.

[Structural diagrams: ammonia, methylamine, diethyl amine]

Amines are sometimes formed as decomposition products of proteins. Diethyl amine occurs in decaying fish and has a particularly unpleasant smell.

On the other hand, the amino group is a component of a group of compounds absolutely essential to life, the *amino acids,* which are the building blocks of protein. The biologically significant amino acids all have the amino ($-NH_2$) group attached to the same carbon to which the carboxyl group is attached, the so-called α carbon. They are called α-amino acids.

[Structural diagrams:
alanine — 2-aminopropanoic acid
glycine — aminoethanoic acid
lysine
tryptophane]

Types of Organic Compounds

$$\text{arginine}$$

$$\text{cysteine}$$

Lysine, tryptophane, and arginine are three of the nine amino acids that are essential to human nutrition but cannot be synthesized by the body. They must be obtained from assorted protein-containing foods. The human body makes about a dozen other amino acids that it also requires.

Esters

Organic acids react with alcohols in a manner analogous in some ways to the reaction of mineral acids with bases, although *esterification*, as the organic reaction is called, is perhaps more different from than similar to neutralization. The common feature is the combination of the acid H and the alcohol —OH to form water:

$$R-C(=O)-OH \;+\; R'-OH \;\rightarrow\; R-C(=O)-O-R' \;+\; H_2O$$

acid + alcohol → ester + water
(R and R' need not be the same alkyl group) (covalent, unlike salts)

Esters are on the whole liquids with pleasant odors and tastes. Many fruit flavors and flower fragrances are esters.

```
        H   O
        |   ‖
    H—C—C        H   H   H   H   H
        |    \   |   |   |   |   |
        H     O—C—C—C—C—C—H
                  |   |   |   |   |
                  H   H   H   H   H
```
amyl acetate
(banana flavor)

methyl salicylate
(oil of wintergreen)

```
    H   H   H      O
    |   |   |     ‖        H   H
H—C—C—C—C—O—C—C—H
    |   |   |              H   H
    H   H   H
```
ethyl butyrate
(pineapple flavor)

Alicyclic Compounds

An old name denoting straight chain single bonded carbon compounds is "aliphatic" (from the Greek word for oil) compounds. If such a straight chain "bites its own tail," as it were, a ring compound of the alicyclic type results. The number of carbon atoms in a ring is determined mainly by the amount of strain due to displacing the carbon bonds from their normal tetrahedral configuration that a ring of a given size introduces.

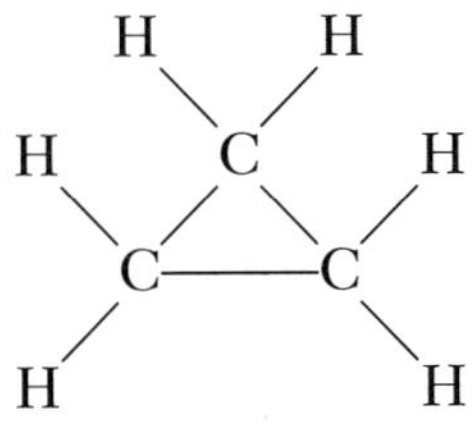

cyclopropane 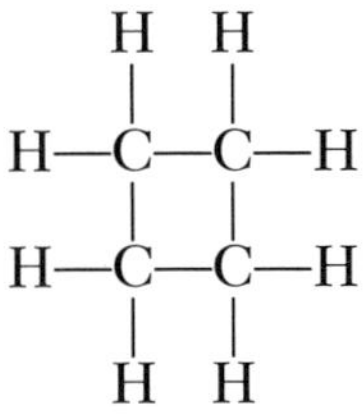   cyclobutane

These rings are strained and unstable and hard to make.

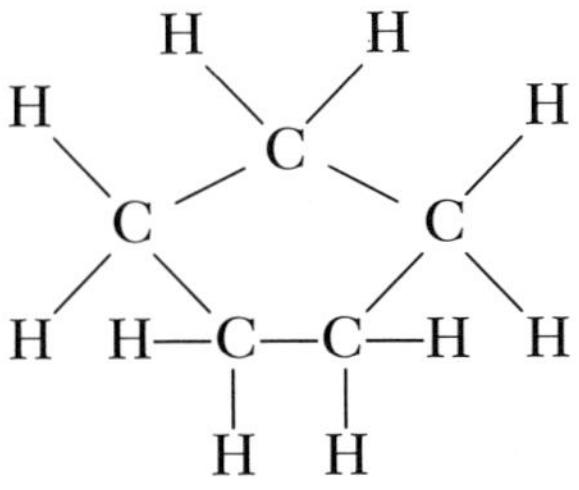

cyclopentane 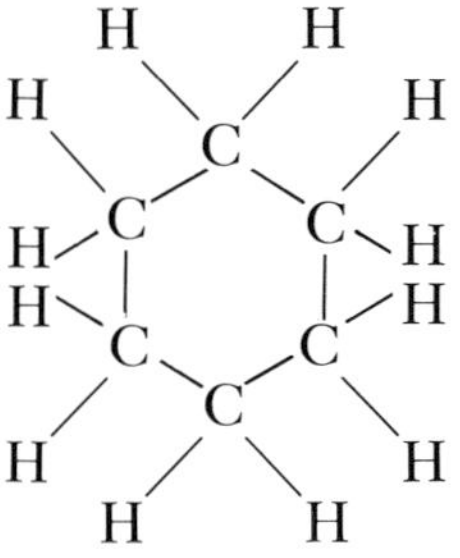   cyclohexane

These rings are stable and occur readily.

Cyclopentane and cyclohexane are useful solvents for nonpolar materials and are intermediates for making various other chemicals.

Aromatic Compounds

Many pleasant-smelling, volatile liquid substances known since the middle of the nineteenth century have presented chemists with a mystifying problem in structure. Benzene, which is typical, is an oily, aromatic, volatile liquid that is a good solvent for nonpolar substances, but immiscible with water. It is very stable and unreactive chemically. Analysis reveals its molecular formula to be C_6H_6. How is it put together? The answer came in one of those moments of inspiration that you don't find listed in the steps of the scientific method, but upon which most creative scientists will admit their dependence. The scientist in this case was Friedrich August Kekulé von Stradonitz, who, in 1865, according to his own account, was dozing on a bus when he "saw" molecules tumbling about and forming the answer to the benzene problem: a ring with alternate double bonds. His solution was not without its problems. Why was not benzene reactive like ethylene or acetylene or other hydrocarbons containing double or triple bonds? The concept of *resonance* was introduced to explain the stability of the benzene ring, which was shown to remain untouched through all kinds of chemical processes by which substituent functional groups could be added and could subsequently react. According to the resonance idea, the moving electrons which form the bonds can "resonate" rapidly between alternate equivalent positions making the ring itself very stable. In the twentieth century, quantum mechanics resolved the benzene question beautifully. If we consider that the electrons in a molecule occupy orbitals that belong to the whole molecule and are different from, although related to, the atomic orbitals from which they are derived, we get a picture of benzene that accounts for the symmetry, stability, and equivalence of position that characterize the benzene ring (Fig. 10.2).

Kekulé structure

resonance in benzene

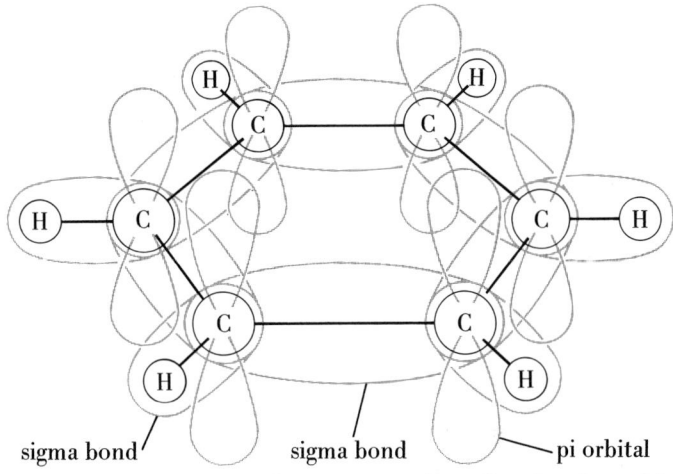

Figure 10.2 Diagrammatic representation of molecular orbitals in benzene. Notice the symmetry and equivalence of all the carbon atoms and the absence of conventional double bonds.

In the case of a 12-atomic compound like benzene, the quantum mechanical calculations are both difficult and approximate, yet they give a better picture of the reasons for the observed chemical and physical properties than any of the simpler assumptions.

The hydrogen atoms of benzene may be replaced by a wide variety of functional groups. The properties of the resulting compounds will be similar, on the one hand, to their aliphatic analogs, because of the properties of the functional group, and dissimilar, on the other hand, because of the influence of the ring. For example, the phenols, having -OH replacing one hydrogen, behave much like alcohols in forming esters, but on the other hand show acidic properties in that the hydrogen of the -OH group actually ionizes off in phenols, whereas it may be replaced by metals in alcohols only with difficulty in most cases.

Many important chemical substances are ring compounds containing the benzene "nucleus," conventionally written

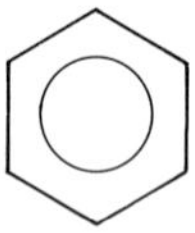

with any vertex assumed to have a hydrogen atom attached unless otherwise specified.

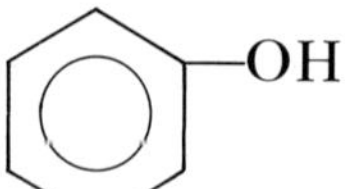

phenol
(carbolic acid)

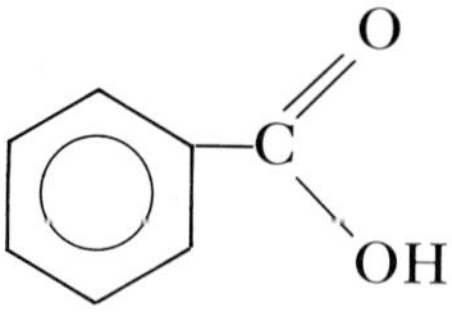

benzoic acid
(its sodium salt
is a preservative)

aniline
(basis of coal
tar dies)

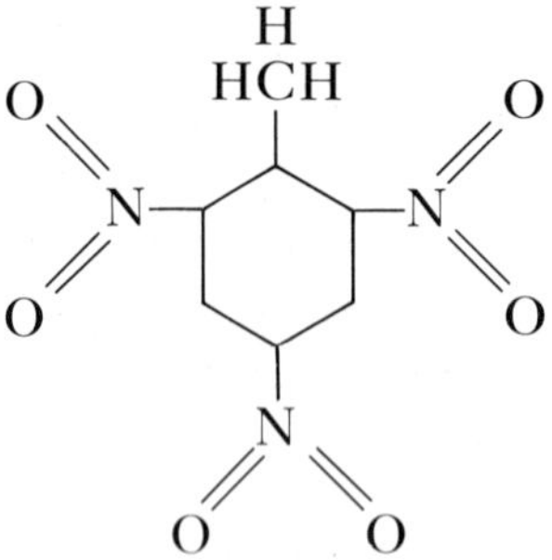

trinitrotoluene
(TNT)

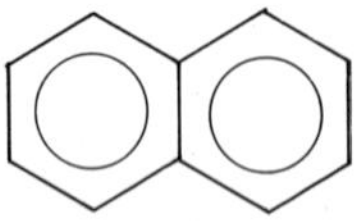

naphthalene
(mothballs)

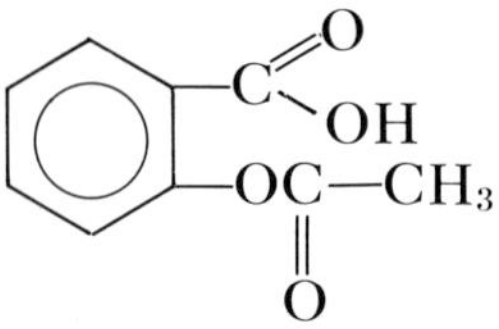

acetylsalicylic acid
(aspirin)

$$\text{sulfanilamide}$$

sulfanilamide

The variety of organic compounds is only suggested by the examples given here. Many of the most useful and important have more than one functional group, which is especially true of the biologically active compounds. Many are very large molecules, polymers built up of various units, often of several types of smaller molecules. Such plastics as polyethylene, fluorinated polyethylene (teflon), and polyvinyl chloride, synthetic fibers such as nylon, acrylics, and polyesters, and even such natural cellulose fibers as cotton and linen and such protein fibers as wool and silk are relatively simple polymers made up of one or a few basic units (monomers). We are only beginning to appreciate the wonderful complexity of the molecular basis of life.

SUMMARY STATEMENTS

1. Some two-and-a-half million organic compounds have been reported in the chemical literature. Those occurring naturally are associated with living systems, past or present, but most organic compounds now known are not natural products, but have been synthesized in laboratories from simpler natural substances.

2. The existence of the large number of organic compounds is due to the ability of carbon to form long straight or branched chains and rings with itself as well as to form covalent bonds with other elements, especially H, O, N, and S.

3. Organic compounds with the same empirical formulas

may be quite different in chemical and physical properties because the same atoms may be arranged in several different ways in a molecule. Molecules having the same numbers and kinds of atoms but different structures are called *isomers*.

4. The chemical characteristics of organic compounds are due in large measure to the functional groups attached to the carbon chain or ring, although rings and chains themselves impart certain properties to compounds. Common functional groups are:

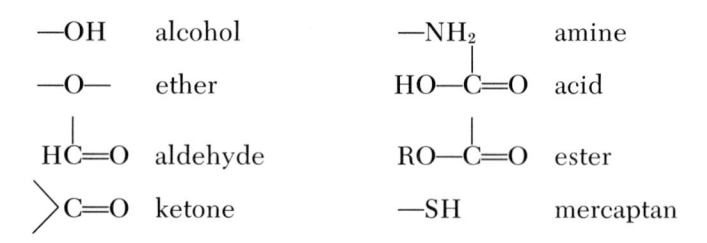

—OH	alcohol	—NH$_2$	amine
—O—	ether	HO—C̣=O	acid
HĊ=O	aldehyde	RO—C̣=O	ester
>C=O	ketone	—SH	mercaptan

5. Most common drugs, fibers, dyes, foodstuffs, and plastics are organic compounds. Many are very complex materials; others, particularly plastics, are polymers made by joining identical simple molecules in long chains.

REVIEW QUESTIONS

1. Why are carbon compounds called *organic compounds*?
2. Who was the first chemist to synthesize a compound recognized as "organic" from starting materials recognized as "inorganic"?
3. Why are there so many carbon compounds?
4. What is chemotherapy? Name some notable discoveries (and discoverers) in this field.
5. Write the structural formulas for the following functional groups:

ether	alkene	phenyl	ketone
alcohol	alkyne	mercaptan	acid (carboxylic)
amine	alkyl	aldehyde	ester

6. Name a familiar or useful compound containing each of the groups listed above (example: alkyne: acetylene).

7. Give systematic names for the following

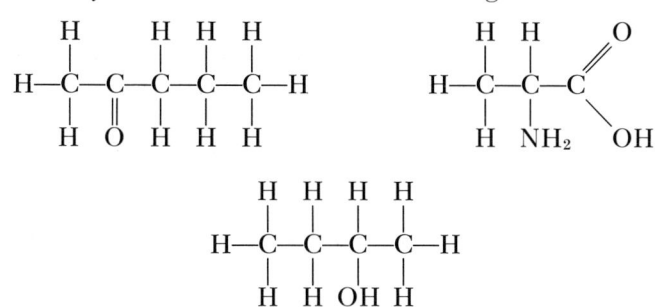

$$\begin{array}{c} H \\ | \\ H-C-C-C-C-C-H \\ | \| | | | \\ H O H H H \end{array} \qquad \begin{array}{c} H H \\ | | \\ H-C-C-C \\ | | \diagdown \\ H NH_2 OH \end{array}$$

with the first structure having H, H, H, H across the top carbons and H, O (double bond), H, H, H below.

$$\begin{array}{c} H H H H \\ | | | | \\ H-C-C-C-C-H \\ | | | | \\ H H OH H \end{array}$$

8. Why do you suppose ethylene,

$$\begin{array}{c} H H \\ | | \\ H-C{=}C-H, \end{array}$$

polymerizes to long chains of polyethylene $H_3C(CH_2)_nCH_3$ so readily?

THOUGHT QUESTIONS

1. Keep a list of the organic compounds you encounter or use in a day. Classify them by functional groups where possible.

2. It is sometimes said that many people today expect "chemical solutions" to all their problems. In what respect is this true? Can you cite cases in which "chemical solutions" themselves become or lead to greater problems? What alternatives would you offer them?

3. Note all the "chemistry," true and false, presented in an evening's worth of television commercials. What misleading statements and out-and-out errors did you spot? Which were most accurate?

4. Look at yourself and your immediate surroundings and list all of the items and materials that could not have been there 150 years ago.

5. Discuss the importance of natural and synthetic polymers in modern life.

READING LIST

Asimov, Isaac, *A Short History of Chemistry*, Doubleday (Anchor), Garden City, N.Y., 1965. A brief, clear account of the history and principles of chemistry by one of the very best writers on science.

Horrigan, Philip A., *The Challenge of Chemistry*, McGraw-Hill, New York, 1970. A topical treatment of chemical principles and their present-day applications in medicine, industry, and agriculture, and of ecology.

Young, Louise (ed.), *The Mystery of Matter*, Oxford University Press, New York, 1965. A collection of readings around the theme of the structure of matter. Part II is of special interest here.

Plate 10 Carbohydrates, such as the sugar in the beaker, contain C, H, O, with a 2:1 ratio of H to O as in water. Adding sulfuric acid irreversibly extracts the water, leaving carbon.

Color illustrations are from the Life Nature Library or the Life Science Library

Plate 11(a) Ethane: the single bo[nd]. When two carbon atoms are joined [by] only one shared pair of electrons, [the] molecule is "saturated." The car[bon] bonds are tetrahedrally arranged [and] very stable.

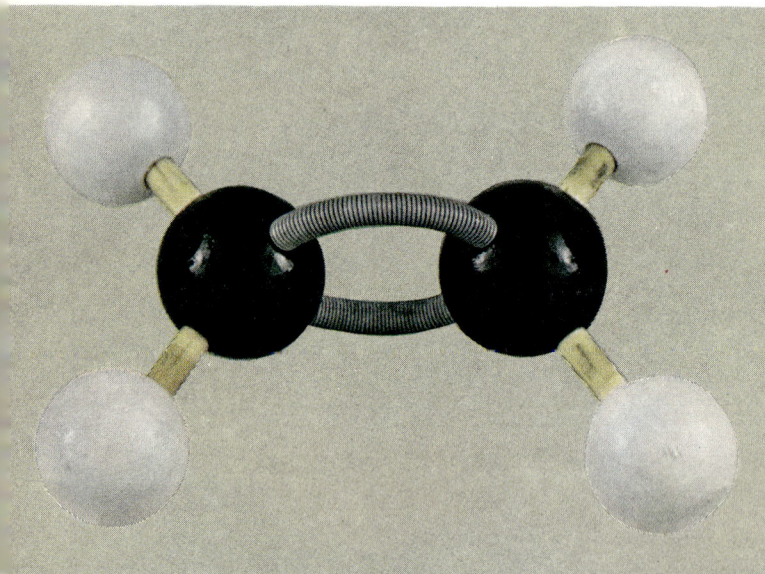

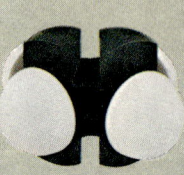

Plate 11(b) Ethylene: the do[uble] bond. When two pairs of elec[trons] are shared between two ca[rbon] atoms, the bonds are strained o[ut of] their normal configuration. Such ["un]saturated" molecules are rathe[r] active, combining with other a[toms] or polymerizing at the double bo[nd].

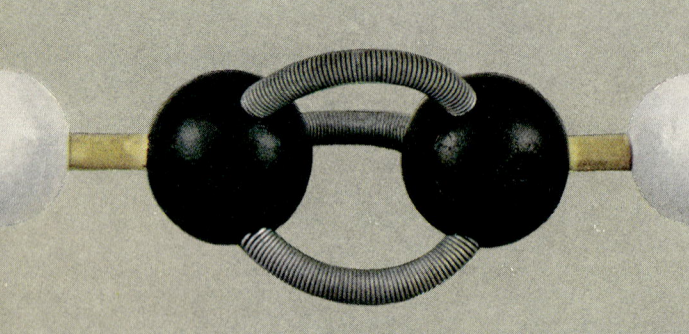

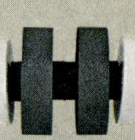

Plate 11(c) Acetylene: the [triple] bond. The sharing of three p[airs of] electrons between two carbon [atoms] represents an extremely strain[ed sit]uation; such molecules are extr[emely] reactive.

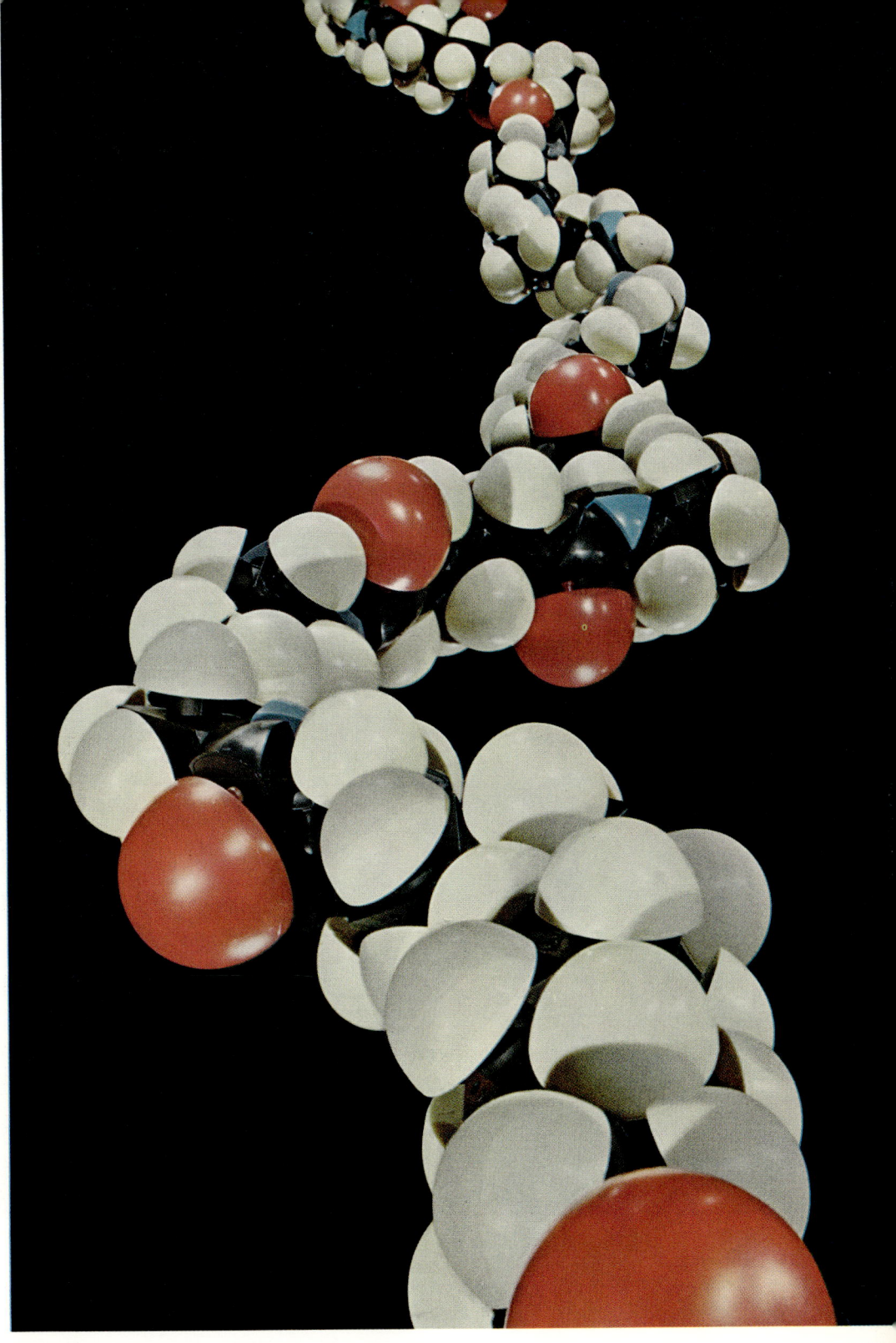

Plate 12 A molecular model of a polymer, nylon.

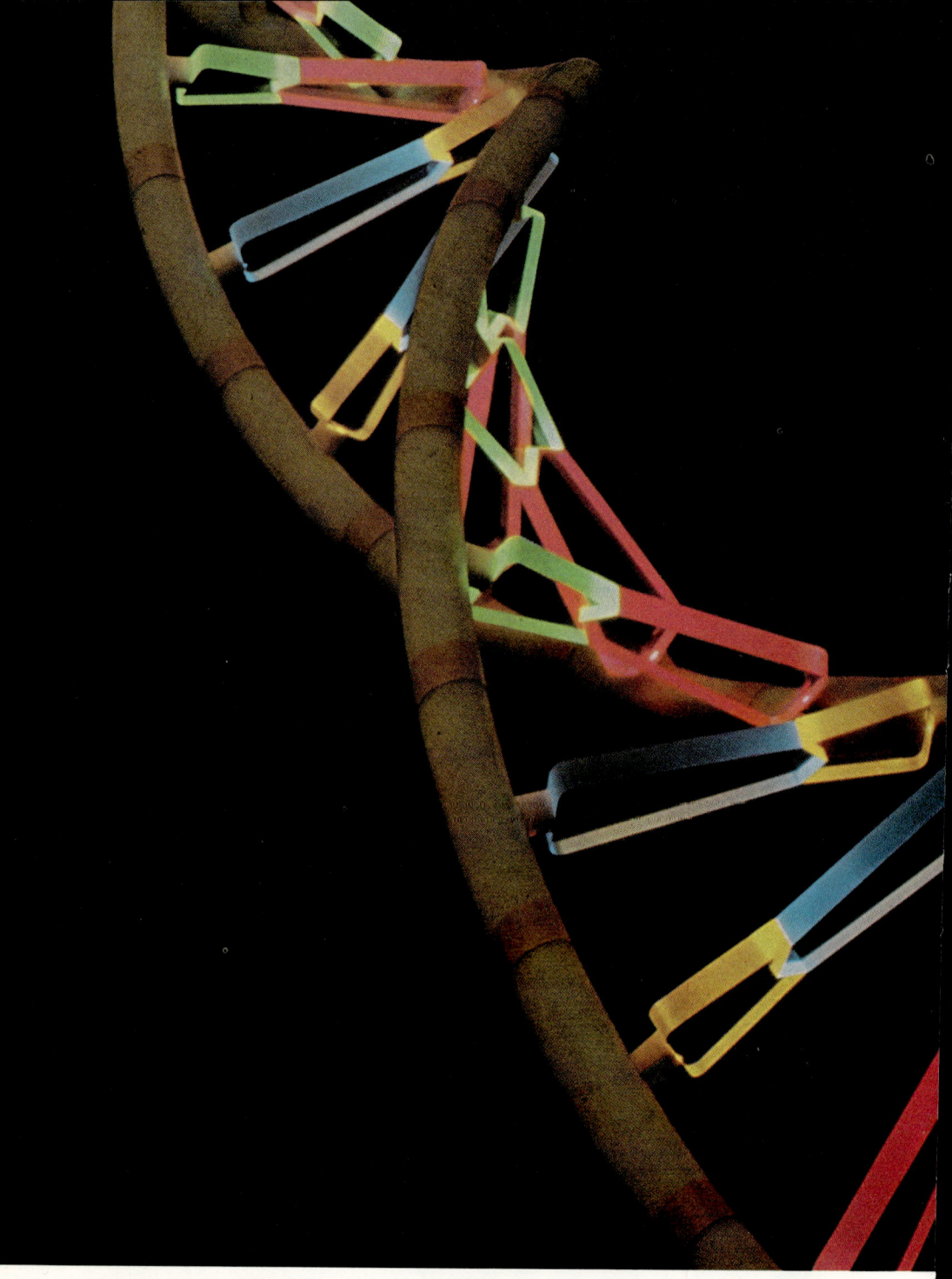

Plate 13 A model of the DNA double helix. The red and blue shapes represent purine bases; the green and yellow, pyrimidines. A red-green pair is an adenine-thymine pair and blue-yellow is guanine-cytosine. The sugar-phosphate "backbone" is made of light and dark tan segments.

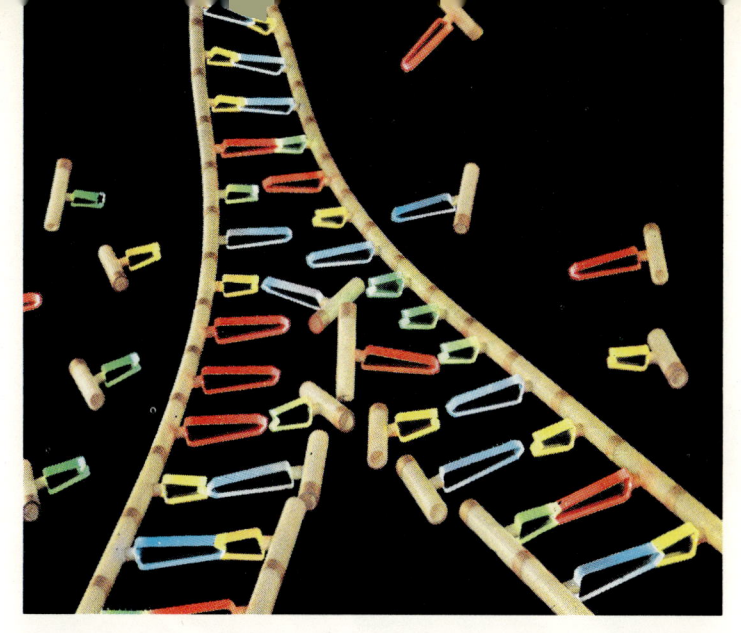

Plate 14(a) The double helix splits and each strand rebuilds a partner identical to its original one by base pairing, using the nucleotides available.

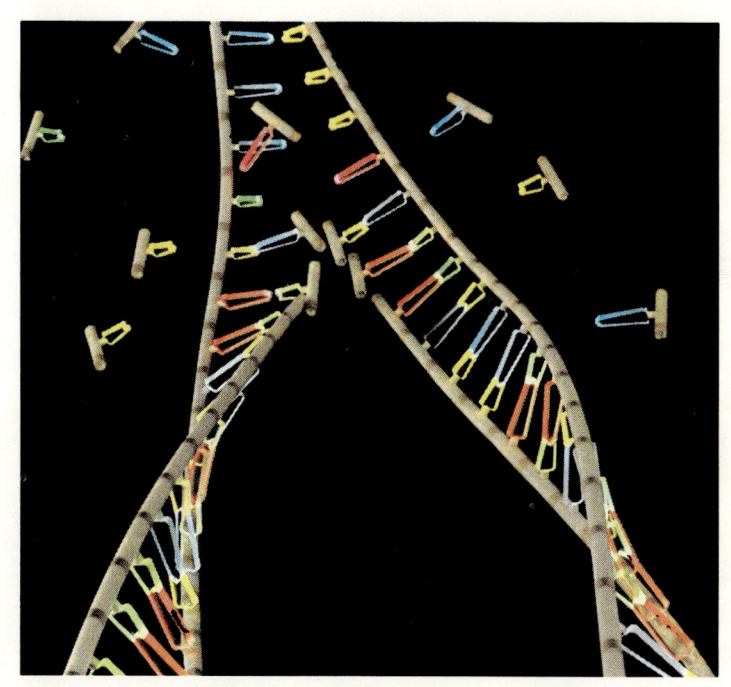

Plate 14(b) Replication proceeds until two complete new double helices are formed.

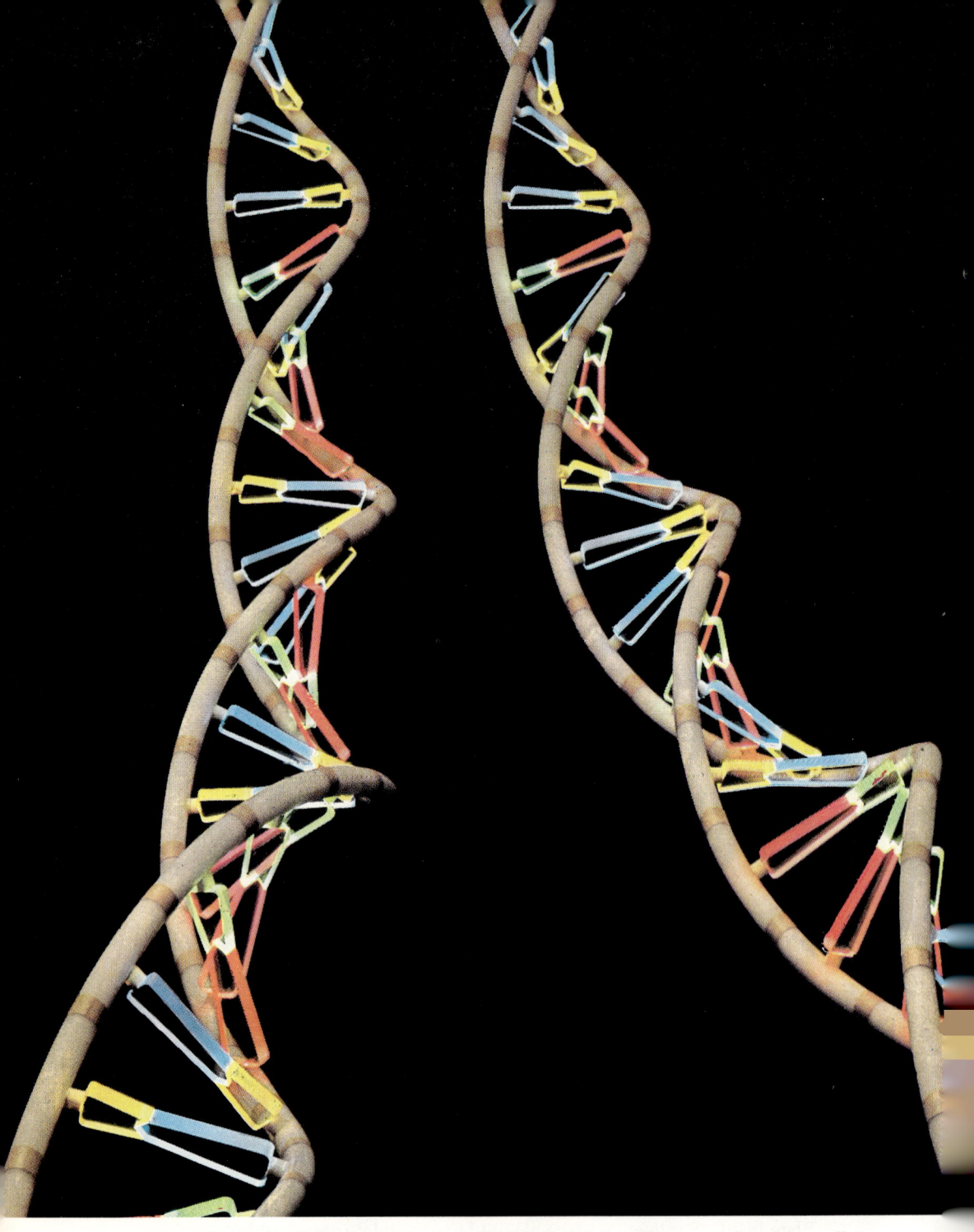

Plate 15 Each of the two identical double helices will go to a daughter cell in mitosis, so that each will have identical genetic information.

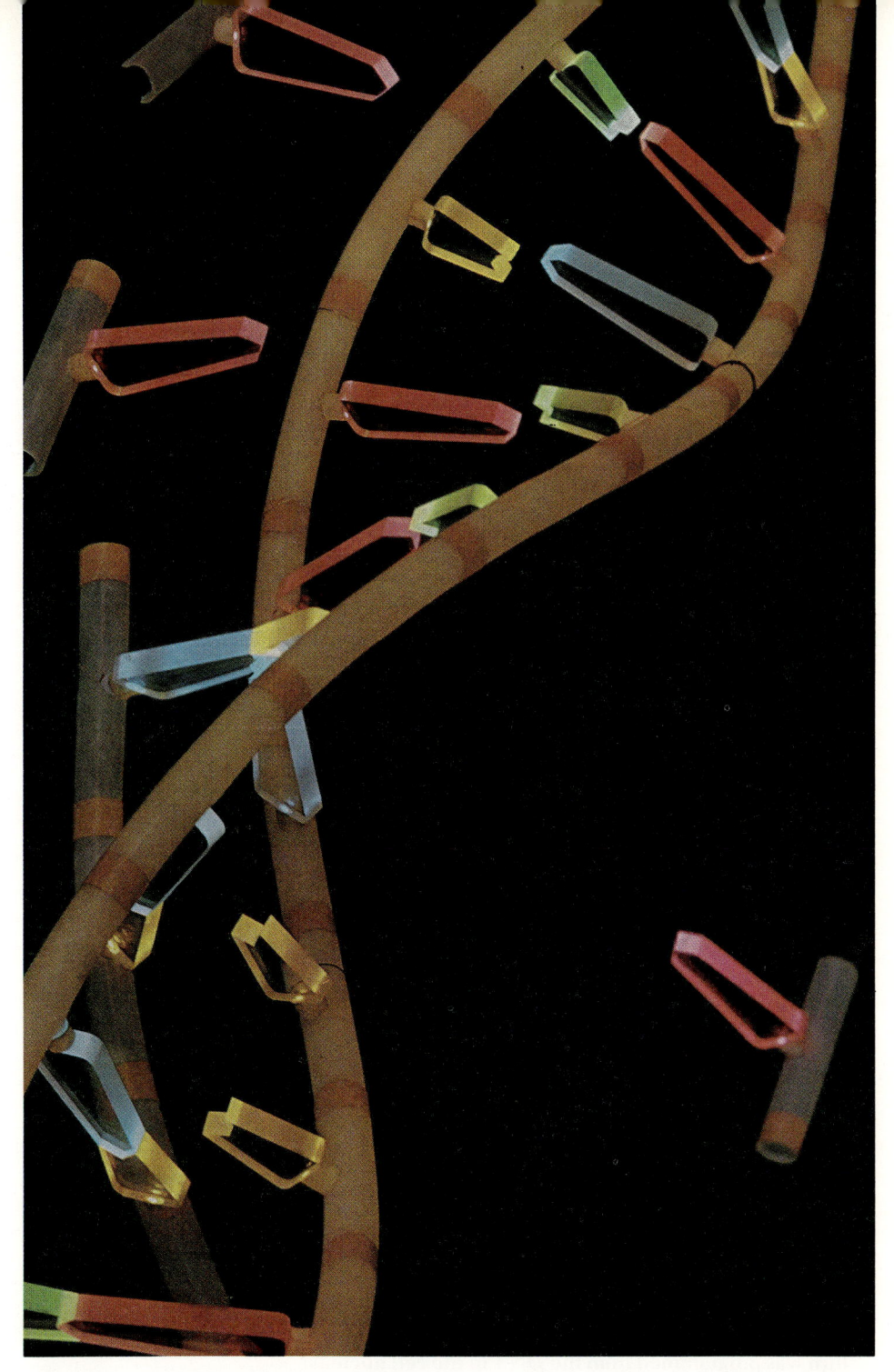

Plate 16 A separated strand of DNA serves as the template for building single strands of RNA, now using nucleotides composed of ribose (gray segments) and uracil instead of thymine.

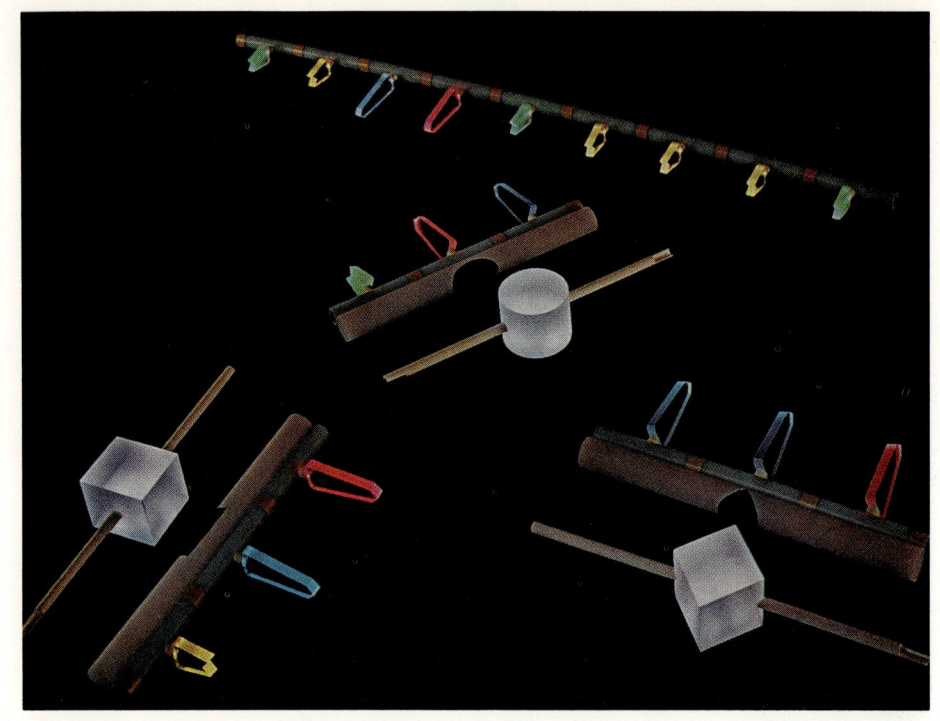

Plate 17(a) Long strands of m-RNA "recognize" short t-RNA strands by base pairing, A-U and G-C. A second code enables t-RNA to "recognize" the amino acid for which it is coded.

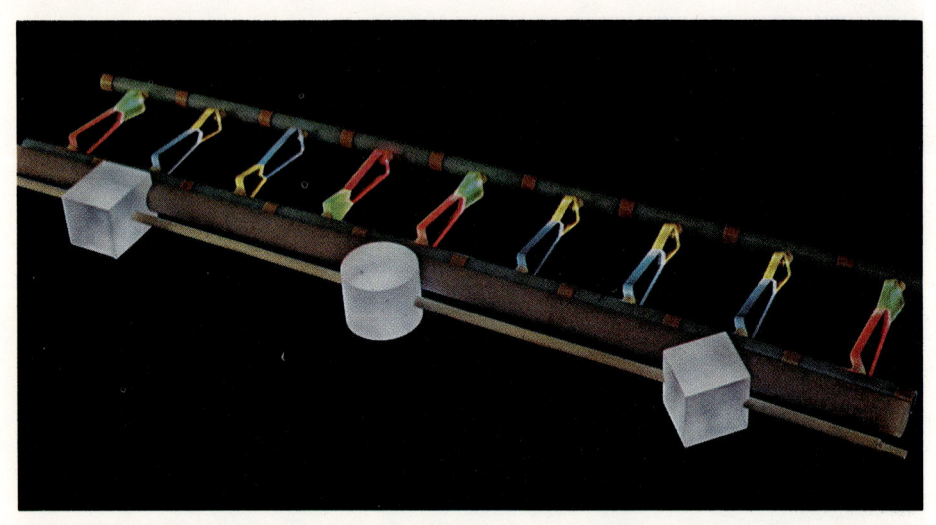

Plate 17(b) Peptide bonds form between adjacent amino acids, producing a protein molecule according to the specifications of the m-RNA template, which in turn was programmed from the DNA in the cell nucleus.

11

Molecular Biology: The Physics and Chemistry of Life

There is a good deal of ad hoc evidence for the various cyclic theories of history and the universe, whether or not any is scientifically justified. Certainly ideas themselves go in cycles. One such concept that goes around and around is vitalism. In prescientific times the ideas that advanced forms of life must come from life (i.e., living parents) and that lower forms were spontaneously generated from nonliving matter existed without apparent conflict side by side. Then in the eighteenth and nineteenth centuries (while Wöhler, Kolbe, and others were disproving vitalism with respect to organic compounds) Spallanzani and Pasteur showed that even the simplest and most minute organisms developed from parent organisms, not from inanimate materials, and the theory of spontaneous generation which had taught that flies came from decaying meat and frogs from mud was laid to rest. Vitalism, the idea that life is unique and must come from living precursors all the way back to some ultimate beginning by special creation, held sway. But along about the same time, the mechanism of creation began to be probed. Must we assume the

Creator did it with a sort of wave of a magic wand, or might He have brought it about by the operation of natural processes? In the phrasing of Loren Eiseley,[1] is life natural or supernatural?

If life is indeed natural, then it must have arisen out of the inanimate matter that preceded it, according to all the evidence of the history of life on Planet Earth. If this is so, we must look for a chemical origin for life, and if it has such an origin, it must still be essentially chemical in nature, at least as far as mechanisms go. In the 1970's, the vitalist-mechanist argument still goes on: When one knows the chemical mechanisms of life does one know all there is to know about it, or is a living cell more than the sum of its chemical parts? The mechanist says, "That's all!" The vitalist says, "There's more!" What does the scientifically accumulated evidence say?

The Chemical Nature of Life

First, the human body, as well as every other living organism (a redundancy! *organism* means living thing), has certain chemical requirements for maintaining life. It must be fueled by food, and food must also provide the raw materials for growth and the production of replacement parts. It must have water as the medium in which its chemical reactions occur, and it must have oxygen to support the combustion of its fuels for energy production. Water and oxygen are plentiful on Earth, although in many places their quality leaves much to be desired, largely because of the activities of that form of life reputed to be the most highly developed on the planet. Technologically, the providing of adequate food supplies may be no more complex than providing pure air and water, providing the world's population does not grow indefinitely, but chemically, food is a good deal more complex.

[1] *The Firmament of Time* (New York: Atheneum, 1962).

Food Chemistry

An organism uses food to produce energy and tissue. Energy is derived in the main by converting compounds containing C—H, C—C, and C—O—H bonds into CO_2 and H_2O. The compounds usually used for this purpose belong to the classes of carbohydrates and fats.

Fats are esters of long chain aliphatic acids called *fatty acids* with glycerol (glycerine), an alcohol containing three —OH groups.

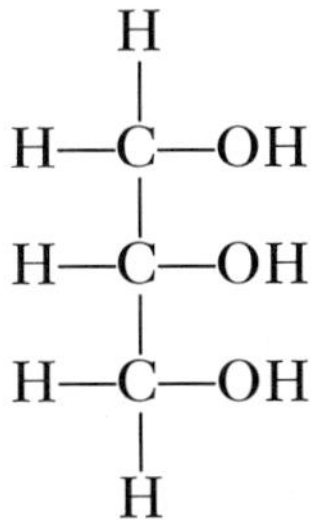

glycerol
(1,2,3 propanetriol)

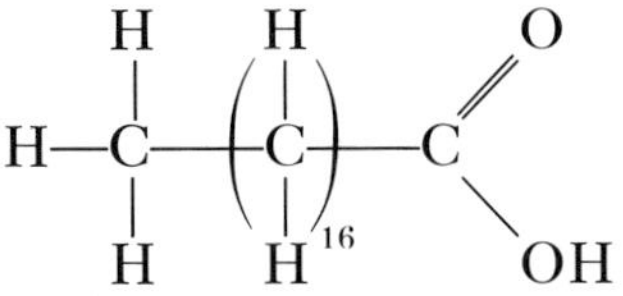

stearic acid

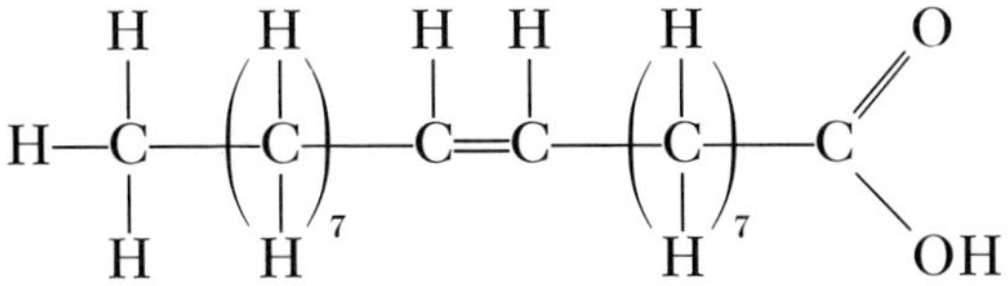

oleic acid

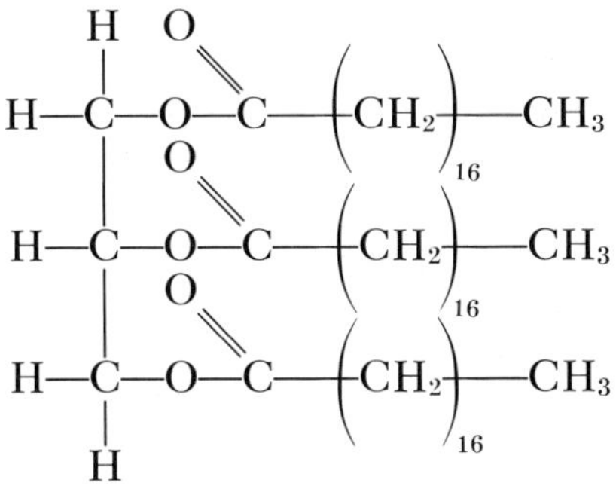

glyceryl tristearate
(stearin)

$$\begin{array}{l} H-\overset{\overset{\displaystyle H}{|}}{C}-O-\overset{\overset{\displaystyle O}{\diagdown}}{C}-(CH_2)_7-\overset{\overset{\displaystyle H}{|}}{C}=\overset{\overset{\displaystyle H}{|}}{C}-(CH_2)_7CH_3 \\ H-\overset{|}{C}-O-\overset{\overset{\displaystyle O}{\diagdown}}{C}-(CH_2)_7-\overset{\overset{\displaystyle H}{|}}{C}=\overset{\overset{\displaystyle H}{|}}{C}-(CH_2)_7CH_3 \\ H-\overset{|}{C}-O-\overset{\overset{\displaystyle O}{\diagdown}}{C}-(CH_2)_7-\overset{\overset{\displaystyle H}{|}}{C}=\overset{\overset{\displaystyle H}{|}}{C}-(CH_2)_7CH_3 \\ \overset{|}{H} \end{array}$$

glyceryl trioleate
(olein)

Stearin and olein are two commonly occurring fats. Stearin is a solid material; olein is a liquid. Generally, esters of saturated acids such as stearic acid are solids, whereas esters of unsaturated acids such as oleic acid are oils. Solid fats are generally of animal origin; oils come from plants. In digestion, fats are hydrolized to glycerol and fatty acids. *Hydrolysis* is the opposite of esterification. In the cells, the esters are reformed.

The body "burns" some fats to produce energy, and stores some. By marking fats with radioactive isotopes such as C^{14}, the radioactive form of carbon, it has been shown that fat reserves are constantly being used up and replaced rather than having the same fats in "permanent" storage without exchange. Fats produce more energy per unit mass than any other foods—about 9.5 kilocalories per gram, compared to a little over 4 kilocalories per gram of sugars and proteins. Kilocalories are the weight-watcher's Calories. Because a little fat goes a long way in energy production, fat eaten over and above current energy needs is stored, usually in noticeable places. Just what it means to say the body "burns" any fuel we will examine later in this chapter.

Carbohydrates, the body's other important energy source, are in composition exactly what their name implies: compounds of carbon and hydrogen and oxygen, the latter two in the same 2:1 ratio as in water. The simplest carbohydrates are the more or less sweet-tasting water-soluble white crystalline solids we call *sugars* (see Plate 10). The simplest natural sugars, or monosaccharides, contain 5- or 6-carbon chains and are called *pentoses* or *hexoses*, respectively.

```
        H                    H                     H
        |                    |                     |
    H—C—OH                  C=O                   C=O
        |                    |                     |
       C=O                HO—C—H                HO—C—H
        |                    |                     |
    H—C—OH                HO—C—H                 H—C—OH
        |                    |                     |
    H—C—OH                 H—C—OH                H—C—OH
        |                    |                     |
    HO—C—H                 H—C—OH                H—C—OH
        |                    |                     |
        H                  H—C—OH                H—C—OH
    d-ribose                 |                     |
                             H                     H
                         d-glucose         possible optical isomer
                         (dextrose)            of d-glucose
```

"d-" refers to a particular configuration of the carbon chain, not to the direction of rotation of polarized light.

As the structural formulas show, the hexoses are isomeric. All are $C_6H_{12}O_6$, and the two shown even have the same systematic name, 2,3,4,5,6-hydroxy-hexanal, because they contain the same groups on the same atoms of their chains. Nevertheless they are separate and distinct compounds because, due to the particular direction in which the groups are bonded to the carbon atoms, the two molecules are not superimposable on each other but are mirror images, like one's right and left hand. Any carbon atom that is bonded to four different groups, no matter what they are, makes this mirror image effect possible, and such a carbon is said to be asymmetric. Any molecule containing one or more asymmetric carbon atoms has the further peculiar property of rotating the plane of

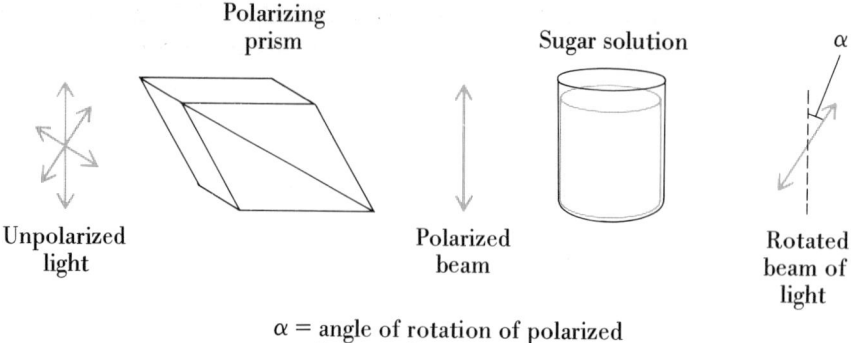

α = angle of rotation of polarized light, a characteristic of the chemical substance.

Figure 11.1 Polarimeter arrangement for measuring the rotation of polarized light by optically active compounds.

polarization of plane polarized light, although it sometimes happens that the effects of two asymmetric carbons in one molecule cancel each other out (Fig. 11.1). Compounds that rotate the plane of polarization of polarized light are said to be optically active, and two molecules that are mirror images will rotate the plane in opposite directions. They are therefore called optical isomers. Perhaps the strangest thing about optical activity is that biologically active molecules, almost without exception, rotate the plane of polarization to the left; that is, they are levorotatory, rather than dextrorotatory. Louis Pasteur discovered optical activity when he observed and separated with tweezers the two kinds of crystals that formed when tartaric acid was synthesized in the laboratory. The two kinds of crystals are mirror images of each other, they rotate the plane of polarization in opposite directions, and only the levorotatory crystals could be assimilated by rats (Fig. 11.2). All tartaric acid from living sources (grapes, for example) is levorotatory. No one yet knows why, but through all the long evolutionary line of both plants and animals, organisms make and use levorotatory compounds almost ex-

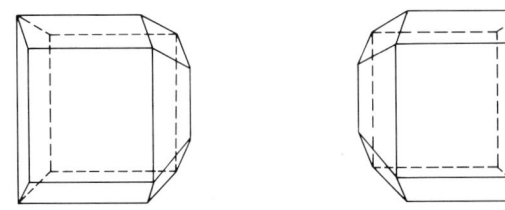

Figure 11.2 Left- and right-handed tartaric acid crystals.

clusively, and cannot metabolize the dextrorotatory isomers.

From the biochemist's point of view, optical activity is an extremely useful property of biological materials, for α, the angle of rotation, is a constant characteristic of the molecule by which an optically active compound can be identified.

Although our digestive system breaks all digestible carbohydrates down to monosaccharides for metabolism in the cells, the bulk of our carbohydrate consumption is in the form of disaccharides (general formula $C_{12}H_{22}O_{11}$) such as cane sugar (sucrose) and polysaccharides, or starches. Cellulose, the main component of wood, cotton fiber, and synthetics such as rayon, is also a polysaccharide, but the hexose units in starch and in cellulose are connected differently. We humans can digest (break down) starch linkages, but not cellulose linkages.

Although we have written the structural formulas of the hexoses as straight chains, chemical evidence indicates that they actually curl around into rings, thus:

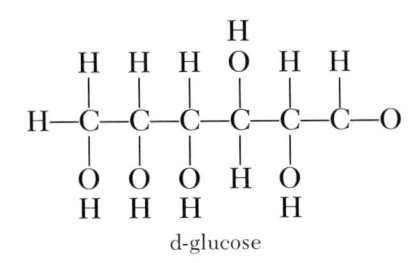

$$H-\underset{\underset{H}{O}}{\overset{H}{\underset{|}{C}}}-\underset{\underset{H}{O}}{\overset{H}{\underset{|}{C}}}-\underset{\underset{H}{O}}{\overset{H}{\underset{|}{C}}}-\underset{\underset{H}{H}}{\overset{H}{\underset{|}{C}}}-\underset{\underset{H}{O}}{\overset{H}{\underset{|}{C}}}-\overset{H}{\underset{|}{C}}=O$$

d-glucose

```
      H H H O   H
      | | | ‖   |
    H-C-C-C-C-C-C-H
      | | | | ‖ |
      O O O H O H
      | | |
      H H H
```
d-fructose

α-d-glucose
(one possible configuration)

α-d-fructose
(one possible configuration)

Glucose, containing an aldehyde group, and fructose, containing a ketone group, are called "reducing sugars" because they react with Fehling's solution.

To form a disaccharide, such as sucrose, two of these hexose rings condense, splitting out a molecule of water. The reverse reaction, hydrolysis of sucrose into glucose and fructose, can easily be carried out by boiling sucrose with hydrochloric acid solution.

sucrose ($C_{12}H_{22}O_{11}$)
(dextrorotatory)

d-glucose γd-fructose
(both levorotatory)

Sugar digestion begins with this hydrolysis reaction, producing levorotatory molecules for the cells' use. Lactose and maltose are the other two naturally occurring and nutritionally useful disaccharides.

Polysaccharides are long polymers made up of repeating identical monosaccharide units, usually glucose.

cellulose

starch

Perhaps the most noteworthy item in all these formulas is that minor structural differences, such as the different arrangements of identical glucose units in cellulose and starch, have tremendous biological effects. The chemical reactions that go on in the human body at the relatively low temperature of about 98.6°F (about 37°C) are nearly all *catalyzed* reactions. A *catalyst* is a substance that increases the rate at which a chemical reaction occurs, without being itself permanently changed. *Enzymes* are body catalysts. No reaction "goes" in the body unless there is a specific enzyme to catalyze it. We have an enzyme to catalyze the hydrolysis of starch. We do not have an enzyme to catalyze the hydrolysis of cellulose. The explanation of our ability to be nourished by potatoes but not by pine logs is as simple as that.

Enzymes

Clearly, then, the real story of the chemistry of life is the story of enzymes, which belong to the third class of compounds our food must contain in quantity, proteins. Proteins are much more complicated, both physically and chemically, than carbohydrates and fats. They are polymers, but not made of identical monomers. The monomers are all of the same chemical type, the alpha-amino acids, but some twenty-odd of these amino acids in varying proportions are assembled to give the enormous variety of proteins that exist in living organisms. Hair, skin, muscle, connective tissue, enzymes—all are proteins. Surely proteins are well-named; the word means "the primary stuff of life."

The amino acid molecules are joined together in protein molecules by a characteristic linkage called the *peptide linkage*. It occurs by splitting out water and amounts to substituting an amino group (however complicated by substituents) for the —OH of a carboxyl group. In general, a compound so formed is called an *acid amide*.

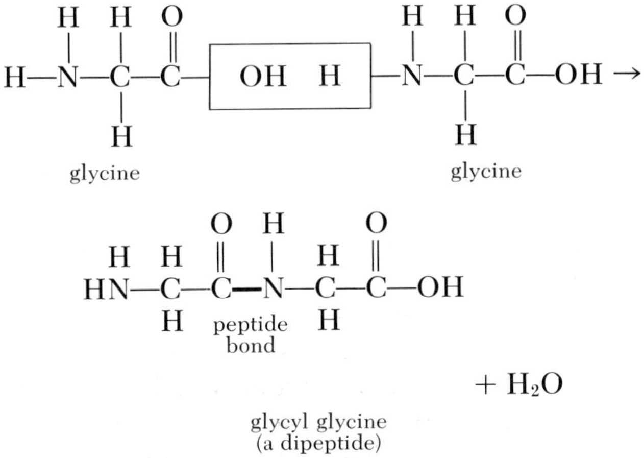

$$H_2N-CH_2-CO-\boxed{OH\ H}-NH-CH_2-COOH \rightarrow$$

glycine   glycine

$$H_2N-CH_2-CO-\underset{\text{peptide bond}}{NH}-CH_2-COOH$$

$$+ H_2O$$

glycyl glycine
(a dipeptide)

When proteins are digested, this reaction is reversed; hydrolysis back to amino acids is the reverse of the condensation by which the peptide linkage is formed.

Every living organism contains protein, but each species has its own characteristic individual proteins. This is as obvious as saying that egg white, chicken, pork, beef, and fish (all largely protein) are different—

in texture, flavor, appearance, and therefore in chemical structure. Analysis of proteins shows that they differ first in the kinds and proportions of the various amino acids they contain, and the much more difficult structural studies now are beginning to show that the arrangement of amino acids is also, as might be expected, unique for each peptide chain. In digestion, each organism hydrolyzes the protein it ingests to amino acids; then in its cells it reassembles the amino acids into its own characteristic protein. (How this is done is a question still not completely answered, but tremendous progress has been made over the last twenty years.)

Vitamins and Minerals

Other essential components of food are chemical substances necessary for life and health but not synthesizable in the body. Nine of the amino acids required for human proteins cannot be made by the human body. That is why a variety of protein foods are necessary to assure that all the essential amino acids are available. The two kinds of chemicals needed in small amounts on a regular basis are vitamins and minerals.

Everyone knows blood contains iron. Iron is, however, only one of the "trace" elements needed in minute quantities to keep the body in working order. We call all these inorganic substances *minerals*. Salt, NaCl, is a mineral needed in significant quantity in the body to maintain its water balance in the cells.

The first of the organic compounds shown to be needed in trace amounts to prevent certain malfunctions (or "deficiency diseases") were amines—hence the name *vit-amine*, for "vital amine." The vitamins now known exhibit a wide variety of structures, some entirely free of nitrogen, but the name sticks. The body requires these substances but cannot synthesize them. A number are known to function as a part of enzyme systems, and because enzymes are specific to organisms it is not surprising to find that different compounds act as vitamins for different organisms. One of the most interesting vitamins is ascorbic acid, Vitamin C. It was known in the 1750's that if fresh fruits such as limes were eaten by sailors on long sea voyages the disease of scurvy could be prevented. (That's why British sailors were called "limeys.") The actual preventive agent was not identified until around 1930, when it was named ascorbic acid. It has this structure:

$$O=C-C=C-C-C-C-OH$$

with OH, OH, H on the first three carbons (in a ring through O), and H, H, H, H on the last two carbons.

Humans are almost the only mammals that get scurvy. Recent investigation has shown that most other mammals can synthesize Vitamin C for themselves. Linus Pauling, Nobel laureate in Chemistry (and Peace, as well), has suggested that this is an example of simplification in the evolutionary process, whereas we usually associate progress with increasing complexity. In this case, however, if the diet of evolving man habitually included adequate supplies of ascorbic acid, the ascorbic acid synthesizing apparatus (doubtless an enzyme system) might have been "selected out" by the operation of natural selection because it was not necessary or advantageous to the organism. Pauling has also suggested that massive doses of Vitamin C can sharpen learning faculties (based on the improved performance of rats in mazes after being fed ascorbic acid) and prevent colds. This potential good news awaits confirmation at this writing, and much evidence has been presented on each side.

The Chemistry of Heredity

How do species maintain their homogeneity both from generation to generation, and from cell to cell in an individual organism as these cells multiply, die, and are replaced? How do individual differences occur within a species? We would expect a chemical answer, that is, an answer in terms of the existence of certain compounds and the occurrence of certain reactions by which proteins are made and by which genetic information is transmitted. Such compounds were identified in the early 1950's, and much information has accumulated during the subsequent years about the chemical mechanisms through which they act.

For many years the focus of biology has been on the cell, which is where our story must start. A cell is a wonderfully complex structure, though microscopic in size (with a few exceptions). It consists of *protoplasm* — living matter — enclosed in a *cell membrane*. The membrane is critically important. It not only keeps the protoplasm in place, but selectively admits the necessary molecules of food, minerals, etc., and selectively lets out wastes or useful products made in the cell for use elsewhere. The effectiveness of penicillin against bacteria lies in its ability to prevent the rebuilding of the bacterial cell membrane when it is opened to allow more space for the growing cell. When the gap in the cell membrane cannot be closed, the cell ceases to function.

The specific structures within a cell, as well as its overall shape and form, depend on its function, but some structures are common to all or nearly all cells. With few exceptions, cells have nuclei. Those exceptional cells that have none cannot divide and reproduce themselves.[2] This is true of red blood cells, which lose their nuclei before they leave the bone marrow where they are produced, and therefore at death must be replaced from the bone marrow construction sites. Thus, cells generally include in their

---

[2] There is an exception to this, too: some very primitive cells have nuclear materials that are not contained within a nuclear membrane.

machinery three structures of particular interest to molecular biology because of their functions in the fundamental biochemical processes of the cell. They are:

1. Nucleus, which reproduces genetic material.
2. Ribosomes, which manufacture proteins.
3. Mitochondria, which produce energy.

Chromosomes

In 1902, Walter S. Sutton suggested that the long tangled threadlike bodies in the nucleus, called *chromosomes* (colored bodies) because they selectively absorbed stains when cells were prepared for microscopic study, might be the carriers of heredity. Thomas Hunt Morgan, acting on the suggestion that the chromosomes were strings of units called *genes*, each of which was assumed to be responsible for a specific hereditary characteristic, began the still-continuing and fruitful work on the tiny fruit fly, *Drosophila melanogaster*, which has only four pairs of chromosomes. These can be altered by radiation or chemicals and the effects studied, and thus genes for specific characteristics can be identified. But this work has not shown exactly what a gene is, structurally, nor how it operates. This turned out to be a chemical rather than a biological problem, at least to the extent that such distinctions can be made.

Chromosomes are made of two kinds of compounds: proteins and nucleic acids. Which one is critical in transmitting genetic information? Around 1950, a group at the California Institute of Technology led by Linus Pauling was concentrating on proteins. At Cambridge's famed Cavendish Laboratory and other British research centers, Francis Crick, James Watson, Maurice Wilkins, and several others were studying the coding possibilities of nucleic acids, especially the predominant form found in chromosomes, DNA. The British group turned out to be on the right track.

Deoxyribonucleic acid, DNA, was known to consist of a long chain of alternating units of a 5-carbon sugar, deoxyribose, and phosphate groups:

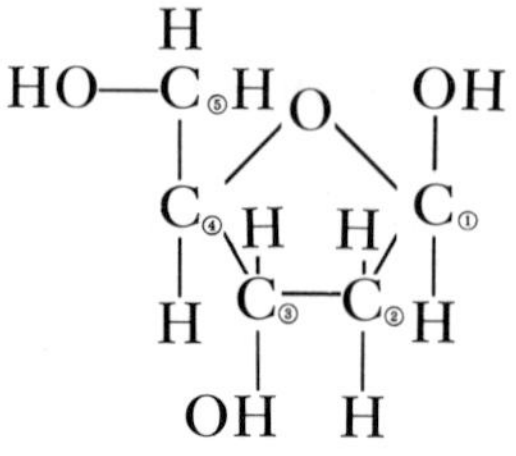

$$\text{deoxyribose} \qquad \text{phosphoric acid}$$

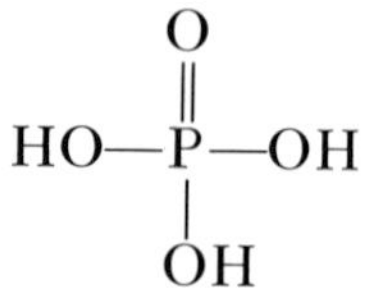

(Notice the absence of oxygen on the second carbon.) These are put together as follows:

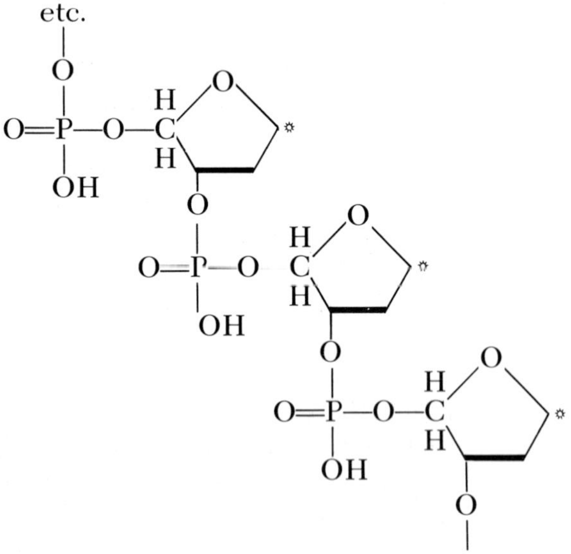

This is the backbone of a DNA molecule. What makes the molecule remarkable is not its backbone but the "side chains" attached to the first carbon atom in each ring—the ones marked with an asterisk in the backbone formula. Each side chain on a DNA sugar-phosphate backbone is one of four cyclic amines often called *nitrogen bases*. These four fall into two groups, the purines and the pyrimidines, respectively.

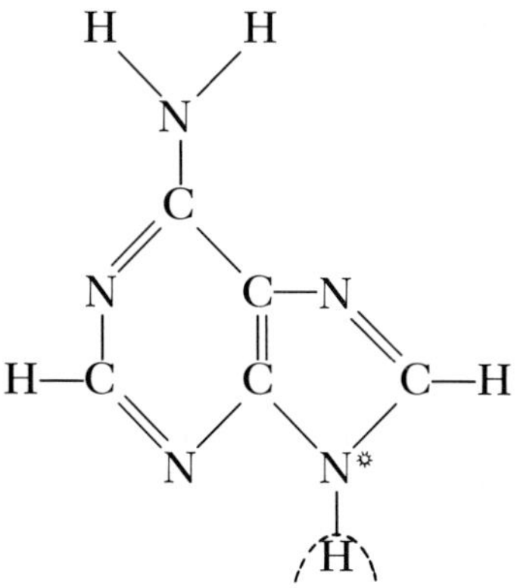

The Chemistry of Heredity

adenine

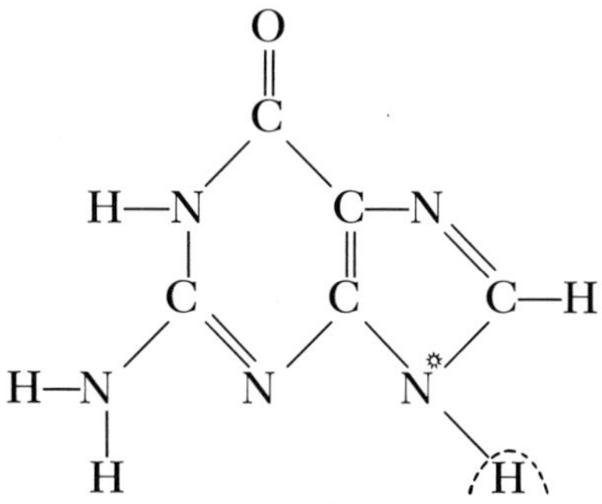

guanine

purines

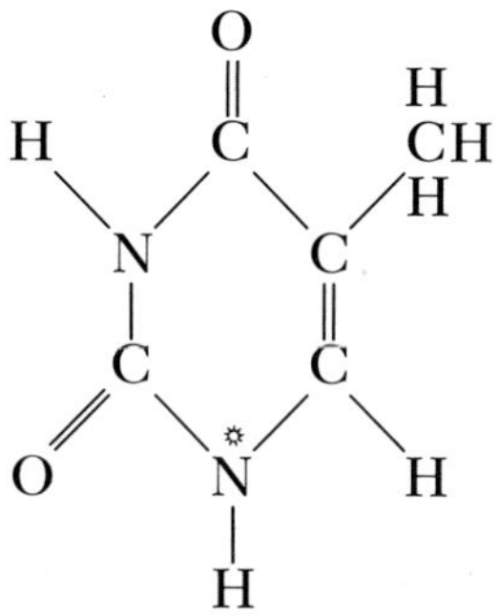

thymine

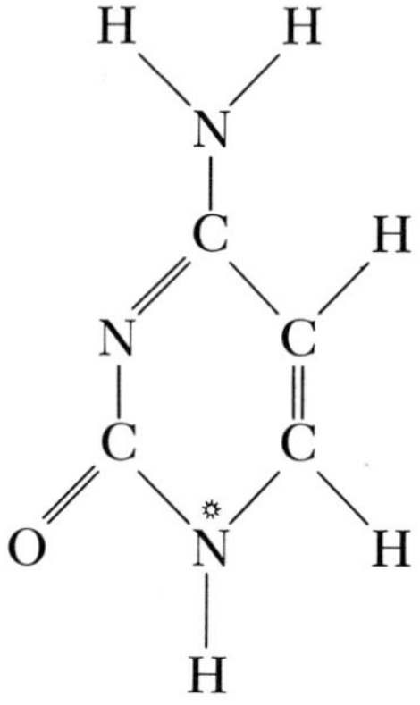

cytosine

pyrimidines

These four units are attached to the first carbon of the sugar by replacing the —OH group by the amine at the starred nitrogen atom, splitting out water, formed from the —OH plus the —H that was attached to the nitrogen atom and marked ⟨H⟩.

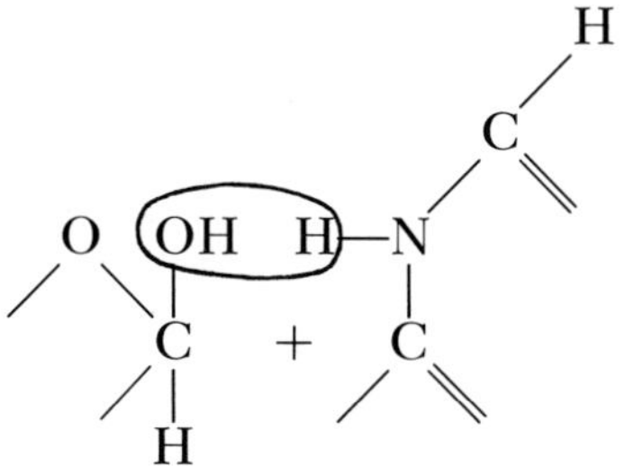

To understand how these bases make DNA remarkable, we shall have to expand our ideas of a chemical bond to accommodate a very special kind called the *hydrogen bond*.

The Hydrogen Bond

Hydrogen, the simplest atom, has only one electron for bonding, and usually shares this electron with oxygen, nitrogen, or carbon to form a covalent bond in organic molecules. The C—H bond is nonpolar and extremely stable, but both N and O are substantially more electronegative than H, so that N—H and OH bonds are polar. If we use dots for electrons we would denote the electron distribution around an ⟩N—H group and an —O—H group thus:

$$: N : H \qquad : O : H$$

Two pairs of the electrons around N and one pair around O are being shared by other atoms, but N still has one unshared pair, and O has two unshared pairs. The hydrogen nucleus, held loosely to an electron pair on one O or N atom can just as easily share a pair of electrons on another O or N that happened to be nearby. That is, if the structure of a molecule is such that —OH and —NH groups on the same or adjacent molecules are in close proximity, the H atoms will form a sort of tentative attachment to both groups. (This is not to suggest that such attachments are of negligible strength. It is rather like the behavior of a girl with two boyfriends who may be quite "attached" to both, and strongly resists giving either up entirely.) This behavior of hydrogen is called the *formation of a hydrogen bond*, and is usually expressed thus:

$$-O-H\text{---}N- \quad \text{or} \quad -N-H\text{---}O-$$

The solid line denotes the original covalent bond; the dotted line is the hydrogen bond. A glance at the structures of the purines and pyrimidines will indicate two or three sites on each molecule where hydrogen bonds might form if other similar molecules were nearby.

The basic problem in understanding the transmission of genetic information on a molecular basis is finding a molecule that can replicate itself. Chromosomes replicate themselves so that each daughter cell, after cell division, contains the same instructions for behavior that the parent cell had. What kind of molecule can do that? We can put together what we know about DNA, plus the great insight of Watson and Crick concerning the structure of the DNA molecule, to answer that question.

1. DNA, on analysis, is always shown to have an equal number of purine and pyrimidine bases.
2. The number of thymine units always equals the number of adenine units.
3. The number of guanine units always equals the number of cytosine units.
4. Crystallographic studies indicate that DNA, like many proteins and other biological polymers, has a helical structure.
5. The nitrogen bases each contain two or three groups that have either unshared electron pairs or polar bonds between H and either N or O, making H-bond formation between pairs of bases both possible and highly probable if the bases are close together in the DNA molecule.

Now comes the insight, intuition, leap-in-the-dark, or whatever it is that the creative scientist has or is able to do that somehow doesn't get listed among the steps of the scientific method:

Suppose the DNA molecule is a *double helix*: two sugar-phosphate backbones, helical in shape, held together by hydrogen bonds between complementary bases, adenine with thymine and guanine with cytosine. Then, if the double helix separates ("unzips"), each half constitutes a template on which another complementary chain could be formed. The molecule could replicate itself from nucleotides (phosphate-sugar base units) in the cell. The nucleotides would, of course, come from food and be reassembled in the cell.

Watson and Crick's proposal, just stated, has become known as the "central dogma" of molecular genetics, a term that is generally offensive to scientists, who prefer to think of themselves as undogmatic and ever open to new ideas and theories. As the years since 1952 have passed, modifications of the dogma have been necessary, and, the name notwithstanding, the Watson-Crick hypothesis has been wonderfully fruitful in leading to increased understanding of the transmission of heredity (Plates 13–17).

The Chemistry of Growth

Of course, understanding the replication of genetic material is of little use in itself. The next question is: how are the genetic instructions put into operation to direct the growth and formation of cells? Because cells are made mostly of protein, and because the enzymes that direct all chemical activity in cells (even DNA replication, it turns out) are protein, a good guess would be that DNA directs the formation of proteins. This turns out to be a process of several steps. A simple outline of these steps, which may actually be rather more complicated than we now appreciate, follows:

1. In the cell nucleus, the DNA double helix acts as a template not only for its own replication but for the production of (usually) single-stranded RNA helices. Ribonucleic acid, RNA, is a molecule with a

sugar-phosphate backbone containing ribose instead of deoxyribose and the four nitrogen bases guanine, cytosine, adenine and uracil. Uracil,

$$\begin{array}{c} \text{structure of uracil} \end{array}$$

always occurs in RNA where thymine would be in DNA. No one knows why, yet.

2. Two main types of RNA molecules are active in protein synthesis. The longer chain molecules, messenger or m-RNA, migrate from the nucleus to the ribosomes, tiny organelles attached to a vast network of internal membranes called the *endoplasmic reticulum.* A strand of m-RNA lines up great numbers of ribosomes along its chain, forming "polysomes."

3. Short RNA chains called transfer or t-RNA, also made in the nucleus on the DNA template, have the function of "recognizing" individual amino acids and lining them up for formation of peptide chains. (A number of other kinds of RNA seem less directly involved in protein synthesis.)

4. The sequence of bases on an m-RNA chain forms a code of three-letter (three-base) words, each of which designates an amino acid. That is to say, the sequence |C—G—G|—A—T—C—|G—A—T—| would represent three amino acids in order in a peptide chain. The so-called genetic code thus is a code for amino acid arrangement in proteins, including the enzymes that direct all the body's processes.

5. Molecules of t-RNA have base sequences complementary to those on m-RNA, as well as a structure or mechanism for recognizing its amino acid. (It was

suggested late in 1970 that a "second genetic code" might exist for this purpose.)

6. In a ribosome the connection between the t-RNA and m-RNA is made (via H-bonding of complementary bases) that aligns amino acids in such a fashion that the peptide linkages can form. The ribosome seems to move along a chain providing sites in which the connection can be made, somewhat as the closing device of a zipper pulls the two rows of teeth together so they can mesh.

7. The m-RNA code apparently contains "punctuation" as well as words. Certain base sequences clearly seem to indicate that the polypeptide chain should end and another begin. A mutation in which the punctuation sequence was removed from an RNA chain produced two complete functional enzymes fused together in one molecule.

In its simplest form, the "central dogma" of molecular biology thus says that DNA replicates itself to transmit the genetic code to new cells and new organisms, and in its own cell serves as the master template from which RNA molecules are formed to serve as templates on which proteins are constructed. Individuality is due to small differences in DNA base sequences among the vast numbers of identical base sequences that account for the identity of a species. The old biological term "gene," for the carrier of heredity, now is understood to be that segment of a DNA chain that codes for a particular enzyme that determines or influences some particular inherited traits.

The beautifully simple Bohr atomic theory had to be refined and amplified into the elegant but no longer simple quantum mechanics. Analogously, the beautifully simple "central dogma" is being refined and amplified to include the fact that even DNA replication, the formation of RNA, and the matching of amino acids to their specific t-RNA are all enzyme-assisted reactions. Just as physicists argued over the implications of quantum theory, so molecular biologists argue the implications of the development from the simple Watson-Crick model: is the living cell no more than the sum of its biochemical components and their reactions, or is the whole greater than its parts? Is the transmission of heredity a DNA function, or a cell function? Stay tuned in. . . .

Energy Production

One other function of every cell is crucial to its life: energy production. The sites of energy production in the cell are organelles ("little organs") called *mitochondria*. What goes on there to extract energy from fuel foods is vastly more complicated, chemically, than the replication of DNA or the assembling of a protein. For example, consider how energy is extracted from a most common cellular fuel, glucose.

If glucose, a white crystalline solid, were burned in a flame until it was completely consumed, we could express what happened in this way:

$$C_6H_{12}O_6 + 6O_2 \rightarrow 6CO_2 + 6H_2O + 690 \text{ kilocalories.}$$

That is, when one mole (180g) of glucose is completely burned to carbon dioxide and water, 690,000 calories (690 Calories or kilocalories) of energy is produced. In the flame, this energy appears as heat. One could, theoretically if not practically, run a heat engine (a steam engine, for instance) on sugar rather than on coal or wood.

The body, however, is not a heat engine. It runs at a constant temperature, and according to the Second Law of Thermodynamics, heat can be converted into useful work in a cyclic process only if the cycle operates between two different temperatures, and its efficiency is proportional to the temperature difference. Instead, the body requires energy in the form of chemical energy to drive biochemical reactions, mechanical energy to contract muscle fibers, and electrical energy to drive nerve impulses. Except for merely maintaining body temperature in warm-blooded animals, organisms do not use heat energy directly at all.

Nevertheless, if in the body 180g of glucose is converted into CO_2 and H_2O, 690 Kcal of energy will be available in some form, simply because energy is conserved. "Respiration" is in the long series of chemical reactions by which this energy is made available in useful forms to the cell (Fig. 11.3). In general, the useful form in which energy is made available is the energy stored in certain high-energy bonds. Several kinds of molecules contain such bonds, but the most important is ATP, adenosine triphosphate, the primary energy carrier of cells ($\sim$ indicates high energy bonds).

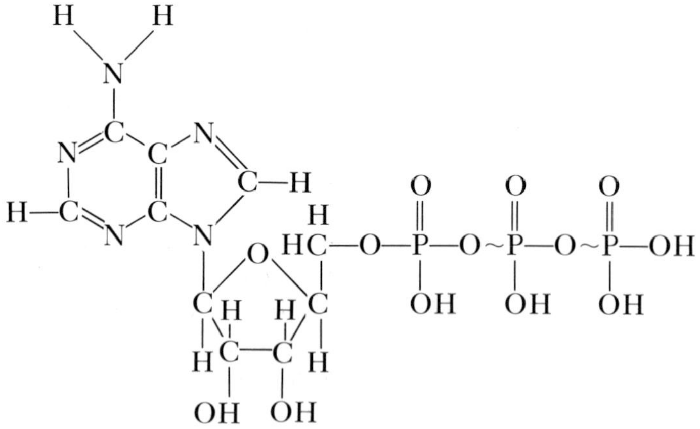

It is very much like our old friends the nucleotides from which DNA and RNA are assembled, in that it contains a 5-carbon sugar, ribose, with a nitrogen base,

---

Figure 11.3 See opposite page. Summary of stages in respiration. Glucose, containing six carbons, is successively oxidized to three carbon (3C) chains, two carbon chains (2C), and CO_2. Redrawn from Nelson, Robinson, and Boolootian, *Fundamental Concepts of Biology*, Second Edition © Copyright 1970, by permission of John Wiley & Sons, Inc.

Energy Production

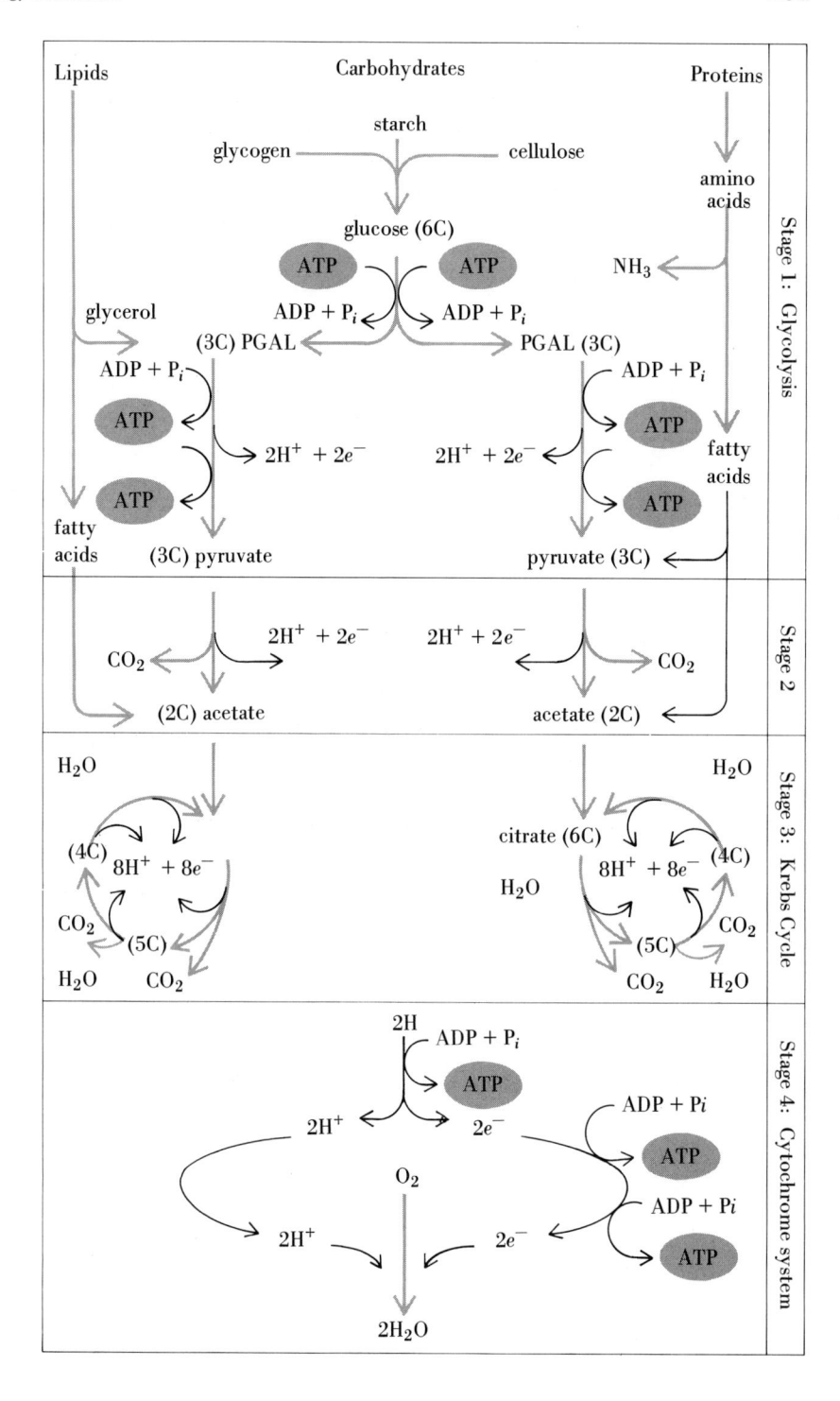

Stage 1: Glycolysis

Lipids → glycerol → (3C) PGAL ... fatty acids
Carbohydrates: glycogen, starch, cellulose → glucose (6C) → PGAL (3C) (using ATP → ADP + P_i)
PGAL → pyruvate (3C), producing ATP and $2H^+ + 2e^-$
Proteins → amino acids → NH_3, fatty acids, pyruvate (3C)

Stage 2

(3C) pyruvate → (2C) acetate + CO_2 + $2H^+ + 2e^-$ (both sides)

Stage 3: Krebs Cycle

acetate (2C) → citrate (6C) → (5C) + CO_2 → (4C) + CO_2
Inputs: H_2O; Outputs: CO_2, H_2O, $8H^+ + 8e^-$

Stage 4: Cytochrome system

$2H \to 2H^+ + 2e^-$; $2e^-$ through chain producing ATP (ADP + P_i → ATP) at three sites
$2H^+ + 2e^- + \frac{1}{2}O_2 \to 2H_2O$

adenine, on the first carbon; the sugar is also attached to a phosphate group. However, that phosphate group is condensed with a second phosphate group, which in turn is condensed with a third phosphate group. (We use the term *condensed* because water is split out between the two phosphate groups in each case.) The P—O bonds that attach the last two phosphate groups are the high energy bonds. When these bonds are broken, energy is made available for some other reaction or process; when glucose is "burned" by many stages, the P—O bonds of ATP are restored, and energy is stored for future use.

Oxidation

The actual chemistry of the extraction of energy from glucose is extremely complex. Remember, first of all, that all the reactions must be carried out at a relatively low temperature, and that energy must be produced mainly in forms other than heat. So "burning" or "oxidation" of glucose is not the simple saturation of all carbon and hydrogen bonds with oxygen, but oxidation must be interpreted in its technical, chemical sense of transfer of electrons and protons. Oxidation is always accompanied by reduction. In the burning of glucose in oxygen we would apply the definitions thus:

$$C_6H_{12}O_6 + 6O_2 \rightarrow 6CO_2 + 6H_2O$$

| This is the reducing agent; it gets oxidized —H is removed from attachment to C and from O atoms. | This is the oxidizing agent; it gets reduced by becoming attached to C and to H atoms. | These contain oxidized forms of C and H because their bonds are saturated with oxygen. They contain oxygen in reduced form, because it is now attached to C or H atoms. |

From another point of view, the oxygen atoms act as receptors for hydrogen atoms, each bringing one electron into the oxygen atom's electron shells, making reduction a gain of electrons. Conversely, the hydrogen has lost at least part interest in an electron, and loss of electrons is oxidation. The process of *respiration,* as energy production is called, is summarized in Figure 11.3

From the viewpoint of thermodynamics, a living organism is a highly improbable state of affairs. Living is an ordering process: a cell takes simple, randomly distributed chemicals and orders them into genetic materials or proteins; it utilizes the energy of a spontaneous, exoenergetic reaction, the oxidation of glucose, to drive a vast number of nonspontaneous, endoenergetic reactions. The whole universe tends spontaneously and irreversibly toward equilibrium, yet every living cell avoids equilibrium with all its resources, for equilibrium is synonymous with death. Surely, as Psalm 139:14 observed, we are "fearfully and wonderfully made"!

This is not to say that life is a contradiction to thermodynamics, nor that the laws of thermodynamics are "broken" by living systems. The laws of thermodynamics are universal, and on the scale of the universe, no exceptions to them have been found. A living organism creates order in itself at the expense of order in its environment. When the organism is human, the price the environment pays is pollution. This observation perhaps suggests that although some disorder in the environment necessarily accompanies life, thinking organisms might be under some obligation to minimize that pollution which is not necessary for life.

SUMMARY
STATEMENTS

1. All living things require food and energy to continue living. Plants containing chlorophyll can use radiant energy from the sun to synthesize sugars and other foods from inorganic constituents; all other forms of life depend ultimately on foods produced by this photosynthetic process in green plants.

2. Animals require carbohydrates and fats for energy and protein for structural and regulatory functions. Certain

minerals and organic compounds called *vitamins* must also be ingested for special uses in the body.

3. Essentially all chemical processes within organisms are catalytic processes. The human body operates at a constant, rather low temperature, and depends on specific protein catalysts called enzymes to make its chemical reactions work.

4. Because enzymes are very specific in their actions, and because enzymes are proteins, all an organism is and does is determined by its proteins.

5. The genetic code carried in every cell of an organism and transmitted to its descendants through its reproductive cells is a code directing the manufacture of that organism's unique set of proteins.

6. The genetic code is carried by molecules of deoxyribonucleic acid (DNA), which have a double helix structure. Each of the two strands consists of a backbone of alternating sugar (deoxyribose) and phosphate units. To each sugar is joined one of four nitrogen bases.

7. The nitrogen bases, adenine (A), cytosine (C), guanine (G), and thymine (T), form hydrogen bonds always in the pattern A—T and G—C. These bonds hold two DNA strands together in the chromosomes of the cell nucleus.

8. When a DNA double helix separates into its two strands, each strand serves as a template for constructing another strand just like its original partner, by base pairing, thus replicating the original DNA molecule.

9. A DNA strand can also serve as a template for constructing RNA, a similar molecule except that its sugar is ribose and it contains uracil (U) in place of thymine (T).

10. RNA, coded from DNA, travels to ribosomes, the cell sites at which protein manufacture occurs. Here transfer RNA (t-RNA) molecules line up amino acids in the order specified by base pairing with the template messenger (m-RNA) molecules. A new protein is built up unit by unit as peptide bonds form between adjacent amino acids. A code word of three bases stands for an amino acid.

11. Energy is produced in cells by the stepwise catalytic oxidation of glucose and other simple sugars and fats produced by digestion of food. The energy so liberated is stored in high energy bonds in a variety of molecules, especially adenosine triphosphate (ATP). Breaking these bonds provides energy for all body processes.

REVIEW QUESTIONS

1. What does "vitalism" mean? How do vitalism and mechanism as theories of life differ?
2. Name the three kinds of food substances the animal body needs in quantity and state the function of each.
3. Name two kinds of substances the animal body needs in minute quantities. Why are they necessary?
4. Mention some chemical characteristics (structure, functional groups) of (1) carbohydrates; (2) fats; (3) proteins.
5. What is optical activity? What must a substance contain to be optically active?
6. What is the principal function of each of these organelles within a cell?
   (a) nucleus   (b) ribosomes   (c) mitochondria
7. What is a peptide linkage?
8. What is a hydrogen bond?
9. Identify DNA; m-RNA; t-RNA; purine; pyrimidine; ribose; nucleotide.
10. Show in outline how a four-letter alphabet of nitrogen bases can form a code for arranging some twenty amino acids into proteins.
11. Define: catalyst; enzyme; ATP.
12. Why must all body processes be catalyzed?

THOUGHT QUESTIONS

1. How do the biological and molecular pictures of genetics fit together? Outline the "whole story" as it is now understood.
2. How is the development and persistence of life accounted for in terms of the second law of thermodynamics?
3. Listed are some of the possibilities for preserving and altering human life as a result of continued progress in the elucidation of the molecular aspects of biology. Discuss them from the point of view of desirability, applicability, control, and responsibility for the decision to apply the available knowledge and techniques:

   a. Replacement of organs as a routine procedure.
   b. Alteration of hereditary material in a fetus.
   c. Asexual human reproduction by "cloning." (In this process, the nucleus of an egg cell is replaced by the nucleus of a nonreproductive cell from the person whose total inheritance the child is to have.)

d. Storage and subsequent use of genetic material from geniuses or persons of special talent.
e. Indefinite postponement of death.
f. Genetic control of whole populations.
g. "Test tube" babies.
h. Routine abortion of defective fetuses.

4. Note in current popular and science-news publications the development of research in molecular biology. How are these investigations related to the fight against cancer and other noninfectious diseases?

5. Actually, much evidence exists that some cancers *are* infectious, being caused by viruses. What are viruses and how do they act?

6. What natural and artificial agents of genetic change (mutagens) are now known? What can a person do to minimize chances of producing defective offspring?

7. The problem of malnutrition is not always one of lack of quantities of food, but of specific deficiencies in kinds of food. If you were going into an area to study nutrition problems for FAO and WHO, what would you look for? List possible nutritional problems you might encounter and suggest solutions.

8. Persons working to improve the nutrition of malnourished groups of people, whether in the jungles of Guatemala or central Africa where Kwashiorkor is endemic or among the recipients of food stamps in Appalachia or the Mississippi Delta, find that it is not sufficient just to provide people with proper foods. Why is it so difficult to change peoples' food habits, especially those in lower income and social groups?

READING LIST

Asimov, Isaac, *The Well Springs of Life*, New American Library (Signet), New York, 1962. The fascinating story of the evolution of life and of our understanding of it.

―――, *The Genetic Code*, Orion Press, New York, 1963. An accurate, well-written account of the genetic code and how it works.

Watson, James D., *The Double Helix*, New American Library (Signet), New York, 1969. The interesting, sometimes irritating, firsthand account of the discovery of the chemical mechanism of heredity.

Adler, Irving, *How Life Began*, New American Library

(Signet), New York, 1959. A simple account of the chemical processes of life.

Scientific American Reprints (W. H. Freeman, San Francisco):

#123 Crick, F. H. C., "The Genetic Code" (October 1962).

#153 Nirenberg, M. W., "The Genetic Code: II" (March 1963).

#1052 Crick, F. H. C., "The Genetic Code: III" (October 1966).

#90 Brachet, Jean, "The Living Cell" (September 1961).

#91 Lehninger, A. L., "How Cells Transform Energy" (September 1961).

#92 Allfrey, V. G., and Mirsky, A. E., "How Cells Make Molecules" (September 1961).

#119 Hurwitz, Jerard, and Furth, J. J., "Messenger RNA" (February 1962).

#121 Kendrew, John C., "The Three Dimensional Structure of a Protein Molecule" (December 1961).

12

The Atomic Nucleus

By 1900 it was fairly clear that the atom, that fundamental unit of matter whose name means "indivisible," was actually the sum of many parts. Electrons, alpha particles, beta particles, and gamma rays all came out of atoms either spontaneously or under conditions of stress such as high electrical potential differences. Rutherford, Bohr, and the "quantum mechanikers" of the 1930's worked out many details of the structure of the simpler atoms, and their theories have been extended to the electronic structure of even complicated molecules. But the heart of the atom, the *nucleus*, remains shrouded in mystery.

Once the general structure of the atom—a nucleus surrounded by orbiting electrons—was known, it became obvious that alpha, beta, and gamma rays were emitted from the nucleus. Alpha rays were identified as helium nuclei, with a mass of four units on the atomic weight scale and a positive charge equal to two electronic charges in magnitude. Beta rays were found to be identical with electrons, although the theory of atomic structure allowed for no electrons in the nucleus. Gamma rays, electromagnetic rays of very high frequency, appeared to be emitted as the nucleus reached a lower energy state, just as visible light emission signaled the dropping of electrons into lower energy states. Chemical analysis of radioactive samples showed that decay by alpha emission left a new nucleus of mass four units less and atomic number two units less than the original nucleus. Beta emission, on the other hand, left a daughter nucleus of substantially the same mass but with an increase of one in nuclear charge, suggesting that the emission of an electron from the nucleus followed the conversion of a neutron to a proton and the emitted beta.

Radioactivity

Natural radioactivity, then, occurs when a nucleus is unstable in either of two ways: if the proton-neutron ratio is too high, the stress is relieved by alpha emission—never the emission of single protons; if the proton-neutron ratio is too low, a neutron would change to a proton and an electron, and the electron is emitted. Gamma rays most often accompany beta emission. Helium nuclei rather than protons are emitted, which suggests that this unit has unusual stability, and any theory of nuclear structure must take account of this. The stability of any nucleus is remarkable in terms of the strength of electric repulsions between protons, and leads to postulating the existence of some kind of nuclear force that at close range is much stronger than electrostatic forces. It is to be expected that the greatest instability would occur in the heaviest nuclei, and we find no stable (i.e., nonradioactive) isotopes among nuclei of atomic number greater than 83 (Bismuth). However, nearly every element has a naturally occurring radioactive isotope: H^3 (tritium), C^{14}, O^{18}, K^{40}, and Fe^{55} are examples of the relatively rare radioactive isotopes of common, stable elements. Potassium-40 decays naturally by a third avenue to stability, less common than alpha and beta decay, called *K-capture*. In this case, an electron from the innermost, or K, shell falls into the nucleus and combines with a proton to change it into a neutron.

The resulting nucleus is argon-40, an isotope of about the same mass as K^{40}, but with one less nuclear charge.

Radioactivity is a property of atomic nuclei. It is normally uninfluenced by the chemical or physical environment and by any stresses—temperature or pressure or electric discharge, etc.—by which we speed up or slow down physical and chemical processes. A radioactive element decays at a steady rate, no matter what else happens to it. However, "steady rate" does not imply a process like the ticking of a clock. If you were to listen to a Geiger counter, which clicks every time a particle from a nuclear disintegration encounters it, you would hear a very irregular pattern of single and multiple clicks. You would conclude that the disintegrations were occurring completely at random, and your conclusion would be correct. However, if you were to "count" for a very long time, you would discover a pattern: the time it takes for one-half of any mass of a radioactive element to decay is always the same. Thus, counting the disintegrations of ten grams of radium over a period of days or weeks will enable you to derive a rate equation from which you could calculate that in 1650 years you would have 5 g of radium left. If you had started with only one gram of radium, however, you would still have 0.5 g 1650 years later. The characteristic time it takes for one-half the atoms of a radioactive element to decay is called its half-life. There is no way to tell which half of radium atoms will disintegrate in a given 1650-year period; there is no way of telling when or if any given atom will decay. It is, as far as we know now, a completely random process, and we do not know why any particular nucleus does—or does not—decay at any given moment. But we can determine half-life with a high degree of accuracy, because enormous total numbers of atoms are involved.[1]

---

[1] One gram of radium contains roughly 3×10^{21} atoms. Although we can make no sure statement about the behavior of any one atom, the laws of probability governing random events make our statement that half of the atoms will decay in any 1650-year period a near certainty.

This constant characteristic half-life of radioactive elements makes them valuable for dating ancient events, from the earth's oldest rocks to charcoal from a "cave man's" campfire. Three isotopes are particularly useful in this application: U^{238}, which decays to Pb^{206} with a very long half-life; K^{40}, which decays to Ar^{40} with a half-life of some tens of thousands of years; and C^{14}, which decays to N^{14} with a half-life of less than 6000 years, and is particularly useful in dating organic remains. Any radioactive dating method depends on the fact that over the long run radioisotopes decay at a constant rate, so that the relative amounts of radioisotope and end product found in association with each other will enable one to calculate how long that rock or that bone has been undisturbed in the place where it was found.

One of the first observations about radioactive "rays" was that they differed greatly in penetrating power, which is to say, in energy and in mass. Recalling that kinetic energy is expressed as $\frac{1}{2}mv^2$, it will be seen that a massive α (alpha) particle would move slower than a light β (beta) particle emitted with the same energy, while two α particles differing in energy would differ in velocity. The measurable quantity by which energy can most easily be measured is *range*. The range of a particle is the distance it can travel before it is stopped. (Of course, the "stopping material" will have something to do with the range, as well as the energy of the particle.) In general betas have longer ranges than alphas, whereas the ranges of particles from a given radioisotope will lie so near their mean value as to be adequate for identifying the isotope.

The detection, production, use, and control of radioactive processes has led to the development of an enormous nuclear technology, which includes detection and counting devices, accelerators, research reactors, and power reactors. Also required are special handling techniques, safety measures, and medical applications equipment.

Detection

The earliest detectors of radiation were sensitive devices called *Lauritsen electroscopes* (Fig. 12.1). All particles emitted by radioisotopes are to some extent ionizing radiations. That is, they knock off electrons from molecules they pass through, producing a trail of ionized particles in their wake. An electroscope exposed to radiation therefore becomes charged and indicates this condition qualitatively by spreading its leaves.

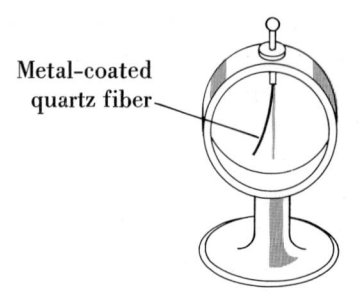

Metal-coated quartz fiber

Figure 12.1 Lauritsen electroscope.

Another qualitative or semiquantitative detector is the cloud chamber, invented by C. T. R. Wilson in 1911. In a cloud chamber the ionized air particles produced by the radiation passing through act as condensation nuclei upon which supercooled alcohol vapor forms droplets which become a visible cloud (Fig. 12.2). Even a small cloud chamber shows the cloud tracks of alpha particles quite clearly. Larger models maintained under carefully controlled conditions have been the detectors in which new particles have first been identified on several occasions.

A refinement of the cloud chamber is the bubble chamber, invented by Donald Glaser in 1952 (Fig. 12.3). The range of high-energy particles is too great for them to be stopped in a cloud chamber of reasonable size, because the cloud chamber contains only

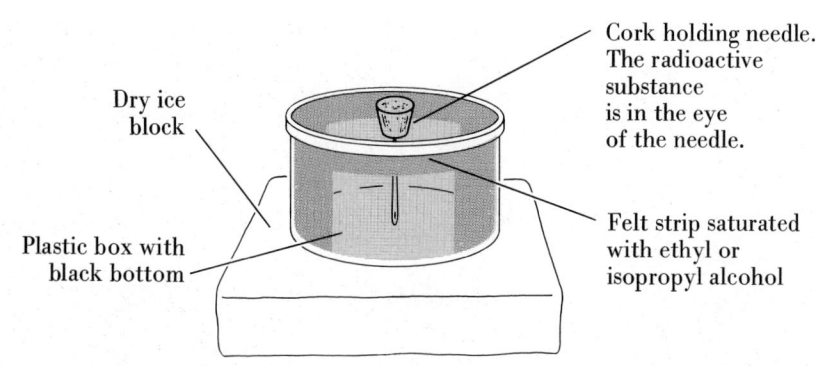

Figure 12.2 Diffusion cloud chamber for alpha particle detection.

gas. A bubble chamber contains a liquid under pressure just below its boiling point. When the pressure is released the liquid will vaporize if there are nuclei on which bubbles can form. These are produced as ions in the liquid by ionizing radiation passing through; thus a track of tiny bubbles that can be photographed by a short-duration flash denotes the passage of a beta or other high-energy particle as well as of the heavier particles, because the liquid has greater stopping power than the gas of a cloud chamber.

Particles are also detected directly by allowing them to fall on photographic emulsions on films or plates, but this solid medium has such great stopping power that it is effective only for high-energy particles.

All these detectors—cloud and bubble chambers and emulsions—provide a photographic record that can be studied at leisure after the instantaneous event. The length of a track (range), its curvature if the chamber was in a magnetic field, and other characteristics of its appearance tell the practiced observer much about the mass, charge, and energy of the particle making the track. Each track is the record of a single event; a branching track indicates that the original particle disintegrated into new particles, so that the range of the original particle also indicates

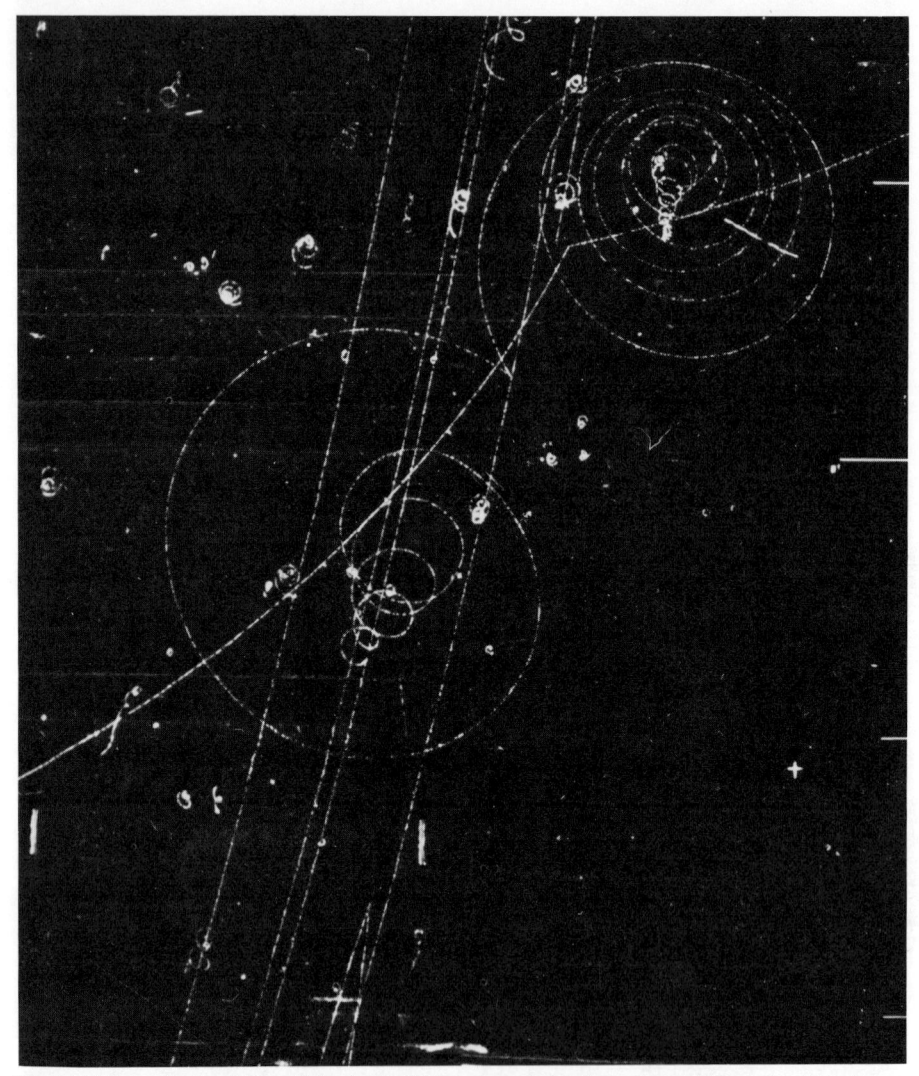

Figure 12.3 Something of the invisible universe of the atom is shown in this photograph that high energy physicists in the University of California Lawrence Radiation Laboratory, Berkeley, use to dig deeper into the ancient puzzle of matter. The photograph was taken in the 72-inch liquid hydrogen bubble chamber at Berkeley, which has accounted for the discovery of a large number of new nuclear particles in recent years. The tracks are of a variety of particles. Studies of these tracks show how particles are created at high energy and how they decay. The spiral tracks are made by electrons slowing down. The tracks running all through the photograph are of K-minus particles generated in the Laboratory's giant 6.2 BeV Bevatron. Lawrence Berkeley Laboratory, University of California, Berkeley

its stability, and several tracks of identical events would permit calculating the half-lives of the particles involved.

Counters

Counters are important for quantitative determination of the activity of radioactive materials for purposes ranging from determining half-lives of new man-made elements to routine monitoring of radioactive areas to ensure the safety of personnel. (The rugged portable Geiger counter is as much the mark of the twentieth-century uranium prospector as the pickaxe and burro were of the gold prospector of the Old West.)

The Geiger-Müller tube (Fig. 12.4) is the most fa-

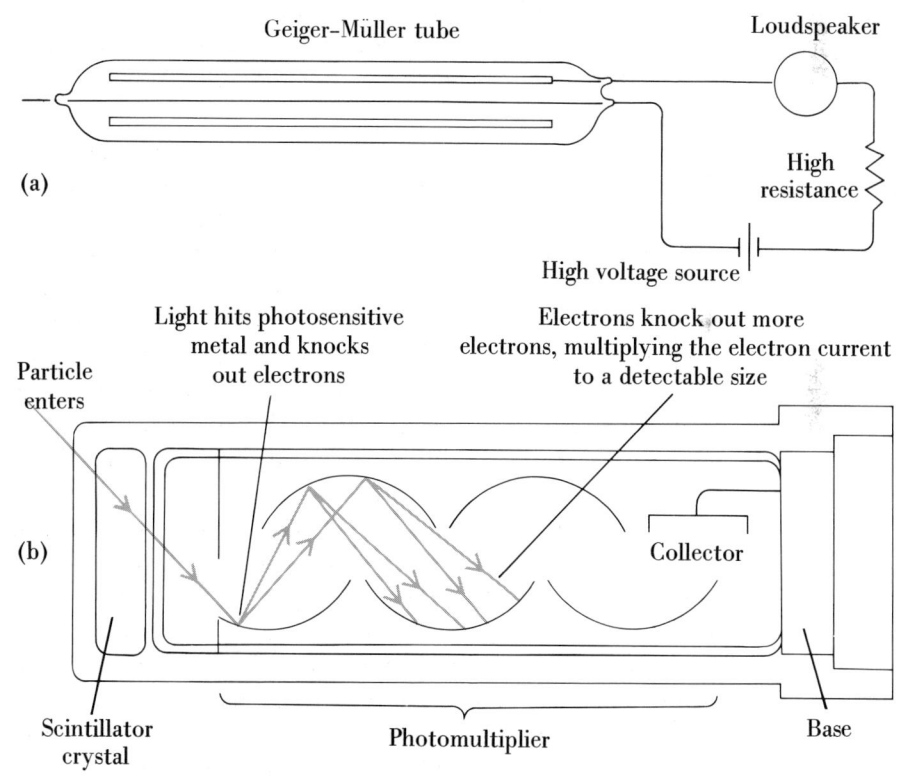

Figure 12.4 Radiation detectors: (a) Geiger-Müller counter tube; (b) scintillation counter tube.

miliar counter because of its wide use in monitoring devices and in uranium prospecting. It is basically a glass tube containing alcohol vapor and argon gas at very low pressure. A potential difference is maintained between a cylindrical metal electrode just inside the glass envelope and a wire electrode running down the center. When an ionizing particle enters the tube a pulse of current flows between the electrodes because of the ionization of the counter gas. This pulse may be amplified in a loudspeaker system to give an audible click, or it may cause a light to flash or a tabulating mechanism to register a number. This basic element can be used in a variety of circuits involving scalers (so that one count in 10, or 100, or 1000, is registered) or coincidence counting, where two events must trigger two G-M tubes simultaneously to be recorded, or proportional counters which are sensitive to very weak responses.

Scintillation counters are generally more sensitive than G-M counters and can be used to detect gammas as well as ionizing particles. Each time a particle or gamma ray strikes the fluorescent crystal, usually sodium iodide containing a little thallium iodide, a tiny flash of light appears. This is amplified by a photomultiplier tube in which light knocks out electrons (photoelectric effect), which in turn knock out more electrons until a measurable "photocurrent" is produced that is proportional to the number and energy of the particles being counted.

Another device, invented by F. J. Aston, a coworker of J. J. Thomson in the early days of electron study, is the mass spectrograph, which has had tremendous usefulness in identifying and determining charge and mass of subatomic particles (Fig. 12.5). Particles, from whatever source—natural radioactive materials or high-energy accelerators—enter a space between the poles of a large magnet. If the particles are uncharged they move straight ahead and record their arrival in the center of the photographic emulsion at the end of

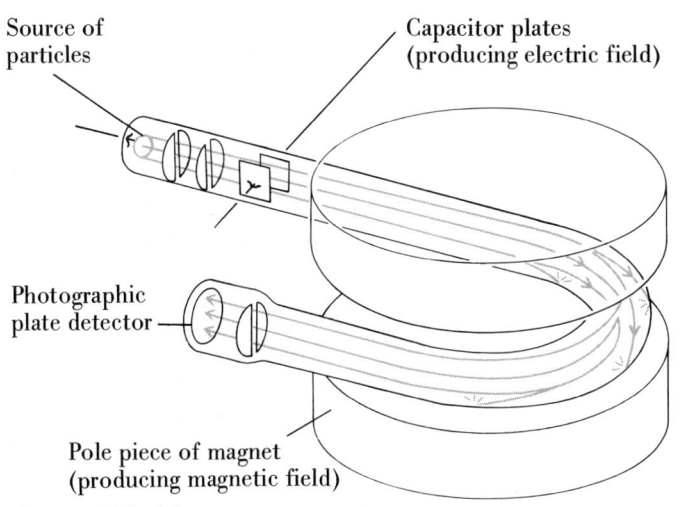

Figure 12.5 Mass spectrometer.

the chamber. If they are charged, their path will be bent in the direction determined by the sign of the charge, and the radius of the curved path they follow in a given magnetic field will be a function of their charge-mass ratio, a quantity that may be sufficient in itself to identify a particle, and is nearly always useful in distinguishing among known possibilities.

Natural radioisotopes kept investigators busy for about the first twenty years after their discovery. Then in 1919 Rutherford identified the long-range cloud chamber tracks, observed occasionally when alpha particles collided with nitrogen atoms, as the tracks of protons knocked out of the nitrogen nucleus. He could summarize the event thus:

$$_7N^{14} + {}_2He^4 \rightarrow {}_1H^1 + {}_8O^{17}.$$

This was the first observed artificial transmutation of one element into another. The dream of medieval alchemists had come true, not through the medium of the philosopher's stone, but of alpha particles. Of course, the product wasn't gold, but if you fired the right nuclear projectile at the right nucleus, gold was sure to turn up, eventually. The actual value of the method has proved to be in making far more precious things than gold, however, chiefly medically significant radioisotopes.

One important aspect of Rutherford's discovery was that it provided a new projectile, the *proton,* for studying further nuclear reactions. A simplified form of writing such reactions was adopted. Rutherford's transmutation came to be written $_7N^{14}$ $(\alpha,p)_8O^{17}$, first the original nucleus, then, in parenthesis, the projectile and the emitted subatomic particle, and to the right of the parenthesis, the product nucleus. Many such reactions were studied in the 1920's and '30's. Then in 1934 another new phenomenon was reported. Irene and Frédéric Joliot-Curie observed that the product of the reaction $_{13}Al^{27}$ $(\alpha,n)_{15}P^{30}$ remained radioactive. The $_{15}P^{30}$ nucleus was thus the first artificial radioactive isotope. It decayed to $_{14}Si^{30}$ by emission of positrons (positive electrons) produced by the conversion of a proton into a neutron with a half-life of 2.5 minutes. Since then, many radioisotopes, including some important in cancer treatment and biological tracer applications, have been produced artificially.

By 1940 several subatomic particles had been identified and others predicted (Table 12.1).

Table 12.1 Subatomic Particles

electron	J. J. Thomson	1897	
proton	E. Rutherford	1914	
neutron	J. Chadwick	1932	
positron	Anderson	1932	predicted by Dirac, 1931
μ meson	Anderson and Street	1939	predicted by Yukawa, 1935
neutrino	proposed by Fermi, (observed in "weak interactions" such as β decay)	1934	to allow for momentum and energy conservation

The first three of these particles describe the structure of atoms in the simplest fashion. The Japanese physicist Yukawa proposed the existence of an intermediate mass "exchange particle" swapped back and forth between neutrons and protons to account for the stability of nuclei, and shortly thereafter the μ meson, or muon, was observed. Since 1940, various mesons have been observed in nuclear reactions. Whether any of these particles have any permanent existence in the nucleus is open to question. P. A. M. Dirac predicted the existence of positive electrons, or positrons, on the basis of symmetry considerations in his relativistic quantum mechanics. He described them as "holes" in the space-time continuum where electrons were not. If an electron encountered a hole, Dirac theorized, it would fall in, and both the electron and the hole would vanish. However, conservation would require that if matter equal to the two electron masses were to vanish, an equivalent amount of energy must appear. This process, the annihilation of an electron-positron pair with the release of two high-energy gamma rays, has been observed, as has its reverse, pair production, the formation of an electron and a positron from gamma rays. If we pursue Dirac's analogy of a positron as a negative energy state of an electron we can see that when a gamma ray knocks an electron out of a "hole" in which it was not detectable, the result would be an observable electron (one with positive kinetic energy) and an observable "hole" or positron.

Since positrons were first identified and have been observed countless times since by their tracks in cloud and bubble chambers, the "hole" picture may not seem completely adequate. Let us accept it simply as the theoretical framework that prepared scientists to recognize the possibility of a whole symmetric system of antiparticles, many of which—the antiproton (charge -1), the antineutron (differing in spin properties), and both positive and negative forms of most mesons—have been observed and identified.

Accelerators

Observing the vast numbers and the numerous varieties of particles that physicists are trying to fit into some coherent nuclear theory today has been made possible by using enormous machines called *accelerators*, which give particles high energies so that they will tear apart nuclei with which they collide. These fragments give us most of our information about the nucleus and its constituent particles.

Particle accelerators range from rather simple electrostatic generators of the Van de Graaff and Cockcroft-Walton type to the 2-mile-long Stanford linear accelerator and such machines as the Brookhaven "bevatron" that are now in operation in several countries around the world. All use strong electric or magnetic fields or both to produce huge potential differences through which charged particles will fall with ever-increasing energies. The usual unit in which these particle energies are measured is the *electron volt*, which is defined as the amount of energy acquired by a particle of one electronic charge (whether + or −) on falling through a potential difference of one volt. (Analogously, we could measure the energy of falling bodies in "tennis ball–feet," the energy acquired by the mass of one tennis ball in falling one foot toward the earth, i.e., through a gravitational potential difference of one foot!)

Two of the earliest accelerators could produce particles of energies of the order of some hundreds of KeV (kiloelectron volts) or even two or three MeV (million electron volts). The Cockcroft-Walton generator, a system of electron tubes and capacitors, could take an input a.c. voltage of 100,000 V from a transformer, rectify it (because electron tubes act as valves they allow current to flow only one way), and build it up to some 800,000 V for accelerating protons (Fig. 12.6). Later Cockcroft-Walton generators produced 3 MeV particles. In the Van de Graaff electrostatic

Radioactivity

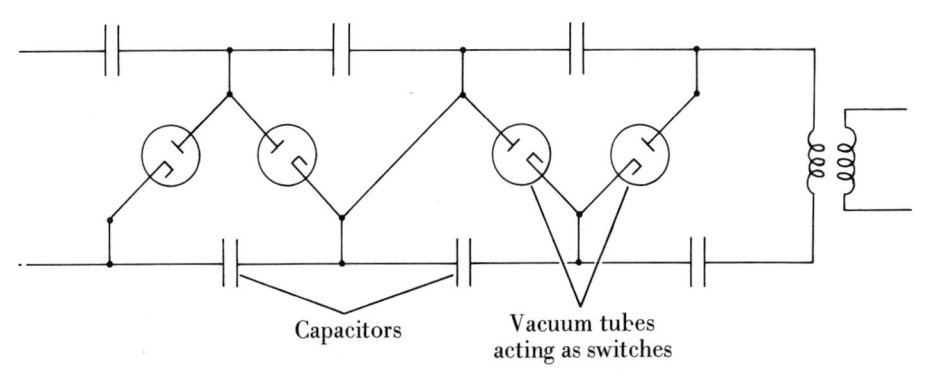

Figure 12.6 Diagrammatic representation of the Cockcroft-Walton generator.

generator, a very high potential can be created on a hollow metal sphere by charging it from a motor driven endless belt on which a static charge accumulates (Fig. 12.7). Particles could be accelerated to several MeV in a tube connecting two such spheres.

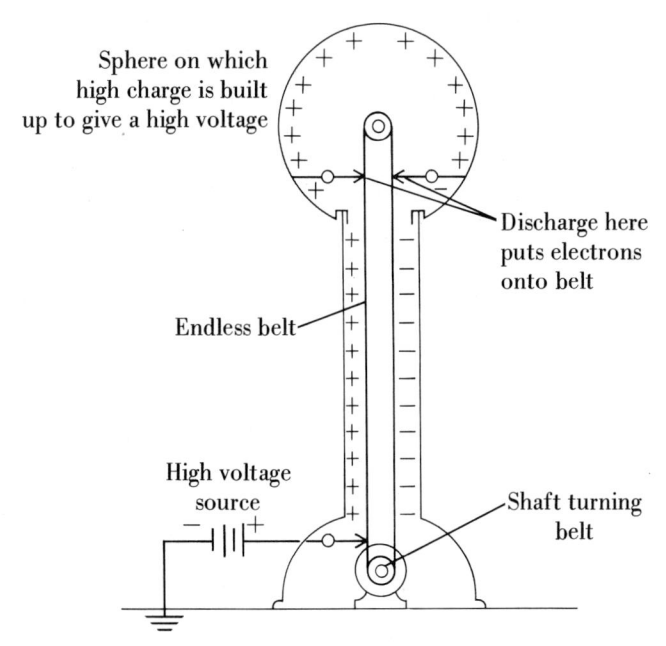

Figure 12.7 Diagrammatic representation of the Van de Graaff generator.

After two decades of concentration on the giant accelerators, interest has been renewed in Van de Graaff generators operating in the 2 to 3 MeV range in the 1970's because small countries can afford them and they can solve research problems in moderate energy ranges that the big machines cannot.

The giant accelerators are of two main designs: linear or circular. The American physicist E. O. Lawrence figures prominently in the development of both types. A linear accelerator is a series of straight tubes of regularly increasing lengths (Fig. 12.8). When a strong alternating electric field is applied to ions from a source, they are accelerated so that they go through the first tube segment in one-half cycle of the a.c. (this is a.c. in the radiofrequency range) and get another accelerating "kick" from the field when they emerge into the gap between the first two tube segments. Each tube segment must be longer than the one before it because the particles are going faster in each tube than in the one before when they enter and they must always emerge into the gap at just the right moment in the cycle to get an accelerating "kick." Linear accelerators can be built to almost any length, and vary from 100-ft models that produce 3 MeV protons to Stanford University's 2-mile-long "Linac" that produces electrons at energies up to 40 BeV (billion electron volts).

The first circular path accelerator was the cyclotron, developed by Lawrence in 1931. A cyclotron

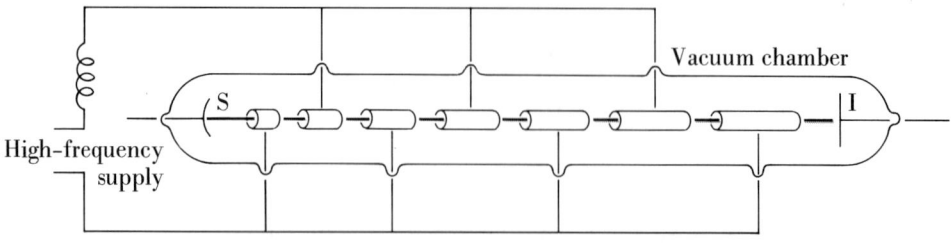

Figure 12.8 Diagrammatic representation of a linear accelerator.

looks rather like a flat cylindrical pillbox cut in half vertically (Fig. 12.9). The two "dees" are between the pole pieces of a powerful magnet, and particles from a source in the center of the dees are accelerated in a spiral path until they come out at the perimeter, where the target they are to strike is placed. Cyclotrons produce heavy ions (protons, alphas, etc.) of energies in the MeV range, depending on the size of the dees and the strength of the magnetic field.

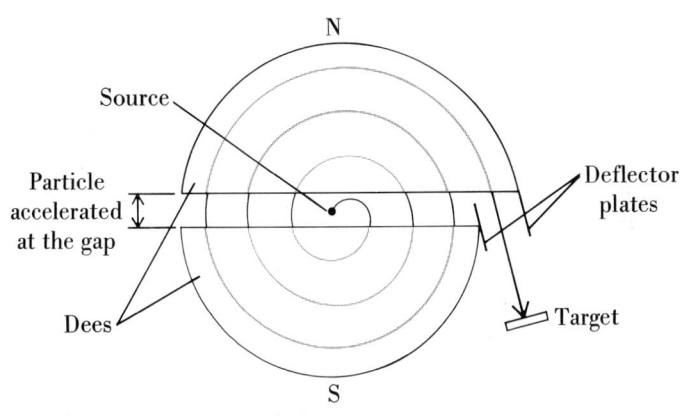

Figure 12.9 Cyclotron (top view).

The behavior of accelerated particles in a cyclotron provided the first experimental evidence for the increase in mass as a particle attained velocities approaching that of light. It was found that particles were arriving "late" at the gap between the dees where the electric field gives them the accelerating "kick" because their mass increase delayed them. This effect on successive spirals would eventually lead to deceleration of the particles when they arrived at the gap when the field was reversed. It was necessary to develop means of synchronizing the alternation of the field with the arrival of the increasingly massive particles at the gap.

Accelerators operating in the BeV range must necessarily take into account the mass-increase effect, and operate on the *synchrotron principle,* which in general involves the acceleration of particles, either electrons or positive ions, in a doughnut-shaped vacuum chamber in a magnetic field that guides the particle in its circular path. The accelerating is done by a radiofrequency electric field, as in a linear accelerator, and synchronization is achieved by adjusting the frequency and varying the magnetic field strength.

The accelerators provide high-energy projectiles to fire at assorted targets to produce nuclear reactions. A study of the masses, charges, and energy spectra of particles emitted by target nuclei gives insight into the structure and composition of nuclei. By 1965, twenty particles of sufficiently long lifetimes to be considered "elementary" had been reported. Such theoreticians as Murray Gell-Mann, attempting to fit all the observed particles into a comprehensive theoretical framework, had found it necessary to postulate the existence of still more particles, some of which were subsequently observed. By 1971 the magazine *Physics Today* considered it newsworthy that a new experiment had been performed without producing any new particles! It began to seem unlikely that all twenty-odd particles were really fundamental or elementary. Some high-energy physicists, including Victor Weisskopf, former director of CERN, the European atomic energy research facility, suggested that perhaps some entities identified as distinct if short-lived particles were really only unusual energy states of stable particles produced under the drastic conditions of bombardment by high-energy particles. The question of what particles are really elementary remains unresolved.

Murray Gell-Mann has attempted to resolve this question by proposing that the truly fundamental particle is a nearly massless entity of one-third electronic charge, called a *quark.* All properties of other particles could be accounted for mathematically in terms of quarks, so the experimental particle physicists began looking for them in their bubble chambers and among the cosmic ray tracks in their photoemulsions. There have been a few false alarms, but to date no quarks have been observed for certain.

The observed particles are classified by mass as *leptons* (very light), *mesons* (middle weight), or *baryons* (heavy). Leptons include the stable particles—electrons ($\pm$) and neutrinos, as well as muons, which have a mean lifetime of about two microseconds and decay into an electron, a neutrino, and an antineutrino. Mesons are all very short lived, decaying into leptons and/or each other. Protons are the only completely stable baryons; free neutrons decay to protons, electrons, and neutrinos after a mean life of about a thousand seconds. Heavier baryons are an assortment of very short-lived particles (mean life $< 10^{-10}$ sec) that decay into other baryons and mesons. Accelerators are not the only source of these "strange" particles. Some are formed by high-energy collisions of particles in remote regions of space and reach earth as a part of the never-ending stream of particles and radiations we call *cosmic rays*. These cosmic-ray particles are traveling at velocities very near that of light, and the lengthened mean lifetimes of muons in cosmic rays afford the best evidence for time dilation.

Nuclear Reactors

In the mid-thirties nuclear physics dominated the world's experimental laboratories. Accelerators were being developed, artificial nuclear reactions were being studied, and natural decay mechanisms were being elucidated (Fig. 12.10). Three radioactive decay chains were identified, beginning with uranium-238, thorium-232, and uranium-235, respectively. A fourth decay chain beginning with plutonium, one of the first man-made transuranium elements, was discovered when it was observed that $_{92}U^{238}$ can absorb a

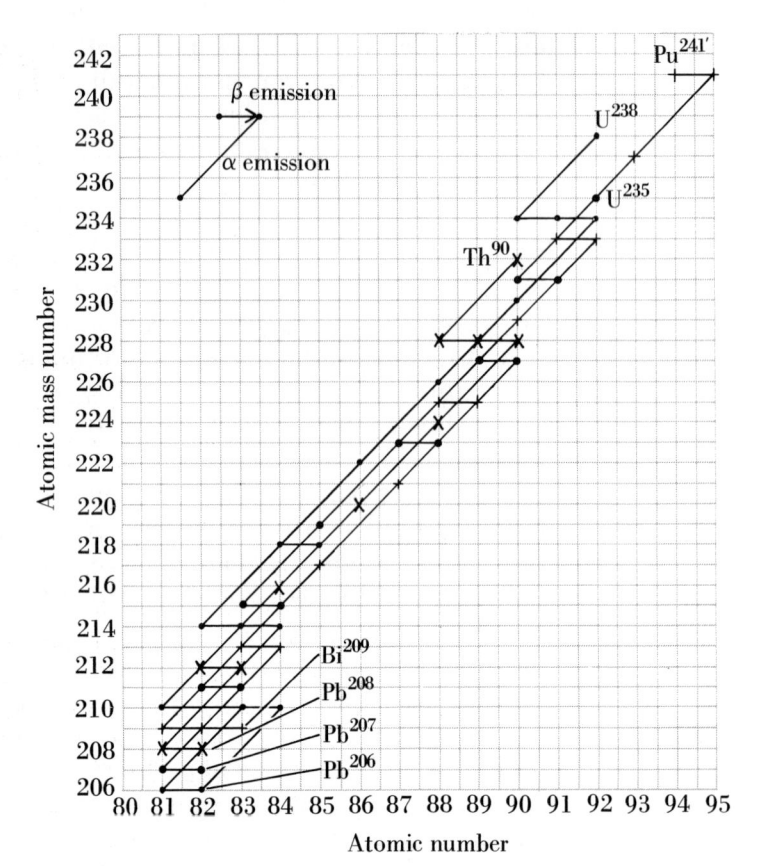

Figure 12.10 Radioactive decay chains.

slow, or thermal, neutron to form $_{92}U^{239}$, which emits a beta particle to become a new element with an atomic number greater than that of uranium. This first transuranium element was named neptunium, for the next planet after Uranus in the solar system. When $_{93}Np^{239}$ proved to be a beta emitter also, the resulting element 94 was named plutonium. Plutonium-239 is a fissionable nucleus, like U^{235}, and now forms the basis for a whole technology of power production by breeder reactors.

Relatively large amounts of energy per particle are associated with radioactive emissions, but because nuclear disintegrations are random and relatively very little of the total mass is converted to energy, radioactivity alone does not lend itself to producing such useful energy as electricity. It was determined early in the study of nuclear disintegrations that nuclei have masses slightly different from the sum of their nucleon masses, and that this "mass defect" constitutes the binding energy that holds the nucleus together. If the mass defect is positive, i.e., if the nucleus has a greater mass than its constituent nucleons, it is unstable, and this excess mass appears as energy when the nucleus disintegrates. On the other hand, an element such as $_{26}Fe^{56}$, which has a large negative mass defect, would require large amounts of energy to break it apart, and such nuclei are extremely stable. The amount of energy equivalent to any mass is, of course, given by $E = mc^2$, and the actuality of the transformation of mass into energy was first demonstrated by nuclear disintegrations.

Fission and Nuclear Power

Useful amounts of energy are produced from the type of nuclear disintegration called *fission*. In this process, a heavy nucleus ($_{92}U^{235}$ or $_{94}Pu^{239}$) splits into two middleweight nuclei and a handful of neutrons. Fission happens spontaneously on a very small scale, but the neutrons produced by a fission can, if they are slow enough, initiate further fissions—the nuclear fission chain reaction. Whenever a sufficiently large mass of U^{235} or of Pu^{239} is brought together, the random spontaneous fissions will initiate the chain reaction, and the release of energy will cause an explosion similar to that of an atomic bomb. Enormous heat and pressure will be developed, along with tremendously high levels of radiation, which result in widespread death and destruction. Radioactive fission fragments, the most dangerous of which is strontium-90, remain in the air at high altitudes, circulate widely over the earth and eventually drift back to the surface as "fallout." Strontium-90 is particularly dangerous because its chemical similarity to calcium leads to its absorption into milk from contaminated grazing areas and its eventual concentration in human bones and teeth where its long-lived beta activity can produce cancer or genetic damage within the body. Useful energy from fission is produced by controlling the rate at which the chain reaction proceeds.

The first self-sustaining chain reaction was produced in uranium slugs embedded in a pile of graphite bricks—whence the name "atomic pile"—in the squash court under the west stands of the University of Chicago's unused football stadium, Stagg Field, under the supervision of the émigré Italian physicist, Enrico Fermi. The U^{235} had been painstakingly concentrated in a months-long process of gaseous diffusion of UF_6 that took advantage of the slightly higher velocities of the molecules of the lighter isotope, U^{235}, to enrich the abundance of U^{235} in natural uranium. On the afternoon of December 2, 1942, the cadmium control rods were withdrawn and the first self-sustaining nuclear chain reaction was initiated.[2]

Graphite pile reactors are still used for research and to produce isotopes for medical and commercial uses (Fig. 12.11). Artificially radioactive materials are prepared by irradiating in the pile. Samples are irradiated, removed, and packaged for shipping in thick lead cylinders, all by remote handling techniques.

The graphite in the pile acts as a moderator. That is, it slows down the emitted fission neutrons to "thermal" speeds so that they can be effectively absorbed to produce additional fissions. The moderator also absorbs some neutrons, thus helping to control the number of fissions. The variable control mechanism

---

[2] For details of this fascinating event, see *The First Atomic Pile: An Eye-Witness Account Revealed by Some of the Participants and Narratively Recorded by Corbin Allardice and Edward R. Trapnell* (1946, published in 1949 by the USAEC).

Fission and Nuclear Power

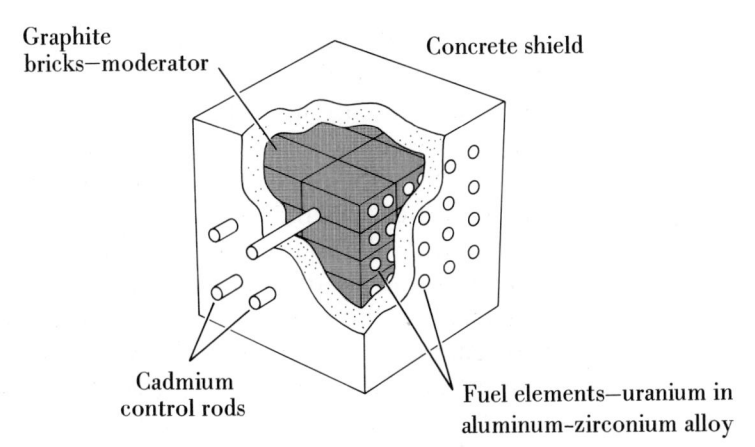

Figure 12.11 Graphite-moderated research reactor or atomic "pile."

in nearly all reactors involves rods of cadmium metal (a heavier chemical cousin of zinc), which is a very efficient absorber of neutrons. Cadmium rods inserted into a reactor between the masses of fissionable material will absorb enough neutrons to stop the chain reaction. Such rods automatically drop into place if the reactor's monitoring system shows an unusual level of radioactivity, and the chain reaction stops.

The basic problems to be solved in constructing nuclear power reactors might be stated as: how to produce electricity from fission energy using natural or moderately enriched uranium fuel in a form that can be conveniently reclaimed with a minimum of radioactive wastes to dispose of and using an efficient heat transfer/moderator material, all operating with a minimum risk of breakdown that would create radioactive contamination over the surrounding area, and competing with conventional power sources on the public market.

In general it is done this way: The fuel elements consist of tubes of an aluminum-zirconium alloy filled with an oxide of uranium, possibly enriched, i.e., containing a higher proportion of U^{235} than naturally occurring uranium. These are arranged in any one of several configurations so that most neutrons emitted from one fuel rod is likely to be absorbed and will produce fissions in another rod. The total amount of uranium will, of course, equal or exceed the "critical mass," the minimum amount of uranium that must be put together to enable the fission chain reaction to maintain itself. Cadmium control rods are so located that they can be dropped down between fuel rods so as to isolate the uranium in subcritical amounts when the reactor is shut down and can be raised or lowered among the fuel rods to keep the chain reaction going at the desired rate. Surrounding the fuel elements and control rods will be a fluid medium, usually water, that may double as moderator and heat exchanger. Some newer reactors are being designed to use liquid sodium; others are gas-cooled. Figure 12.12 represents a basic design.

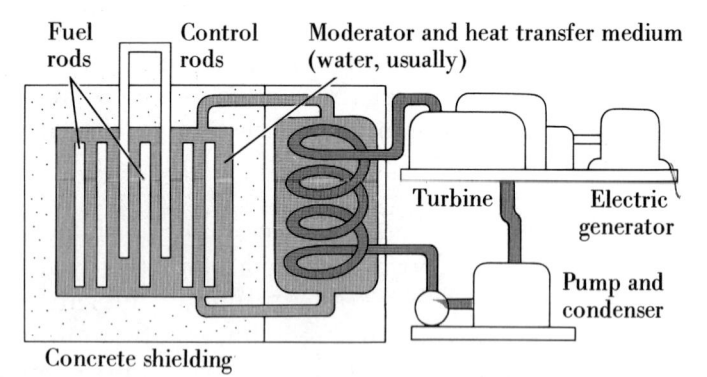

Figure 12.12 Diagrammatic representation of a nuclear power reactor.

Because U^{238} can absorb neutrons to produce fissionable Pu^{239} thus:

$$_{92}U^{238} + n \rightarrow {_{92}}U^{239} \xrightarrow{\beta} {_{93}}Np^{239} \xrightarrow{\beta} {_{94}}Pu^{239},$$

it is possible to take advantage of this reaction to produce new fuel (Pu^{239}) as the U^{235} is used up. A "breeder" reactor is designed to produce and use

Pu^{239}. From time to time fuel rods become depleted and must be removed, dissolved in acid, the uranium and plutonium chemically separated from the fission products, the fissionable material reclaimed and the rods reassembled. A major problem in nuclear reactor technology is the safe disposal of fission products, which are usually highly radioactive. The usual method is to store the solutions of waste material in concrete and lead tanks until the activity diminishes to "safe" levels, then to bury the solutions or discharge them into large bodies of water. There is, of course, some finite chance of leakage from tanks, and no one really knows what a "safe" level of radioactivity is because the problem simply has not existed long enough to assess its long-range effects on human and other life. People who work in radioactive areas wear film badges or other devices designed to monitor the total radiation dosage received. The United States Atomic Energy Commission sets up allowable dosages for a month or a year, based on the best available data, and a worker who absorbs his allowable monthly dose in a week will have to be transferred to a radiation-free area for the rest of the month, or other appropriate period, because radiation effects are cumulative.

Radiation may affect living cells by either destroying them outright by disrupting their vital metabolic functions or changing the genetic material by breaking the DNA chains at random. If the chains recombine in a different sequence, as is likely, the genetic message is changed, and the offspring will be a mutant that will differ in some more or less drastic way from the parent. An overwhelming majority of such mutations are harmful or fatal. The unknown and almost unassessable danger from radioactivity is that done to germ cells which leads to mutation. Of course, the natural radioactivity amidst which we all inevitably live, many of the chemical substances we ingest knowingly or unknowingly, and the cosmic rays that bombard us from outer space all have a finite possibility of causing mutations. This is one reason why older mothers are more likely than young ones to produce defective babies—their germ cells have simply had more chances to be altered. The fact that we unavoidably are subjected to mutagens all our lives is all the more reason to avoid those mutagens that are avoidable, such as needless exposure to radiation, including X rays or drugs of suspected mutagenic activity.

Radiation Uses

Although radiation has great potential for doing damage to life, in its medical application it is a great life saver. Radioisotopes of elements associated with the normal function of some bodily organs, such as iodine in the thyroid gland, play valuable roles in diagnosing disorders of those organs, because the presence and behavior of such radioisotopes in the body can be traced by counting techniques or analyzing body fluids. (Radioactive tracers are used in industrial processes as well as biological processes.) Better known to most people is that radiation can kill cancer cells selectively. Radiation treatment of cancer is particularly valuable where the malignant cells cannot be reached by surgery. Probably the most widely used isotope in cancer therapy is Co^{60}, which decays to Ni^{60} by beta emission accompanied by high-energy gamma rays that kill the malignant cells.

For better or worse, the world has entered the nuclear age. Since 1968 more nuclear power plants have been built each year than conventional steam or hydroelectric power plants. The dangers inherent in the disposal of radioactive wastes, the chance, however remote, of mechanical failure that would turn a power reactor into an "atomic bomb," the thermal pollution of rivers by the discharge of large volumes of cooling water into them so that their overall temperature is raised enough to upset their ecological balance—these are problems to be solved and risks that must be weighed against the need for energy for modern living. Conventional steam plants produce pollution, too, pollution of the conventional kind—smoke and smog. The choices are seldom clear, which is why every citizen needs to know both the problems and possibilities inherent in modern technology, for it is the lay majority rather than the scientific minority which ultimately decides the course of society. And if that majority is thoughtful and well informed it is probably a good thing that the decision is theirs.

SUMMARY STATEMENTS

1. The discovery of natural radioactivity led to the conclusion that the atomic nucleus, like the atom itself, has particles and structure.

2. The radioactive emissions, α, β, and γ rays, were identified as helium nuclei, electrons, and very short X rays, respectively.

3. Radioactive decay proceeds in completely random fashion, but where large numbers of atoms are involved it can be determined with great precision how long it will take for half the atoms in a radioactive sample to decay. This time is called the radioactive element's *half-life*.

4. Radioactivity may be produced artificially by bombarding stable isotopes with subatomic particles. The new nucleus formed by absorbing the bombarding particle may emit a different particle. Such a nuclear reaction is written, for example, this way: $_{13}Al^{27}(\alpha,n)_{15}P^{30}$.

5. Some nuclei, notably U^{235} and Pu^{239}, when bombarded by slow neutrons, split into two middleweight elements and several neutrons. The neutrons so produced can split other nuclei, leading to the establishment of a fission chain reaction if enough fissionable atoms, i.e., a critical mass, is present. This is the source of useful nuclear energy.

6. A vast nuclear technology has developed in the last forty years. Its devices include detectors such as cloud chambers and Geiger and scintillation counters; accelerators ranging from Van de Graaff and Cockcroft-Walton

generators to giant synchrotrons and linear accelerators; nuclear reactors for research and power production; and radiotherapy units for treatment of cancer.

7. Nuclear power production brings hazards as well as a plentiful energy supply, and every precaution for safe operation of nuclear power plants and disposal of radioactive wastes must be taken, because long-range effects of radiation on living things is not yet known.

8. Nuclear research has presented physicists with a bewildering array of subatomic particles and many still unanswered questions about the structure of the nucleus.

REVIEW QUESTIONS

1. Identify: alpha rays; beta rays; gamma rays.
2. Where do radioactive "rays" come from? Under what circumstances is each kind produced?
3. Complete the following nuclear equations:

   a. $_6C^{14} \rightarrow$ _____ $+ \, _7N^{14}$
   b. $_{92}U^{238} +$ _____ $\rightarrow \, _{92}U^{239} \rightarrow \, _{93}Np^{239} +$ _____
   c. $_{92}U^{238} \rightarrow \, _2He^4 +$ _____

4. What is meant by half-life?
5. Name and describe the operation of three types of radiation detectors.
6. Name and describe the operation of three types of accelerators for subatomic particles.
7. Define: (1) positron; (2) annihilation; (3) pair production; (4) meson.
8. Sketch the essential features of a nuclear power reactor using uranium fuel.
9. Define: control rods; breeder reactor; fission; critical mass; chain reaction.
10. Why is it so difficult to set definite safe or allowable levels of radiation for humans?

THOUGHT QUESTIONS

1. If radioactive elements in the earth have been decaying since the beginning of earth, why was radioactivity so long in being discovered?
2. Enumerate the various applications of radioactivity in fields other than physics.
3. Why are gamma rays so dangerous?
4. How can you justify the enormous expenditures of

money necessary to build giant accelerators in government laboratories?

5. The term *fundamental particle* becomes questionable when applied to twenty or thirty particles, and one wonders if they can all be truly fundamental. Watch for developments in this area as reported in the scientific press. What criteria would you expect for a "fundamental" particle?

6. What facts does the general public need to know about nuclear power plants, medical use of radioisotopes and X rays, and other aspects of radioactivity?

7. How do you decide what risks are tolerable when such decisions as the location of a nuclear power plant (or the use of a food additive, or a drug, etc.) must be made?

READING LIST

Curie, Eve, *Madam Curie*, Doubleday, Garden City, N.Y., 1940. The biography of Marie Curie by her daughter, beautifully told.

Gamow, George, *The Atom and Its Nucleus*, Prentice-Hall, Inc., Englewood Cliffs, N.J., 1961. Interestingly written and accurate, as are all Gamow's books.

Wilson, Robert R., and Littauer, Ralph, *Accelerators*, Doubleday (Anchor), Garden City, N.Y., 1960. A description of the construction and use of the big machines of nuclear physics by two scientists now working in the field.

Cohen, Bernard L., *The Heart of the Atom*, Doubleday (Anchor), Garden City, N.Y., 1967. An understandable account of the current theories of the nucleus.

Glasstone, Samuel, *Sourcebook on Atomic Energy*, Van Nostrand, New York, 1958. This book tells everything about atomic energy—history, theory, and practice—in very clear and readable language.

Smyth, Henry DeWolf, *Atomic Energy for Military Purposes*, Princeton University Press, Princeton, N.J., 1948. The first officially released account of the Manhattan Project. Interesting both as history and as science.

13

Stars and Galaxies

Our earliest scientific forebears directed their observations and questions to the heavens above them. And twentieth-century man still finds an excitement about the stars. Our universe is far grander and vaster, both better understood and more mysterious than that of the Egyptian priests or the Babylonian astrologers or the first scientific philosophers of Greece.

The ancients found astronomy a source of both fun and profit. As we have seen, the regular movements of the stars provided a calendar and marked the seasons for agriculture — a most practical service of astronomical observation. But the stars were also the picture book of ancient times. Lonely shepherds on midnight hillsides entertained themselves by illustrating the traditional tales of gods and heroes with patterns in the stars. The fact that the stargazers of Mesopotamia saw the same pictures (although their stories differed) as the North American Indians says something about the fundamental unity of the human family.

The pictures in the stars, which we call *constellations,* may not be as clear or obvious to us as they were to those who named them long ago, but they are still useful in mapping the sky. Modern astronomers divide the whole sky into areas dominated by prominent constellations, much as the map of the United States is marked off in states. As the earth orbits the sun, the constellations near or along the *ecliptic* — the plane containing the earth's orbit, infinitely extended — change with the seasons (Figs. 13.1, 13.2). Those toward which the earth's axis point remain visible all year — one set to dwellers in the northern hemisphere and another set to dwellers below the equator. The earth's north pole points almost at a star (therefore called Polaris, the Pole Star, or the North Star), which is the end star in the handle of the Little Dipper, part of the constellation Ursa Minor (little

Figure 13.1 The winter constellations.

Figure 13.2   The summer constellations.

bear). There is no corresponding south pole star. Events and "moving" objects such as planets are located in the sky according to the constellation forming their background. To say that at a certain season "Jupiter is in Scorpius" means that the planet is viewed in the southern summer sky in the area named for that constellation. Learning to recognize the more familiar star patterns such as Orion, Scorpius, Taurus, the Dippers, the Royal Family, Leo, Virgo, and so on can add to anyone's stargazing enjoyment. Unfortunately, stargazing is becoming one of the less generally available recreations for amateurs, and city smog in California is even interfering with the operation of the great Wilson and Palomar observatories.

Except for such rare cases as the Pleiades (Plate 23) and Taurus (consisting mainly of a star cluster called the Hyades), which are really groups of neighboring stars moving together in the galaxy, most constellations are strictly fortuitous temporary configurations of stars that are actually widely separated and moving essentially independently of each other. The most familiar and easily recognized constellation, the Big Dipper, won't even look like a dipper in another 100,000 years (Fig. 13.3). It is like seeing three trees aligned from one point along a road and looking back from a few yards farther on to see that they are quite widely and randomly separated.

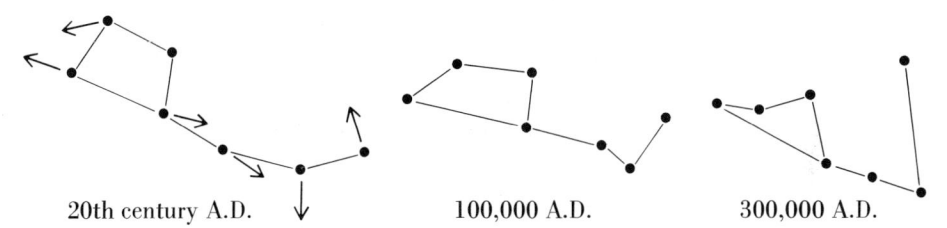

20th century A.D.   100,000 A.D.   300,000 A.D.

Figure 13.3 Projected changes in the Big Dipper resulting from proper motions of the individual stars.

For mapping purposes there are eighty-eight constellations. Those longest recognized represent the heroes of tales dating from the classic days of Greece. Many constellations of the southern hemisphere, however, were named by seventeenth- and eighteenth-century seafarers, and bear such names as the *Telescope* and the *Forge*. Although most bright stars have individual names, each star is also named systematically according to its visual magnitude (apparent brightness) within its constellation. The North Star is thus α Ursae Minoris—the brightest (α) star in the constellation Ursa Minor. We shall follow the astronomers' convention of identifying stars and sky areas by their constellations as we attempt to see something of structure and change in the universe beginning with our own solar system.

Kepler's laws made the universe simple; the law of gravitation made it understandable. The quantum theory, relativity, and the discovery of nuclear reactions helped to solve many remaining mysteries in the twentieth century. At the same time the building of bigger and better optical telescopes and the development of receiving equipment responsive to many ranges of the electromagnetic spectrum other than visible light revealed many other objects and phenomena in the universe yet to be understood.

Astronomy

The history of astronomy from the first telescopic discoveries of Galileo to the present is fascinating. We have already seen that the discovery of the planet Neptune from its gravitational perturbations of the orbit of Uranus was one of the great triumphs of the theory of gravitation. Before long, however, it became apparent that another body was perturbing both Uranus and Neptune. The American astronomer, Percival Lowell, made very detailed calculations and predictions as to its nature, but died before they were fulfilled. His assistant, Clyde Tombaugh, continued the search and in 1930 took the series of photographs showing a tiny, dim body moving perceptibly against the background of "fixed" stars that demonstrated the existence of a ninth planet. It was named Pluto for the god of the dark underworld, whose name happened to start with Percival Lowell's initials. For the forty-odd years since Tombaugh's discovery, the controversy has simmered over whether or not Pluto really meets Lowell's requirements. Many astronomers feel it is too small to be the cause of all the perturbation that must be accounted for. Perhaps there is a tenth planet; if so, it is even farther away, and so much dimmer even than Pluto that we have no means by which to detect it today.

Comets

Other members of the solar system spend most of their time even farther away from the sun than Pluto, but at the other end of their orbits come nearer the sun than any other bodies. These are the comets. A large number belong to the solar system, traveling around the sun in orbits that are very elongated, flattened ellipses. Whether these bodies were always integral parts of the system with a common origin with the sun and planets or whether they have been captured from beyond the system by gravitation when they strayed into the sun's neighborhood is not known. They are smallish bodies that have been described as "dirty snowballs"—chunks of rocky material stuck together with frozen water, ammonia, methane, and the like. These frozen substances vaporize when the comet comes near the sun, and the pressure of the sun's radiation pushes them out in a long tail that reflects the sunlight like streaming hair, from whence the name: *coma*, Latin "hair."

The most famous comet is undoubtedly that named for Edmund Halley, friend of Newton and British Astronomer Royal. He identified as a single body with an orbital period of seventy-five or -six years the comets reported at that interval since very early times. When the comet returned in 1758, sixteen years after Halley's death, exactly as he had predicted, it came to be known as Halley's Comet. Aristotle had taught that comets, being other than perfect spheres, were atmospheric rather than celestial phenomena. Because they appeared unexpectedly—and because there is always some disaster somewhere coincident with almost anything—they came to be considered portents of evil. As late as 1910 many citizens of the enlightened United States of America panicked at the news that the tail of Halley's Comet would envelop the earth on its close approach to the sun that year. In fact, the tail was so sparse no one could tell it was there except by very accurate spectral analysis that would identify trace constituents not usually detectable in the earth's atmosphere.

One group of comets has short periods of roughly three years. These comets have been "captured" by the giant planet Jupiter so that they travel elliptical orbits with the sun and Jupiter as foci. New comets are reported frequently. Some return; others go so far out in space that they are captured by some other system and never return, or their orbits are parabolic and never close up and repeat themselves (Fig. 13.4).

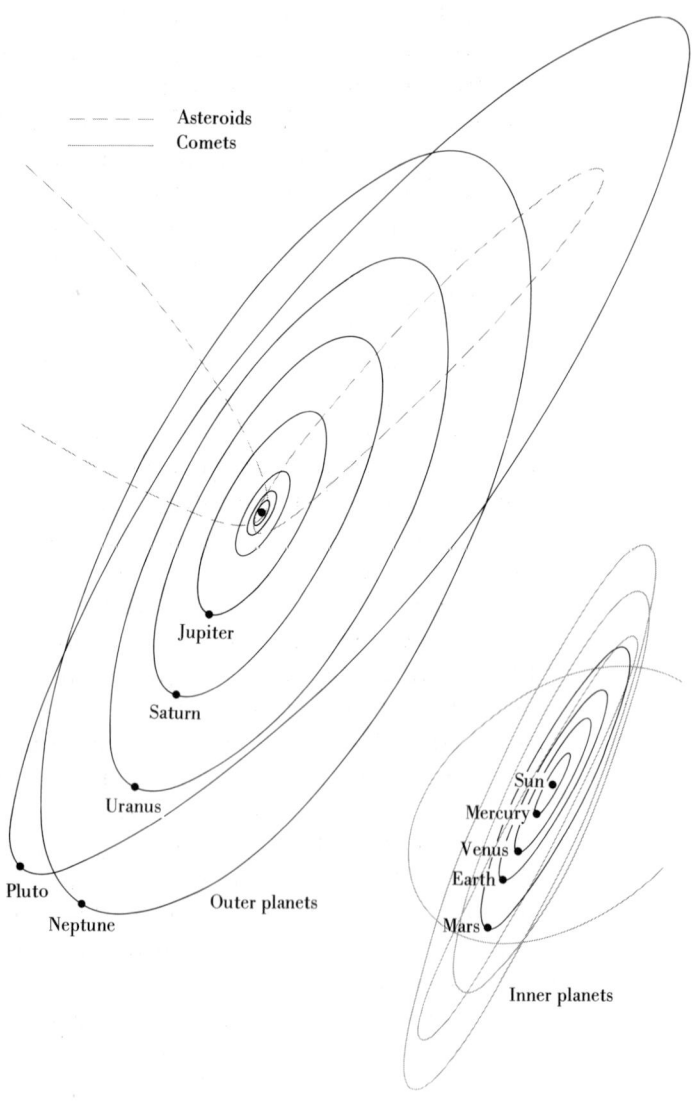

Figure 13.4 Comet orbits.

Planets and Planetoids

After Neptune was discovered in the 1840's every astronomer became a planet-hunter. And a large number of them were successful. A thousand planets were discovered, but they were not the cause of irregularities in Uranus' orbit. They were very small, from large boulders to small islands in size, and they all whizzed around in orbits of varying eccentricity between the orbits of Mars and Jupiter. This flying brickyard came to be known as the asteroid belt, although these boulders are not asteroids—baby stars, but planetoids—baby planets. In fact, one of the most acceptable explanations for planetoids is that simply not enough material was available in this part of the solar system to get itself together gravitationally to form a planet even as big as Mercury or Mars. The alternate theory—that a planet disintegrated—seems less likely from the quantity of scraps.

One or two of these planetoids come closer to the earth than any other body except the moon. Careful measurements of a planetoid's distance on these occasions of nearest approach give astronomers a reliable standard by which to refine other distance measurements. The solar system as we now know it is pictured in Plate 18.

All but three major planets are orbited by smaller "satellite" bodies, or moons. The giant planets, Jupiter and Saturn, have a few large satellites each, and numerous small ones, and more may remain to be discovered. In fact, they may even acquire more moons by capturing, gravitationally, close-approaching planetoids. Probably several of the present moons of Jupiter and possibly of Saturn are captured planetoids. On the other hand Neptune may have lost a moon by having it captured by the sun. Pluto not only is very much smaller than its neighbor planets, but also part of its orbit lies nearer the sun than the orbit of Neptune. This suggests that at one time Pluto may have been a satellite of Neptune, but on its nearest approach to the sun, once upon a time, it was deflected by the sun's gravity into a solar orbit. Pluto will spend the remainder of the twentieth century (until 2004) nearer the sun than Neptune is.

Earth's natural satellite is unique in at least one respect. It is much larger in proportion to its planet than any other satellite. Several moons of Jupiter and Saturn are larger than Luna, but they are still dwarfed by the giant size of their planets. It has been suggested that the earth-moon combination is really a double planet. The distinction is not really important.

The gravitational capture of planetoids is not restricted to the giant planets. Our pock-marked moon is evidence that it has captured numerous unburned remains of meteors that we call *meteorites*, and Earth, too, has some obvious meteor craters. Meteors have two probable sources. Some are planetoids whose orbits intersect the earth. They burn because of frictional heating in Earth's atmosphere: we see them as "shooting stars." At various times of the year there are showers of shooting stars, dozens or hundreds visible per hour. The timing and location of these meteor showers suggest that these meteors are actually fragments of comets remaining in orbit after the sun has vaporized all the ices that were holding them together. When Earth's orbit intersects the old comet orbit, many of the fragments fall into the atmosphere and burn up.

In 1908 in an uninhabited area of Siberia an enormous explosion was heard and seen by villagers many miles away. The crater, the charring of trees, and the mineral fragments found suggest that this "meteor" must actually have been a comet that crashed to Earth.

Creation Theories

Perhaps the first question man asked after the fundamental query "What's for supper?" was answered was "Where did I come from?"—where "I" includes not only the person but all his world. The oldest myths of every civilization are those that account for creation. The universe as conceived by ancient man was not nearly as vast as we now know it to be. Nevertheless, it was so overwhelming to him that he could only imagine it spoken into being by the Creator. Only in the last century has scientific theory been sufficiently advanced to permit man to envision a mechanism by which the universe might have come into being.

Just how vast is the universe? An attempt to locate ourselves in it may give us some insight. We live on a planet roughly 8000 mi in diameter, 93,000,000 mi from the center of our solar system, the sun. The sun, in turn, is some 30,000 light years[1] from the center of our Galaxy, the Milky Way, one of a local group of galaxies some few million light years apart. (Andromeda, our nearest large neighboring galaxy, is 2,000,000 light years away.) The farthest objects in the visible universe lie some two or three billion light years (1.2 to 1.8×10^{22} mi) away. We do not know what may lie beyond our present sight.

How, then, are we to account for a rocky, watery planet a comfortable distance from a moderate sized, steady second-generation star, one of a hundred billion stars, some bigger, some smaller, some hotter, some cooler, some with planetary systems like its own, in a huge rotating spiral moving through space in company with other similar spirals and smaller irregular galaxies, rushing away from ever more distant galaxies at ever-increasing speeds? Two main theories merit consideration. One presents an evolutionary view of the universe, having a beginning, a progressive development, and a conclusion—but with the possibility of recycling the whole process. The other—the "Steady State" theory of the universe—

[1] A light year is the distance light travels in one year, just under six trillion miles.

might be said to adopt the philosophy that "the more things change, the more they stay the same."

Any acceptable theory must explain several well-known features of the universe. First, there is the relatively uniform distribution of galaxies in space. "Galaxy" is defined as a large group of stars moving together in space. Many galaxies have the structure characteristic of our own Milky Way—a dense nucleus of bright reddish stars (Population II) from which two spiral arms emerge. These arms contain stars that are generally bluer (i.e., have higher surface temperatures) and are embedded in vast clouds of gas and dust. These so-called Population I stars are considered younger than the Population II stars, although they appear to be in a wide variety of stages of evolutionary development. In fact, new stars now seem to be forming in the spiral arms of our galaxy. Other galaxies are ellipsoidal and composed entirely of Population II stars. Still others, most of them small, have no obvious structure at all, and are classed as "irregular." The ratio of the distance between galaxies to the diameter of a galaxy is actually smaller than the distance-to-diameter ratio for stars within a galaxy. That is to say, galaxies are more densely scattered through space than stars through the galaxy.

Rotation must also be accounted for. That is, a theory of origins, either of the galaxy or the solar system, must account for their angular momentum, because this quantity—the product of the mass, velocity, and radius of rotation of any rotating body—is quite rigorously conserved.

There is considerable evidence that the universe is expanding. Every other galaxy in space is rushing away from ours, and the farther away they are, the faster they are receding. Presumably, this would be our observation no matter which galaxy we called home. The situation can be pictured thus: suppose you put tiny ink dots on an empty balloon and then blow it up. The dots will "rush away" from each other, and each dot will disappear from the view of others as the balloon expands. Though the universe is expanding at a great rate, the density of galaxies does not seem to change appreciably. Of course the whole period of human observation is but the wink of an eye in the lifetime of the universe. It would take many astronomers' lifetimes to establish the reality of the pattern of change on a universal scale.

The "Steady State" and "Big Bang" Theories

The basic premise of the "Steady State" theory, proposed in the 1950's by three astronomers, Fred Hoyle, Hermann Bondi, and Thomas Gold, is that there is no net change in the universe. As galaxies expand "over the edge," as it were, out of sight, new galaxies must form to maintain a uniform density in space. Because hydrogen is the ultimate element of the universe, this theory requires the continuous creation of hydrogen, from which stars and galaxies are presumed to form by all the current theories. There are, of course, troublesome questions about where the disappearing galaxies go and how to account for the appearance of new matter, because both violate the traditional idea of conservation. Some rather esoteric experiments done in the late 1960's with high-energy particles were expected to confirm or deny the validity of the Steady State theory. They did not produce the hoped-for confirmation, and the Steady State theory has fallen out of favor.

The "Big Bang" or evolutionary theory has its difficulties, too, some of them in common with the Steady State theory. It suggests that the universe had a beginning in a superdense plasma of fundamental particles (protons and electrons, at least) and that some disequilibrium at some point in time and space caused this plasma (Gamow called it "ylem") to expand. In the first few minutes of expansion, hydrogen, some helium, and lesser amounts of other elements formed from the protons and electrons, and through all subsequent time, stars and galaxies have formed from this primordial hydrogen-helium cloud and have lived out their lives over millions or billions of years, creating other heavier elements from hydrogen and helium during their evolution. This theory does not say where the ylem came from, nor does it say where or if the expansion will end. It has been built into a cyclic theory of a pulsating universe by some cosmologists, with eternal cycles of expansion and contraction.

Evolutionary Development

Whether the ultimate behavior of the universe as a whole is best described in evolutionary, cyclic, or steady state terms, there is ample evidence for the evolutionary development of given stars and galaxies. A star is born in a swirl or eddy in a cloud of cosmic gases, and it contracts into a sphere by the gravitational interaction of the gas atoms. Gravitational potential energy is converted into heat, heat at an intensity of 20 million degrees at the center of the mass. At this temperature the newborn star starts to generate radiant energy at visible wavelengths. The star has begun to shine.

The source of energy that enables a star such as the sun to shine for billions of years with a steady outpouring of electromagnetic energy remained a mystery until the 1940's. Just after World War II, Hans Bethe suggested that the nuclear fusion reaction, in which four hydrogen atoms are converted into a helium atom of sufficiently smaller mass to account for the enormous production of energy, is the source of the sun's — and most other stars' — radiation. At last, here was a theory that could account for the vast amounts of energy produced over vast periods of time, something impossible to explain in terms of any known or imaginable chemical process of combustion. Although Bethe's original mechanism for hydrogen fusion in the sun has been modified, a fusion reaction of one kind or another is believed to operate in all stars. The star we call the sun, and on whose energy all life on Earth is directly dependent, is a medium- to small-sized, moderately hot, stable star, roughly 5 billion years old and in its prime, confidently believed able to continue its steady functioning for another 10 to 30 billion years. Yet the sun is converting mass into energy at the rate of 4,600,000 tons per second by fusing four hydrogen nuclei at 1.00797 atomic mass units (amu) each into a helium nucleus of 4.0026 amu. Thus: $4 \times 1.00797 = 4.03188 - 4.0026 = 0.0293$ amu converted to energy. One amu is $1/N$ grams, where N is Avogadro's number, the number of atoms in one gram atomic weight of an element. Thus

$$0.0293 \text{ amu} = \frac{0.0293}{6.023 \times 10^{23}} = 4.86 \times 10^{-26} \text{g}.$$

This amounts to about 0.73% of the hydrogen mass being converted to energy. By Einstein's equation, $E = mc^2$, the energy per fusion is

$$4.86 \times 10^{-26} \text{g} \times 9 \times 10^{20} \text{cm}^2/\text{sec}^2 = 4.37 \times 10^{-5} \text{erg}.$$

This alone is a minute amount of energy. But the same equation applies to the conversion of the 4,600,000 tons of matter to energy per second. This amounts to energy production at the rate of 3.75×10^{25} kilowatts.[2] The sun's total weight is some 2.2×10^{27} tons. If it were originally all hydrogen it would take 100 eons (100 billion years) to use up all of it at the present rate.

Now remember, these figures apply only to a

---

[2] A kilowatt is a unit of power, or rate of energy production, which is equal to 1000 joules per second, or about 250 calories per second.

mediocre sort of star like our sun. Numerous stars are many times as massive as the sun. The more massive a star, the faster its fusion reactions proceed, so that a blue-white giant like Rigel (Orion's left knee in the familiar constellation) may run its whole course in the (astronomically speaking) short time of a billion years.

Measuring the Stars

At about this point the question of how well we know all these things may arise. We have stated these astronomical figures with apparent confidence, yet the stars, if they are that big, must be very far away. We can hardly have weighed or measured them directly. How can we know with such certainty what we are talking about? Fair enough. Let us stop here and see what we can measure about the bodies in the universe, and how.

Until very recent years, all the information we could get about bodies beyond the earth came in one form only: the *light* from them that reaches our eyes or our camera film. The eye is the most precise and sensitive receptor of light, but film doesn't get tired of looking, and astronomers have been photographing the heavens for a century, now. We can put a clock drive on our telescope so that it turns oppositely to the earth and stays focused on a single patch of sky all night, collecting on film light from stars the eye cannot see because the effect of light on the eye is not cumulative. The size of a telescope determines its light-gathering power. Magnification is important when we are studying our own solar system to show us the detailed features of the moon and neighboring planets, but no existing (or proposed) telescope can magnify any star sufficiently to make it more than a point of light. No star except the sun is near enough to "show a disk" to even the largest telescope.

Light from the stars tells us where they are in relation to each other, their brightness or apparent magnitude, and their color. Comparison of many observations can also call attention to changes in any of these properties with time. From these things that we or our instruments can see, we must figure out all the other things we want to know: the distance to the stars, their actual brightness (absolute magnitude), their diameters and masses, their temperature, their chemical composition, and even their life histories.

Distance

This seems a lot to ask from so little direct information, but it can be done. Brightness can mean several things. A star may be very bright because it is really emitting a lot of light, or it may appear very bright simply because it is close to us. How do we decide which? We utilize a phenomenon that many 5-year-olds will have noticed. If you are riding along a country road on a moonlight night you notice that the moon "keeps pace" with you, although the trees and telephone poles along the roadside appear to be passing you in the opposite direction. This apparent relative motion or *parallax* of near and far objects enables us to distinguish between these objects, and even to determine distances to the near ones. Another illustration of parallax will demonstrate the possibility: hold one index finger as far from your eye as the length of your arm permits, and, closing your left eye, align that finger with a distant object—a tree outside the window or the farthest corner of the room. Now close your right eye and look at the finger with your left eye. It is no longer aligned with the tree or corner. You see it from a different viewpoint against a different background. Thus you can measure the angle between your eye and your finger by comparison with that differing background, and if you can measure the distance between your eyes, you can calculate the length of your arm. (It might be easier to measure your arm directly with a tape measure, but no tape measure will reach to the stars!)

Figure 13.5 is terribly distorted. Even with a base line equal to the diameter of the earth's orbit the largest angle of parallax (θ) we ever observe is less than 2 seconds of arc. The stars are very far away, but if we know angles α and β and the length of the base line we can calculate the distance AS or OS (which are not very different from each other).

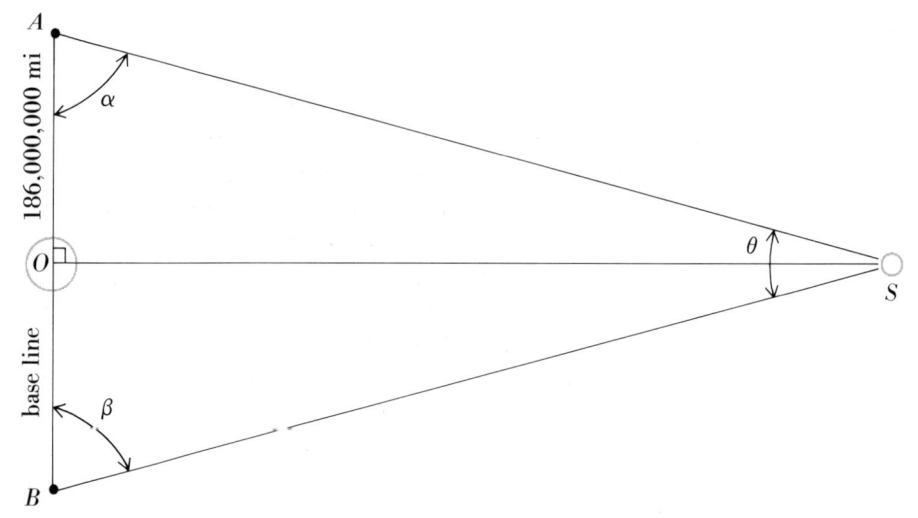

Figure 13.5 Parallax method for determining distances to the nearer stars by observing their changing background of "fixed" stars.

However, only about 10,000 stars have parallax angles large enough to be measured; this includes stars within a radius of 300 light years of the sun. But these few give us a foothold, because once we know the actual distances of these stars we can interpret their brightness in absolute terms[3] by the inverse square law of light intensity. If we apply Occam's Razor, we will assume that all the stars that look alike are alike, so another star of the same color is likely

[3] That is, the intrinsic brightness is a measure of the actual amount of energy the star radiates.

to be of the same intrinsic brightness as our measured star, so its apparent brightness gives us a clue to its distance. On this basis, astronomers confidently make tables of stars showing their distances and intrinsic brightness.

Mass

We weigh stars by exactly the same principle—Universal Gravitation—we use to weigh ourselves, although we use different equipment. Fortunately for the weigher of stars, many of the easily observable stars are double, i.e., many clouds of gas and dust are condensed into two protostars instead of one protostar and a flock of protoplanets, as with the solar system. A pair of stars constituting a "double star" rotates around the center of gravity of the double star system. If their distance from the earth can be measured or estimated, their distance from each other can also be calculated by observing their maximum separation. The rotation period can also be observed, and from these data the force of gravitation can be calculated as a centripetal force. Only the product of the masses remains unknown, but it can be calculated, too. To determine the individual masses we would observe the pattern of rotation to locate the center of gravity (c.g.). For rotating bodies the mass times distance from c.g. is the same for each, so the ratio of these distances, or radii of rotation, enables us to obtain also the ratio of masses. (Now we have two relationships—one for the product of the two star masses and one for their ratio, and it is possible to solve for two unknowns, given two equations.)

But what about determining the masses of stars that are single (or multiple)? We fall back on the information found in the stars' light in terms of the "mass-luminosity" relationship. But to determine luminosity, or intrinsic brightness, we need to know distances, which, as we have seen, are hard to come by, except for the handful of stars near enough to exhibit parallax. Fortunately, another relationship allows us to estimate even very great distances with considerable accuracy. About 1912, American astronomer Henrietta Leavitt began a systematic study of short period variable stars called *Cepheid variables*, after the archetypical member of the group, Delta Cephei. These stars dim and brighten in a regular cycle of from two to forty-five days and are distributed throughout the sky in the Milky Way as well as in other galaxies our largest telescopes can resolve into stars, particularly our nearest neighbors, the two small irregular galaxies called the Clouds of Magellan. Miss Leavitt's crucial discovery was that there is a direct proportionality between the period and the luminosity of Cepheids, so that any two Cepheids having the same period of variation could be assumed to have the same intrinsic brightness. Because the brightness of a light source is inversely proportional to the square of its distance from the observer, the period-luminosity relation provided astronomers with a very long measuring stick.

As it turned out, the mass-luminosity and period-luminosity relations are very closely associated. A star is a system in delicate balance. As it condenses under gravitational forces it gets hotter and begins to radiate energy, which exerts an outward pressure. In a stable star like the sun the inward pressure due to gravity and the outward radiation pressure remain in balance over long periods. In variable stars, some instability requires continual adjustments. These adjustment cycles are longer the more massive the star is, so the longer the period of a Cepheid, not only the brighter it is, but the heavier. Putting all these clues together, we can thus estimate distance, mass, and intrinsic luminosity for a large number of stars.

Temperature and Composition

The most direct information starlight gives the astronomer is about surface temperature and chemical composition. To process this information we run the light through a prism or a diffraction grating to spread out its spectrum. A stellar spectrum is a continuous rainbow crossed by numerous vertical dark lines (Plate 20). The dark lines can be matched with the bright lines emitted by elements that are heated in arc lamps or excited by high potentials; in other words, stellar spectra show absorption lines corresponding to atomic line spectra. The dark line is where energy is missing because quanta of a certain size (frequency, wavelength) have been absorbed to promote electrons to a higher energy state. Absorption lines make as good fingerprints of elements as emission lines. A star contains those elements whose fingerprints it displays.

It may seem far-fetched, but if we consider a star to be a "blackbody," a perfect radiator, a well-known law relating temperature and wavelength of maximum intensity of blackbodies can give us the star's surface temperature. This is the Wien Displacement Law, which says the absolute temperature is inversely proportional to the wavelength of maximum intensity, or

$$T\lambda_{(max)} = k.$$

Now if we know the temperature and the brightness of a star we can figure out how much surface is required to produce that much light, and we will have another item on our list, the diameter of the star.

How do we tell where the stars are going when they appear "fixed" relative to each other, night after night and age after age? Constant observation would soon convince us that some stars, at least, are not fixed. Halley first demonstrated this in 1718 by showing certain bright stars, including Arcturus, Sirius, and Procyon, were a full degree of arc from their positions reported by Greek observers two millennia before. This "proper motion" of stars is naturally more noticeable for nearer stars, such as the three mentioned, but others also show it, though perhaps only to the extent of less than a second of arc per year. Again, however, a great many small observations point to a trend, so we may say in general that our galaxy is rotating, and that our sun, some 30,000 light years (30,000 × 6 trillion miles) from the center of the galaxy, completes one revolution in 230,000,000 years.

The Doppler Effect

We have also noted that every other galaxy in space appears to be receding from us, leading us to the conclusion that the universe is expanding. The evidence for this expansion, as well as a method for calculating its rate, is provided by the absorption lines in galactic spectra. Hydrogen, that most abundant stuff of the universe, has two characteristic lines, one blue and one red. Now let us talk about train whistles.

Anyone who has waited, patiently or otherwise, at a railroad crossing while a fast freight went through has observed the Doppler effect. As the train approached, its whistle or horn rose in pitch *(frequency)*, but as it passed the pitch dropped suddenly and continued to get lower as the train thundered off into the distance. Actually, the frequency emitted by the whistle did not change, but the crowding together of the sound waves as the source of the sound approached effected an increase in the heard frequency, just as their spreading out from the receding source effected a decrease. The Doppler effect occurs with all kinds of waves (Plate 19). Applied to radar, it provides the police with a means of detecting speeding motorists.

Now back to hydrogen absorption lines. They are easy to pick out in stellar spectra, but don't always come at the "right" place. In rare cases, their frequencies appear a little too high, i.e., shifted toward the blue end of the spectrum. In most cases they are more or less low, or shifted toward the red end. Such a blue or red shift is simply a Doppler effect (Plate

19). A blue shift means that a star is approaching us (for example, a neighboring star that is revolving a little faster than the sun around the galactic center might be "catching up" with the sun, or the sun might similarly be closing in on a slower star in a more or less parallel orbit, no danger of collision!). A red shift means the star is receding, and the farther away it is, the faster it is going. (Remember the polka-dot balloon experiment?) Again we can put two and two together and get velocities and distances for the receding galaxies. The farthest observable ones are some 12.5 billion light years away, and receding at speeds very close to the speed of light. If there are more distant galaxies we won't know it, because their light will never reach us.

Let us pause here for an intriguing thought. A "light year" is the distance light travels in one year. At 186,000 miles per second, that comes to just under 6 trillion miles. When we say that the Andromeda Galaxy (Plate 22) is approximately 1,500,000 light years away from us, we are saying the light we see tonight from the fuzzy spot near the great square of Pegasus left its source, a galaxy very like our Milky Way, only a bit bigger, 1.5 million years ago. Whenever we look at the heavens we are looking back in time. Even sunlight is eight minutes old when it reaches us, and the very nearest star, Alpha Centauri (not visible above 30° north latitude), appears to southern skywatchers as it was 4.29 years ago.

A Model of the Universe

With all these astronomical numbers and these bewildering details, based on nothing but painstaking observation of rays of light, it is really no wonder that a coherent model of the universe is only now beginning to emerge. A very convenient preliminary sketch around which our model can be built, however, was

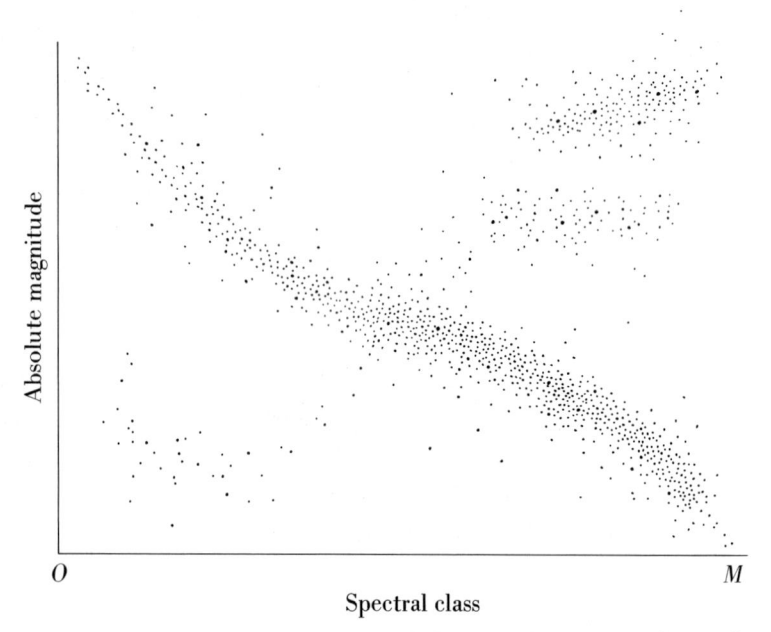

Figure 13.6 Hertzsprung-Russell diagram for stars in the sun's neighborhood in the Milky Way.

made by Ejnar Hertzsprung and independently by H. N. Russell around 1913. It was a graph, that data-displaying device so much loved by scientists, in which the absolute magnitude (a measure of intrinsic luminosity) was plotted against *spectral class*,[4] which is simply a way of specifying temperature, which, as we have seen, is denoted by the color or wavelength of maximum intensity of the star. The H-R diagram plotted for a large number of stars in the sun's neighborhood, i.e., a spiral arm of the Milky Way, looks something like the drawing in Figure 13.6. If we draw an enclosing curve around the greatest concentrations of dots (locations of individual stars by brightness and temperatures) we get the shape in Figure 13.7.

Most stars in the sun's neighborhood lie in the area

---

[4] Spectral classes, from hottest and bluest to coolest and reddest, are designated O, B, A, F, G, K, and M, which astronomers presumably remember by the mnemonic: "Oh, be a fine girl, kiss me."

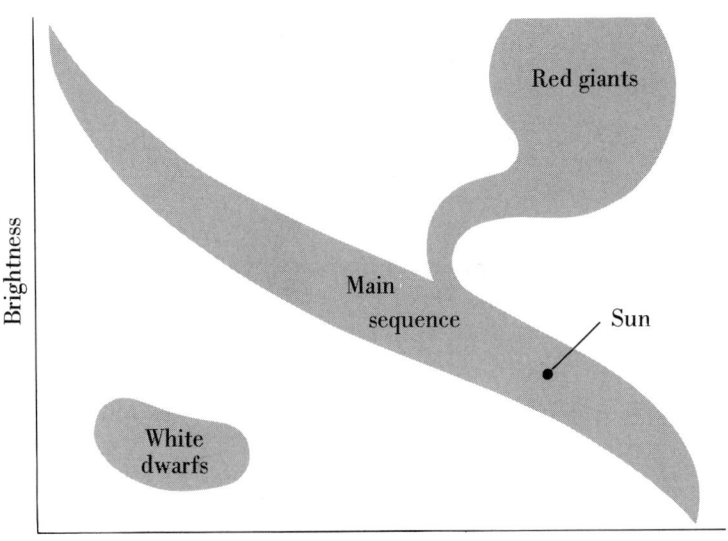

Figure 13.7 H-R diagram showing relative locations of star types.

labeled "main sequence." Apparently these are "normal" stars. They vary from dim reddish stars, not very hot and not very bright, and, as it turns out, not very massive, to very hot, bright, massive stars at the upper left. The mass-luminosity relationship was a sort of bonus from the Hertzsprung-Russell diagram. As we have noticed, only double stars can be "weighed." But the arrangement of weighed stars in the main sequence showed that luminosity is proportional to mass. The relation is not a simple linear one. The brightness of stars varies over a much wider range than do their masses, but the two quantities do vary in the same direction. It seems that a small increase in mass increases the energy production by a large factor.

There is a considerable density of stellar population in two other areas of the H-R diagram. In the upper right corner are the red giants, stars of high luminosity and low temperature. This can only mean that these stars are very, very large in surface area. Antares, a super giant, is some 320 times as large in diameter as the sun. Aldebaran, a red giant, is 57 sun-diameters across. These giant stars, however, are only five to ten times as massive as the sun, so they have very low density. In fact a space the density of Antares would be considered a very good vacuum on earth!

In the lower left corner are stars so dim for their temperatures that their surface areas must be extremely small. Although their masses may be of the same order of magnitude as the sun, their diameters are only a few hundredths or thousandths that of the sun. In a typical white dwarf, matter is so compressed that a teaspoonful of it would weigh tons.

Why such a vast range of temperatures, luminosities, and diameters among stars whose masses lie within the small range of less than two powers of ten? When we consider the nuclear reactions by which a star makes its energy, and the delicate balance between radiation pressure and gravitational pressure, the idea occurs that these various kinds of stars reveal the stages through which most stars pass. So we read from the H-R diagram the life line of a typical star.

The story begins, as we have already seen, with condensation of clouds of dust and gas under gravitational forces. The compression produces heat; heat tears the atoms apart into plasma of nuclei (mainly single protons) and electrons; and when the plasma gets hot enough, nuclei begin to fuse, generating energy in the process. Just what would such an "infant" star, just getting started, look like? It seems quite possible that the infant star is a red supergiant. That is, the low-density supergiant stars such as Antares (in Scorpius) and Betelgeuse (Orion's right shoulder) are at the beginning of their careers, still in the early stages of condensation. The constellations containing these two stars have been familiar to stargazers for some five thousand years. Stars have a long infancy.

As a star continues to condense, it gets hotter, both inside where the nuclear reactions are taking place, and outside, where the color and brightness indicate its surface temperature and size. Presently the star will join the main sequence. Just where it fits into this sequence is mainly determined by its mass, which determines its luminosity and its temperature. Mass also determines how long a star stays in the main sequence. The sun is about a third of the way up the main sequence band. Its mass is fairly small, so that gravitational compression has produced a core temperature of probably no more than 20,000,000°K. At this "modest" plasma temperature, hydrogen fusion proceeds at a moderate steady rate that will take some 10 billion years in all to fuse the hydrogen of its core to helium. The sun has spent about half its life expectancy on the main sequence so far. A much more massive star, such as the blue giant Rigel, may spend as little as a billion years on the main sequence, because its nuclear furnace operates at a much higher rate than that of the sun.

Stars on the main sequence are hydrogen burners. They make their energy almost exclusively by fusing hydrogen to helium by any one of several possible mechanisms, depending on the elemental content of the particular star. The sun does not fuse hydrogen nuclei directly, but by means of a cycle in which carbon, nitrogen, and oxygen nuclei are intermediates. A star with little or no carbon to begin with would use the more direct mechanism. When the hydrogen supply is exhausted, the star's radiation pressure can no longer counterbalance gravitation, so the star collapses. But the heat generated in collapsing ignites a new fusion process in the core: helium nuclei begin to fuse into carbon and/or oxygen nuclei, generating even more energy than before. Radiation pressure causes expansion of the star, which is likely to grow far beyond its original size. Until a new equilibrium is reached, the star is likely to pulsate.

Variable Stars

The sky is full of stars with variable brightness. Some vary because they are double and the brighter is regularly eclipsed by its dimmer partner. Tens of thousands of observed stars, however, are intrinsically variable. Some dim and brighten again over periods of several months; others, the Cepheids, have periods measured in days; still others, the RR Lyrae stars, vary from maximum to minimum brightness and back again in a matter of hours. In the last few years high frequency variables that emit mainly radio frequencies — although some emit ultraviolet or X radiation — called *pulsars* have been discovered. How can we account for this widespread and wide-ranging instability among the stars? One way to account for the longer period variables is in terms of alternate gravitational collapse and renewed bursts of radiation pressure, indicating a flickering of the original nuclear fires of hydrogen fusion as the hydrogen supply approaches some lower limit. Apparently such a state can last a very long time, for some variable stars have been observed for centuries.

If this is an accurate explanation of intrinsic variability, it would seem to be a fate in store for every star on the main sequence. Sooner or later the hydrogen supply must be exhausted, and the star must move off the main sequence into a variable stage, the length of which is doubtless related to its mass, just as is its tenure on the main sequence.

Novae and Supernovae

A fair number of stars undergo a briefer but more violent transition period than the regular variable stars. Such a star increases suddenly so rapidly in magnitude that the naked eye observer, one night, sees a bright star where none was visible before. It appears to be a "new" star, a "nova." Comparison of photographs taken through large telescopes, however, will reveal that the star was there all along—its brightness has just suddenly increased by several magnitudes. In this case, a sudden enormous burst of radiation pressure, probably due to the ignition of some new fusion process, has blown off an outer layer of the star's substance. This rapidly expanding envelope becomes visible, then fades over a period of months or years as it disperses. It is estimated that only about 1% of the star's mass is lost in such an upheaval, and it may happen again and again to the same star at intervals of ten years or so.

In any case, every star must progress through successive stages of nuclear fusion processes, producing successivly heavier nuclei, until no other energy-producing processes are available to it. Radiation pressure diminishes as the nuclear fires die out, and gravitational collapse crushes the remains into a white dwarf. The star will shine on, feebly, for a long time, emitting the energy released as its atoms are pressed more and more tightly together until even atoms as such cease to exist. Electrons, crushed into the nuclei, combine with protons to form neutrons. *Pulsars*, which have periods of variation not of days but of fractions of a second, appear to be neutron stars emitting their last bit of gravitational energy.

Once in a while—about three times per millennium—a really violent stellar event occurs. This is a *supernova*. In 1054, Chinese astronomers recorded their observation of a star so bright it could be seen by day as well as by night. (This is true of the planet Venus occasionally, but these observers had perhaps two thousand years of records of Venus observation behind them, and Venus isn't what they watched in 1054 for some months.) Stargazing was not much in vogue in Europe in 1054, and anyone who did notice the phenomenon probably kept mum about it. No good could come of such a thing! (And it didn't: Britain was conquered in 1066. At least such was the reasoning of the day.) Modern astronomers looking toward the constellation Taurus see a wispy cloud of colored gases, somewhat elliptical in shape, with appendages all around, which is called the Crab Nebula (Plate 24). It is just where the bright star of 1054 was located. Putting eleventh-century and twentieth-century clues together, we come up with this story.

Once in a while, a very massive star becomes so unstable that it blows itself up completely. This is absolutely the end of the nuclear fusion road for such a star. Not only its hydrogen, but its helium and perhaps other light elements are all but gone. It has collapsed and its core has been heated to the higher and higher temperatures (around 3,000,000,000°K) at which heavy nuclei can be formed, and it has extracted the last bit of energy from the formation, not only of stable iron and other middleweight elements, but even inherently unstable atoms like radium and thorium and uranium. This last bit of energy takes it out in the blaze of glory we call a *supernova,* and the exploding cloud expands for thousands of years.

When a supernova explodes, is anything left behind? Yes. The remains of a supernova is a white dwarf or neutron star, a tiny, hot, dim, incredibly dense object located in the lower left hand corner of the H-R diagram (Fig. 13.6). The feeble glow is due to energy liberated in further gravitational collapse. When this process also ends, as it must, what is left then? A burnt-out ash of a star so dense that, even if it were producing light it would never be seen, for such concentrated mass would so warp the space around it (remember General Relativity) that light would be bent back into the "black hole" that was once a blue-white giant at the top of the main sequence.

Heavy particles—protons and neutrons, and to a great extent, the nuclei they form—persist, and so the materials expelled by a supernova are swept up in gravitational eddies eventually, and used again. You can identify a "second-generation" star by its content of heavy and medium weight elements that were manufactured in a supernova. The sun and its planets (especially the inner four) contain such elements. The ingredients of the solid earth were made in a star that lived out its tumultuous life well over five billion years ago. So goes the cyclic story of the lives of the stars.

Galactic Evolution

One thing remains: if stars evolve, what about galaxies? Must not they, too, have a beginning, a developmental stage, a fading away, and an end? If star lifetimes are measured in billions of years, galaxies must have even longer histories. We have no reason to think that we can observe galaxies in every stage, as we can observe stars, but we do see enough variety among the millions of telescopically visible galaxies to get at least an inkling of how the story goes.

The three main types of galaxies, within which all sorts of variations and gradations are found, are irregular, spiral, and elliptical. Irregular galaxies in general are very fuzzy objects. They appear to be composed of the large, hot blue stars we consider youthful, and much of their irregular shape is due to the clouds of bright gases surrounding these stars and illuminated by them. It is at least reasonable to think of these irregular galaxies as young galaxies, born of enormous gas clouds in which billions of stars are condensing.

The Doppler shifts observable from the opposite edges of galaxies indicate that they are all rotating. (One edge is thus showing a blue shift as it turns toward us, as the other edge, turning away, exhibits a red shift.) Rotation alters the shape of a rotating mass, as is evidenced by the equatorial bulge of the spinning earth. So rotation ought eventually to organize irregular galaxies into ellipsoidal shapes, even though wisps of matter might trail off from the central cores for a very long time. That is, an irregular galaxy ought to evolve into a spiral galaxy (Plate 22), which is dense with stars in the middle, and still rich in uncondensed gas clouds in its trailing spiral arms. The Milky Way is such a galaxy.

Eventually, except for occasional replenishing during supernova explosions, all the gases in a galaxy's spiral arms must be used up, and the trailing stars themselves wound up into an ellipsoidal shape. The elliptical galaxies we see appear to consist almost entirely of old, red, first-generation stars that have followed an evolutionary course similar to that predicted for the sun. These stars contain no heavy elements; they are of the first generation, and were made of hydrogen and helium, and they have expanded and reddened at a relatively steady rate as they have aged. The typical Population II star never becomes a supernova. It lives a long, relatively uneventful life, although novae are observed to occur among Population II stars.

The time scale for galactic evolution is necessarily longer than that for many, if not most stars, and because all the evidence, from the ages of stars to the ages of the chemical elements to the time of expansion of the universe, points to a unique beginning and therefore a finite age for the universe as we observe it, we are obviously not in a position to see the whole history of a galaxy from beginning to end. At least at this writing nothing has been observed that has been interpreted as a galaxy at the end of its evolution.

Then what seems to be the most likely course of evolution in the universe?

Begin with a "Big Bang." A dense mass of plasma, protons and electrons and perhaps neutrons, origin unknown, begins to expand. In minutes hydrogen and helium atoms form, perhaps even traces of heavier atoms. The expanding gas whirls and eddies, and gravitational forces take over in areas of greater density to produce local contractions. ("Local" means an area some hundreds of thousands of light years across!) This contracting cloud is going to become a galaxy. Within its dense central area, stars form at a great rate, forming the core of hydrogen (up to 90%) and helium (virtually all the rest of their mass) stars that we call *Population II.* In the less dense outer regions, the gas is condensed much more slowly. In our own 10-billion-year-old galaxy, stars are still forming in the dusty, gas-rich spiral arms. Eventually the raw materials for star-making run out, the spiral arms wind up, and a mature, elliptical galaxy endures for eons with little overall change.

Within any galaxy, the individual stars live out their lives according to patterns predestined largely by their masses. The first energy-producing nuclear reaction, once the star has condensed and heated up enough to ignite its nuclear furnace core, is the fusion of hydrogen to helium. This may take from a very few to tens of billions of years to exhaust the hydrogen in their cores; this is their lifetime on the main sequence. As core hydrogen is exhausted, gravitational collapse will produce heating of layers of hydrogen farther out from the star's center; fusion begins again; the star expands, perhaps even returning to the main sequence. Alternatively it may pulsate with long or short periods (months or days) for a long, long time before all the hydrogen is gone. Then gravitational collapse produces another (perhaps very rapid) temperature rise in the helium core; a new series of fusions begins, producing carbon, oxygen, magnesium, and other nuclei of moderate mass. Or, the star's instability before this happens may be so great that an outer layer or envelope of gas is blown off entirely in the event we call a *nova.* When the helium supply, too, is exhausted, gravitational collapse takes over to produce high core temperatures, on the order of 3 billion degrees. Then any element may be formed, but in the process, the whole star, if it is sufficiently massive, may fly apart as a supernova, pouring out the atoms it has manufactured into the intragalactic spaces where another generation of stars will form. Whether with a supernova bang or the merest pulsating whimper, every star must eventually reach the end of its nuclear resources. The only remaining source of energy is gravitational collapse. The star becomes more and more dense. Its surface temperature is high, but its surface area, and hence its luminosity, is very small. It is a white dwarf. As it continues to collapse, the plasma itself condenses from protons and electrons to neutrons. Pulsars, discovered in the late 1960's to be small dense sources of radiation ranging from X rays to radio waves pulsing at very high frequencies, are now thought to be neutron stars. If evolution continues, its product disappears from observation. It becomes a "black hole," bending any further radiation back in on itself.

Is there an end to the story? Was the Big Bang unique, and will the universe expand forever? Or will it at some point begin to contract and return to a dense plasma from which the whole story begins again . . . and again . . . and again. . . .

Since about 1960, a large number of strong emitters of radio waves have been located by radio astronomers, and many of them have been identified with small, dim visual objects on the photographs taken by optical telescopes. If their red shifts are to be interpreted in the usual way as "cosmological," that is, due to the expansion of the universe, these objects are very far away — 2 to 3 billion light years. And they must be emitting enormous amounts of energy to be detectable at all. Surely, they must be galaxies at the very edge of the visible universe. But here the trouble begins. They are not spread-out, fuzzy objects. They are points of light. So they are called quasi-stellar objects or *quasars*. After a decade of study, quasars remain a mystery. If they are really so far away, they are very small; how can they be emitting so much energy? If they are really large and near, in our own galaxy, what is the explanation for their very large red shifts?

In 1971, for the first time, a few quasars were shown to have the same red shift as galaxies that appeared to be near them, establishing that these are, at least, distant objects. Perhaps quasars fit into our scheme of galactic evolution in some manner yet unknown.

SUMMARY STATEMENTS

1. Constellations, the more or less fortuitous arrangement of stars that were seen by ancient peoples as pictures of their legendary heroes, are useful to modern astronomers in mapping the skies and locating celestial objects and events.

2. All that we can know about bodies beyond our immediate neighborhood in the solar system to date comes from the light and other radiations they emit. These radiations cover the whole electromagnetic spectrum.

3. By indirect methods based on the observed location, brightness, and relative motion of stars we can determine sizes, masses, distances, and intrinsic luminosities; by spectroscopic analysis of their light we can determine the temperatures and chemical composition of the stars.

4. Plotting temperature vs. intrinsic luminosity for a large number of stars in the Milky Way by Hertzsprung and Russell led to a theory of evolution of stars.

5. The life history of a star may be traced as follows:
   a. A cloud of dust and gases (mainly hydrogen) collapses and contracts because of gravitational forces among the particles.
   b. The gravitational energy released in contraction heats the mass until it exists as a plasma of electrons and nuclei at several million degrees centigrade. Then hydrogen nuclei begin to fuse together, directly or indirectly, to yield helium. The energy released in the fusion process causes the star to shine.
   c. A balance between gravitational pressure inward and radiation outward is maintained as long as the hydrogen supply lasts. When it is gone, the star collapses and heats up until a new fusion process, such as helium to carbon, is ignited. A star may go through a long period of pulsation as successive equilibrium stages are achieved.

d. As the energy-producing resources of a star are used up it may, in seeking new equilibrium states, blow off its outer shells (the nova process); explode completely, creating debris of heavy elements and leaving behind a pulsing neutron star (supernova process); or it may slowly fade into a white dwarf, a star of superdense matter, perhaps neutrons, that glows feebly from energy produced by continuing gravitational collapse after all fusion processes cease.

e. The ultimate fate of all stars appears to be the "black hole," matter so degenerate and dense that any radiation it produces would be bent back into itself.

6. Similarly, it may be hypothesized that galaxies evolve from vast clouds of hydrogen and traces of other elements through an irregular stage, to spirals like the Milky Way, and finally to elliptical galaxies that contain no more star-forming clouds.

7. At the present, the evolutionary picture of the universe, starting with a Big Bang, appears to have more supporting evidence than does the alternate Steady State Theory.

REVIEW QUESTIONS

1. Sketch the solar system, drawing both sizes of bodies and sizes of orbits to scale. Is it feasible on one diagram to draw planets and their orbits to the *same* scale?

2. Sketch the shapes (using dots for stars) of the following familiar constellations and explain how they may be located:

Big Dipper	Cassiopeia	Taurus	Scorpius
Little Dipper	Orion	Leo	Cygnus

Natural History magazine's monthly star maps will help.

3. Why can you see some constellations every night of the year while others are seasonal?

4. Why will the shape of the Big Dipper change markedly over a period of thousands of years while Taurus will change hardly at all?

5. Identify: comets; meteorites; asteroids (planetoids); meteors.

6. What are the main ideas of the Big Bang Theory and of the Steady State Theory? Cite evidence for each.

7. What does "expanding universe" mean?

8. How do stars like the sun produce energy?

9. How does the Hertzsprung-Russell diagram suggest a possible course of stellar evolution? Trace the probable life history of a typical star (such as the sun) in terms of the H-R diagram. Do the same for a white giant such as Rigel.

10. Outline briefly how we can measure the following:
   a. distance to near stars
   b. distance to far stars and galaxies
   c. mass of relatively near double stars
   d. mass of distant stars
   e. surface temperatures of stars
   f. chemical composition of stars
   g. diameter of stars
   h. intrinsic brightness
   i. stellar movements

11. What does it mean to say the sun is a second generation star?

12. Trace the possible course of evolution of galaxies.

THOUGHT QUESTIONS

1. You have been invited to address the British Flat Earth Society on the subject "Why I am convinced that the earth is spherical." You will have a very skeptical audience. What evidence for your point of view will you present?

2. Why was it easier for primitive people to see pictures in the stars than it is for most moderns?

3. Enumerate all the ways in which the "fixed stars" are not "fixed" (i.e., how they actually move).

4. What are the conditions under which life comparable to earth life could develop elsewhere in the universe? What odds would you give on its having happened? What are the odds on our ever finding out? Could there be numerous worlds like ours without our knowledge? Why or why not?

5. On reading the writings of various proponents of alternate theories of the universe, it often appears that preferences for one theory or another are based as much on philosophic as on scientific grounds. How do the philosophic bases of the Steady State theory and the Big Bang theory differ?

6. Is astronomy a useful science? Does a science have to be utilitarian to justify time and money being spent on it? (Do the arts?)

7. Will it do you any good to study the stars? Will you miss anything if you don't?

READING LIST

Asimov, Isaac, *The Universe*, Walker, New York, 1966. One of the best books on astronomy for the nontechnical reader.

Adler, Irving, *The Stars*, New American Library (Signet), New York, 1958. An excellent guide for beginning students.

Bondi, Hermann, *The Universe at Large*, Doubleday (Anchor), Garden City, N.Y., 1960. Cosmology for the nontechnical reader, by a leading cosmologist.

Bova, Ben, *The Milky Way Galaxy*, Holt, Rinehart, and Winston, New York, 1961. Well-written, with clear explanations of how as well as what we know about astronomy.

Hoyle, Fred, *Frontiers of Astronomy*, New American Library (Mentor), New York, 1957. Cosmology from the point of view of the leading "steady state" theorist.

Plate 18 Model of the solar system showing relative sizes and locations of sun and planets and their moons. The colors and markings shown are visible through large telescopes.

Color illustrations are from the Life Nature Library or the Life Science Library.

Plate 19 The Doppler effect. Sound waves produced by the moving car are crowded together ahead of the car, producing a higher frequency, and stretched out behind it, so that the frequency is lower. All kinds of waves show the Doppler effect. The frequency of waves from an approaching source rises; from a receding source, it decreases.

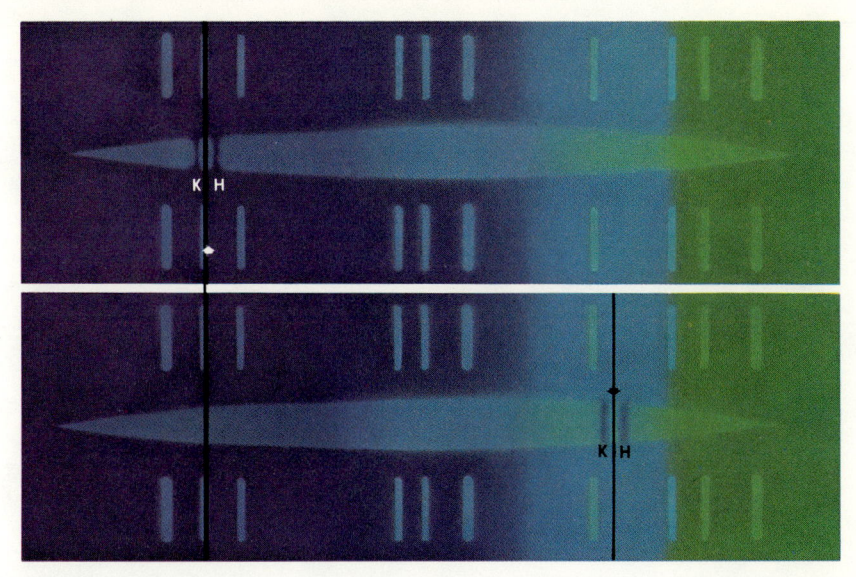

Plate 20 Doppler effect in spectral lines from a star. The red shift (toward lower frequencies) indicates the source is receding.

Plate 21 The layers of the earth. Under a thin rocky crust is an 1800 mile thick layer of mantle, surrounding an iron-nickel core roughly 4000 miles in diameter, liquid except at its center, where the extreme pressure makes it solid in spite of the high temperature.

Plate 22 Andromeda, a spiral galaxy similar in shape to the Milky Way but somewhat larger. It is about 2,000,000 light years from earth.

Plate 23 The Pleiades as seen through a large telescope. The clouds of dust and gas indicate this is an area where stars are still forming.

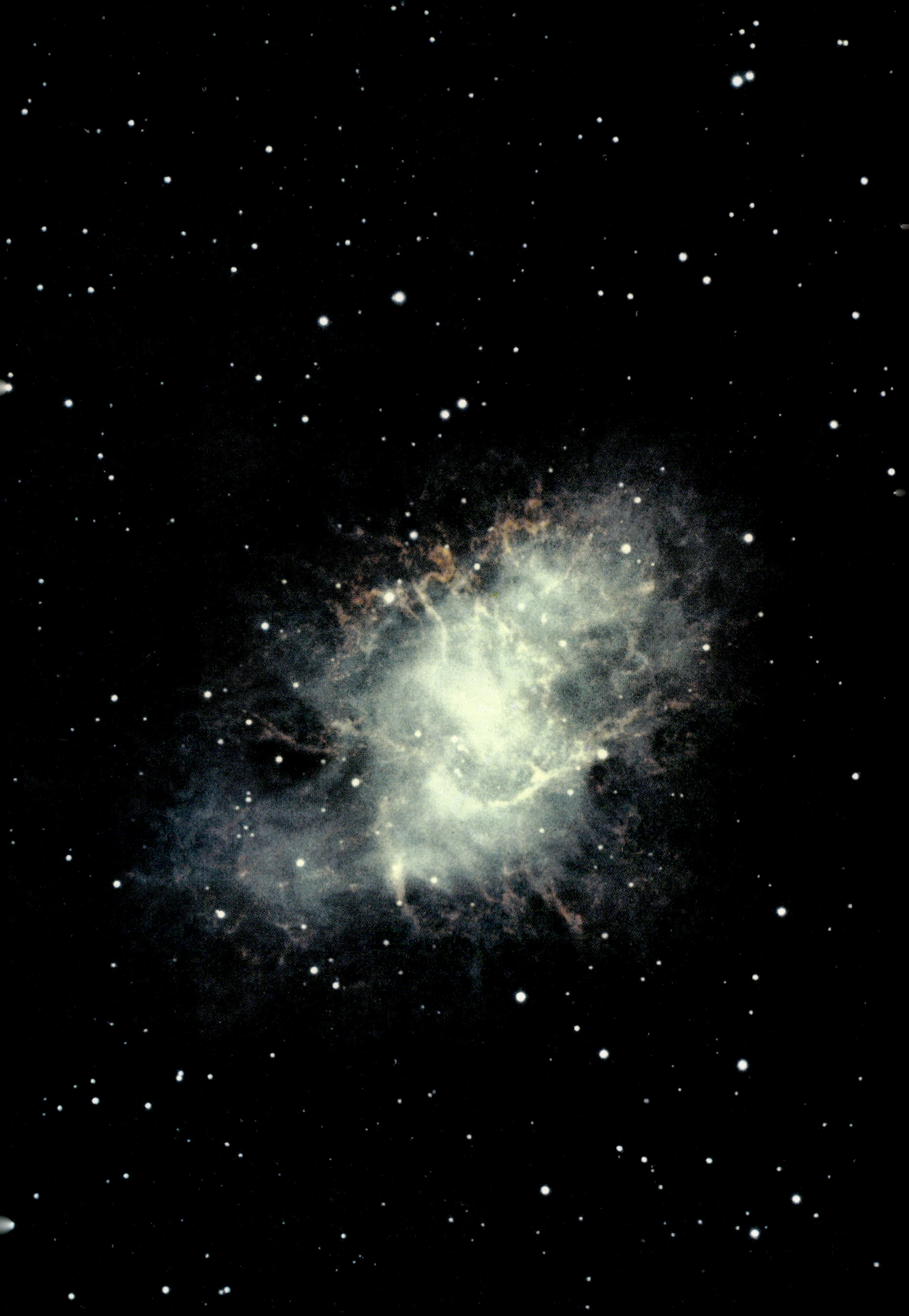

Plate 25 Battlements of resistant rock, probably of volcanic origin, stand above the wind-eroded desert landscape.

Plate 24 The Crab Nebula, the expanding cloud of gases from the supernova of 1054 A.D.

Plate 26 The entrenched meanders of an ancient river cut through a rejuvenated landscape. General uplift of the valley through which the river wound its meandering way gave the stream a new supply of energy by which to cut its way down toward sea level, creating the present-day canyon.

Plate 27 A river of ice reaches down from the mountains like the paw of some snowbound beast of titanic proportions. Such a glacier carves a **U** shaped valley, and builds low hills called moraines at its terminus from the material it has ground out and carried along. Melt water flows away in streams from the glacier's end.

Plate 28 A lake of fire and molten lava fills the crater of an active volcano. Magma from the deep crust or upper mantle melts when a volcano vent opens, releasing the pressure, and comes out either as a violent eruption or a steady flow.

Plate 29 Carbonates dissolved and recrystallized provide the interior decoration of limestone caves.

Plate 30 A page of earth history, written in the lines of strata.

Plate 31 A hurricane, earth's mightiest heat engine.

Plate 32   The separation of Pangaea, 3 billion years ago, showing the schism between Laurasia to the north and Gondwana to the south (above). The configuration of the continents today (facing page, top). By 10,000,000 A.D. the Atlantic will have widened, with a corresponding closing of the Pacific. Africa will be further separated from Eurasia along the Great Rift, and Australia will be reunited with Southeast Asia (facing page, bottom).

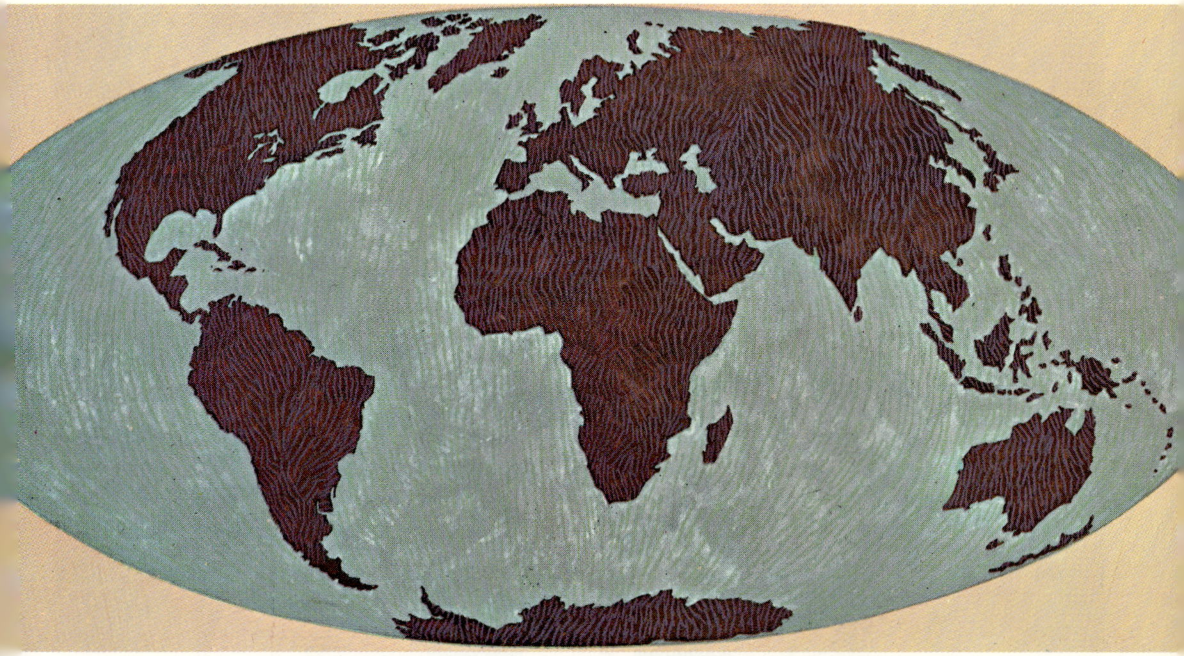

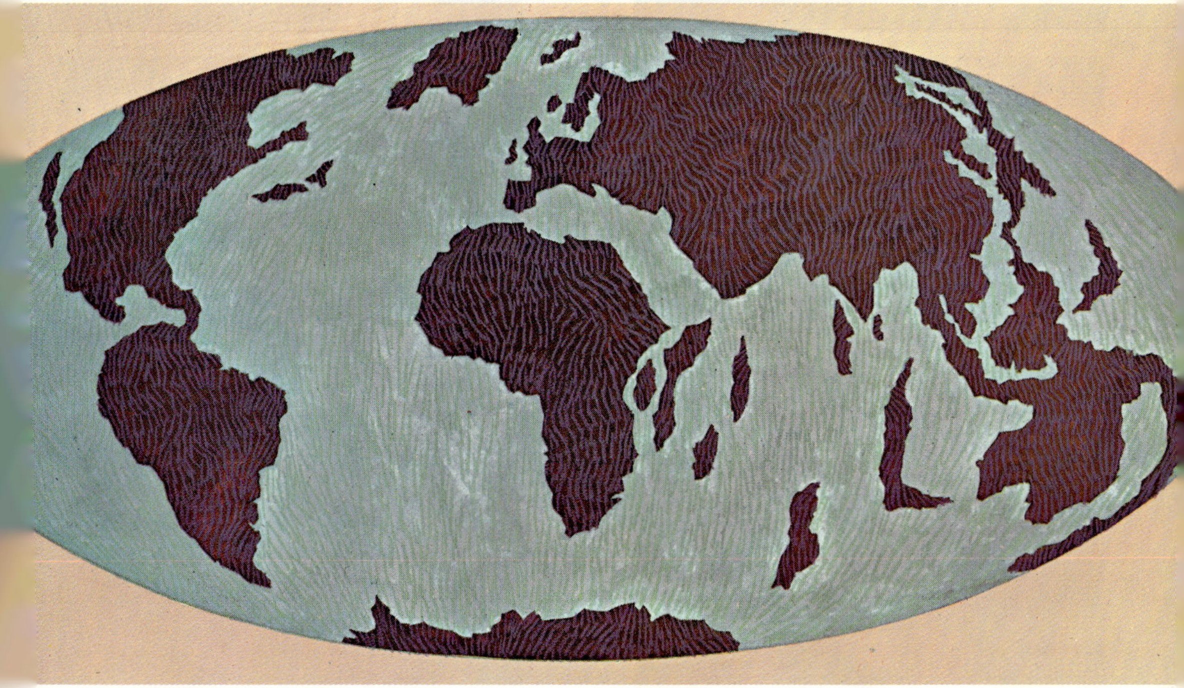

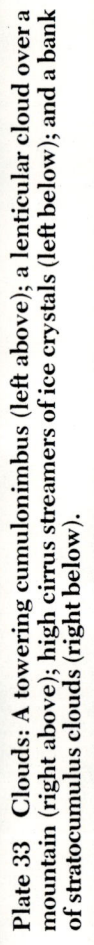

Plate 33 Clouds: A towering cumulonimbus (left above); a lenticular cloud over a mountain (right above); high cirrus streamers of ice crystals (left below); and a bank of stratocumulus clouds (right below).

14

The Evolving Earth

"Some say the world will end in fire,/Some say in ice," the poet Robert Frost wrote, but he didn't mention that the same argument goes on about its beginning. Although the current best theory of the formation of the solar system proposes that planet formation occurs by the agglomeration of cool fragments left over from the sun's condensation, the earth's own internal evidence indicates that it reached a very high temperature somewhere near its beginning. This is to be expected, because compressing these fragments together by the force of gravity produces heat, which will raise the temperature of the forming planet above the melting point of most, if not all, of the materials it contains. From this plastic, if not fluid, mass, the central force of gravitation would form a sphere. When a cloud of dust and gas begins to condense it may form only one body, a single star, or it may form two or three or four bodies of nearly the same size, a double or multiple star system. Such multiple stars are quite common within the galaxy. Perhaps it is just as common for a third alternative to occur: the formation of one large body, massive enough to become self-luminous by nuclear fusion, and a number of much smaller bodies that can never get that hot by gravitational collapse. Because such planetary bodies are not self-luminous we cannot see them; because they comprise so small a fraction of the total mass of the system, their gravitational effects on the central star are almost imperceptible. (Astronomers suspect one or two stars of having large planets orbiting them because of tiny irregularities in their motion that could be explained in terms of planets' gravitational attraction, but we can't tell for sure, yet.)

Solar System Formation Theories

Any theory of formation of a solar system must account for sizes, distribution of materials, orbital distances, and, quantitatively, the angular momentum of the system. The most successful theory to date is that proposed in the late 1940's by Carl F. von Weiszäcker (Fig. 14.1).

As we have seen, there was a variety of raw materials from which our solar system could form—not only the primordial hydrogen and helium, but iron, carbon, oxygen, silicon, calcium, aluminum, nitrogen, and even the very heavy unstable elements of Period 7. The distribution of these elements was determined mainly by the two factors of gravitation and temperature. So we have, in the third orbit out from the sun, a small rocky planet whose gravity could hold only relatively dense materials against the twin effects of the kinetic energy of the light gases heated by the sun, giving them velocities far in excess of their escape velocities from the earth's gravity, and the enormous pull of the sun's gravitational field.

Because the earth started hot, its primitive atmosphere was almost certainly entirely different from the atmosphere today. Oxygen could not have been present. It is a very active element, and at the prevailing temperature it would certainly not have remained

Solar System Formation Theories

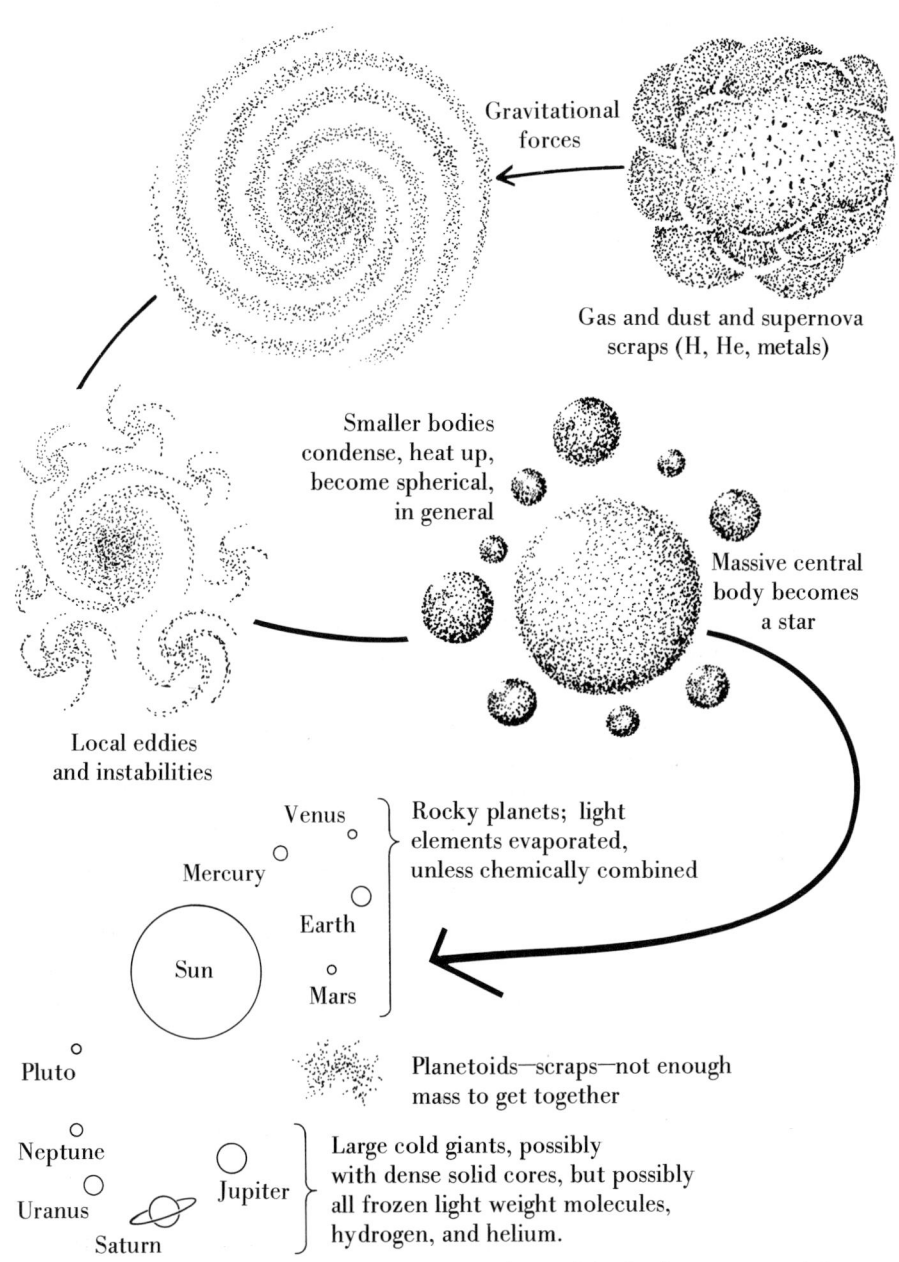

Figure 14.1 Von Weiszäcker's model of solar system origins.

uncombined. On the other hand, nitrogen, carbon, and other less reactive substances at today's "room temperature" would have been reactive at the higher temperatures then prevalent. For this reason, as well as because spectroscopic analysis shows the presence of such things today in the atmospheres of the cold giant planets Jupiter and Saturn, we believe that the primitive atmosphere of earth contained primarily oxides of carbon, CO and CO_2, and hydrogen compounds such as water (H_2O), ammonia (NH_3), methane (CH_4), and even hydrogen cyanide (HCN).

Oxygen is indeed earth's most abundant crustal element, but most of it is chemically tied up in minerals, mixtures of which we call *rocks*. The next two most abundant elements in order are silicon and aluminum, and most rocks are made up largely of complex silicate ($SiO_3^=$) and aluminate ($AlO_3^{\equiv}$) salts. The cations may be Mg^{++}, Fe^{++} or Fe^{+++}, Al^{+++}, Na^+, or any other metal, or mixtures of two metals. Those mentioned are the most common. The earth's crustal rocks are not very dense, on the whole—a range of 2 to 4 grams/cm³ would include most of their densities. This gives us additional clues to the earth's formation. We can imagine the agglomeration of matter destined to be earth ("protoearth," we shall call it) seething with chemical reactions among the gravitationally heated elements that were to produce the various components of the planet. The energy of gravitational collapse of such a small mass as a planet (even a giant planet like Jupiter) could never produce temperatures sufficiently high to trigger nuclear fusion. At those temperatures matter exists as plasma, not as atoms and molecules. On earth, the temperatures were such as to promote virtually total reaction among all but the inert elements, at least as long as the supply of reactants held out. The most abundant element in the whole sphere of earth is iron, and as the protoearth cooled, the light, high-melting oxides, silicates, aluminates, and other salts, solidified and floated to the "top," really the outside, of the mass, while the unreacted molten iron, with substantial amounts of its close relative, nickel, sank toward the center. In between is a layer of denser rocks and minerals similar in composition to the crustal rocks. A hard-boiled egg halved crosswise is an excellent model of the earth (Fig. 14.2; see also Plate 21).

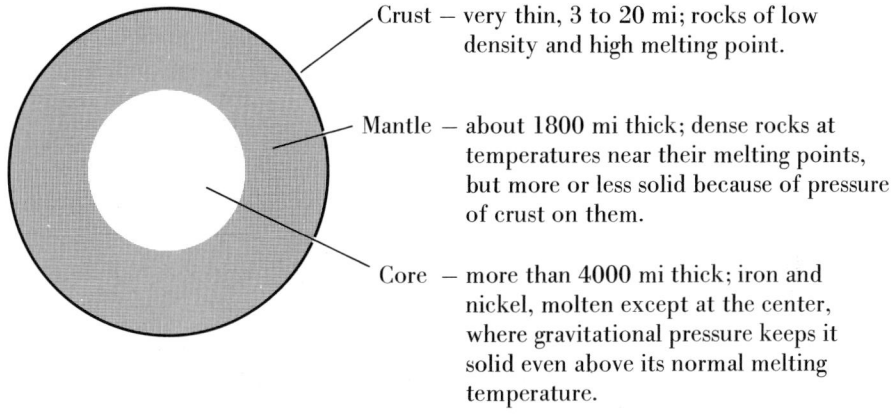

Crust — very thin, 3 to 20 mi; rocks of low density and high melting point.

Mantle — about 1800 mi thick; dense rocks at temperatures near their melting points, but more or less solid because of pressure of crust on them.

Core — more than 4000 mi thick; iron and nickel, molten except at the center, where gravitational pressure keeps it solid even above its normal melting temperature.

Figure 14.2 Cross-section of the earth.

In the 1960's there was an unsuccessful attempt to drill through the earth's crust down to the mantle. It was called "Project Mohole," after the Yugoslavian geologist Mohorovicic, who first noted the location of a discontinuity in the properties of rocks that marked the boundary between the crust and mantle. The Mohole was to be dug through the sea bottom where the crustal rocks are only three miles thick. But the difficulty of anchoring a stable drilling platform several miles up on the surface of the sea outweighed the advantage of having a thin solid layer to drill through. Project Mohole was poorly organized and inadequately funded, and came at a time when the public imagination had already been captured by plans for a trip to the moon. Perhaps as interest in our own planet supersedes our interest in space exploration, Mohole's day may yet come.

If we have not actually explored even the depths of the earth's thin crustal shell, how can we say what's inside? Several lines of evidence are available.

1. We can "weigh" the earth and calculate its volume. So we can then calculate its density, or mass per unit volume. We assume earth contains no "elements" inside other than those listed in the Periodic Table, so we can make good guesses as to the composition of materials that would have the measured density.

2. The earth has a strong magnetic field having a roughly north-south direction. Iron is a strongly magnetic element; iron is an extremely stable middle-weight atom likely to be a main product of stellar nuclear reactions — and the density is right. It is likely that the earth has an iron core.

3. Gravitational sorting would shift the heavier materials to the center of the earth. If, as is indeed the case, the lightest materials were also those of highest melting point, these would solidify first as the earth cooled, and float on top of the still molten, heavier materials. Fitting together a solid crust over the whole earth would produce pressure on the denser materials underneath so that they, though still hot, would to some extent solidify.

4. The crust is irregular, and weaker in some places than others. Sometimes cracks (faults) occur that reach all the way through the crust and the blocks of crust on the two sides of the fault move up or down or sideways relative to each other. This movement is an earthquake. It produces three different kinds of waves, each with its own characteristics of propagation. The behavior of these waves is our chief source of information about the earth's inner structure (Fig. 14.3).

5. When pressure is released by faulting some material from deep in the crust or the mantle will melt and be pushed up through the fault. If this happens rather quietly, it is called a lava flow. Often it happens violently as ground water is turned to steam by the hot

The Earth's Crust

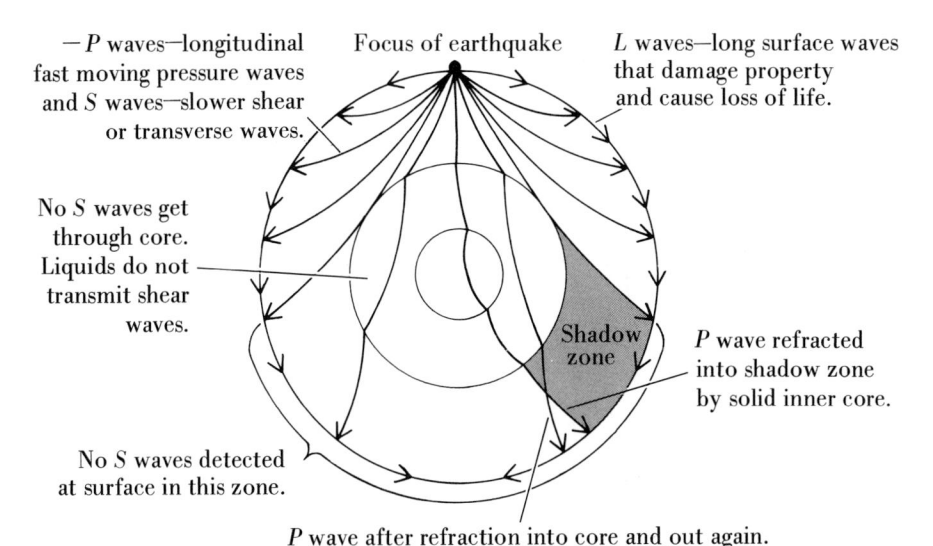

Figure 14.3 How earthquake waves contribute to our knowledge of the earth's interior.

Labels on figure:
- *P* waves—longitudinal fast moving pressure waves and *S* waves—slower shear or transverse waves.
- Focus of earthquake
- *L* waves—long surface waves that damage property and cause loss of life.
- No *S* waves get through core. Liquids do not transmit shear waves.
- Shadow zone
- *P* wave refracted into shadow zone by solid inner core.
- No *S* waves detected at surface in this zone.
- *P* wave after refraction into core and out again.

lava and other gases are formed or released so that the whole mass shoots out with great speed and force. The lava solidifies quickly on reaching the surface, the cinder falls to earth, and the visible result is a cone-shaped mountain, the familiar volcano. The vent of a volcano may remain open or be reopened over a period of many years, and the volcano may erupt repeatedly at long or short intervals, producing, as a sort of scientific by-product, mantle materials for study.

The Earth's Crust

At first, some 4.5 to 5 billion years ago, the crust was hot. It thrust irregularly into an atmosphere containing such gases as earth's gravitational field had been able to hold against the nearly overwhelming pull of the sun and the random expansion of the hot gases themselves. This primitive atmosphere was likely, as we have seen, to contain CO_2, CH_4, NH_3, HCN, and above all, H_2O (Fig. 14.4). When the crust had cooled suffi-

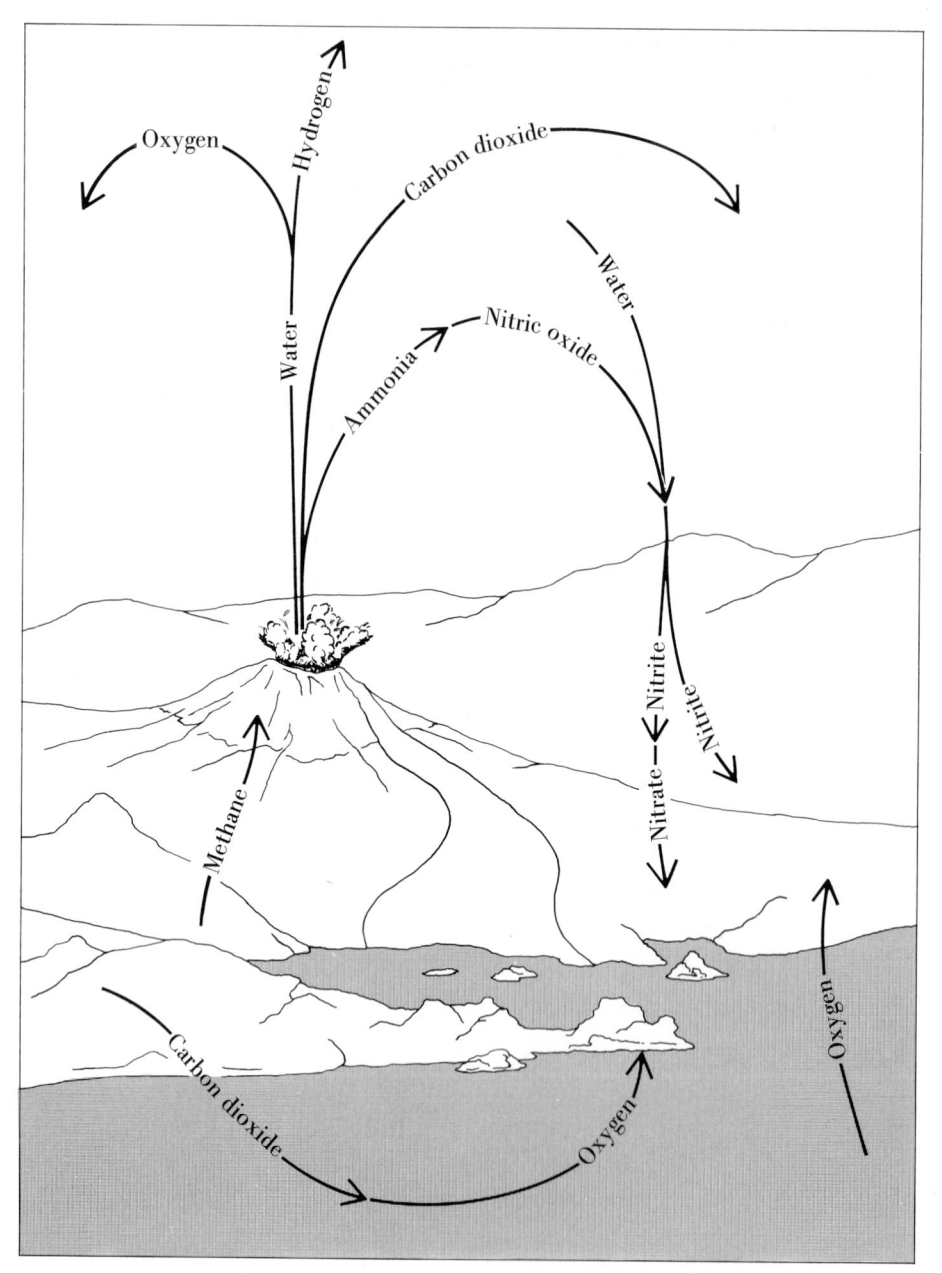

Figure 14.4 The atmosphere of the primitive earth.

ciently, the water vapor began to condense out of the atmosphere and to fall as rain. Some historians of earth envision the cooling crust inundated by a million years of rain. When the rain was over, there was the sea, and the dry land. There were the cycling seasons, too, and wind, and rain again, leaching out the soluble minerals from the rocks, making the sea salt. But the land and seas were barren of life.

A far greater wonder than the condensing of gas clouds into galaxies is the development of the simplest living cell. Every other phenomenon in the universe is understandable in terms of the obvious interpretations of the thermodynamic laws: systems tend to equilibrium, to the lowest energy state, to randomness. Then along comes life, creating order out of disorder, shunning equilibrium, for equilibrium is death. How did such an unlikely event as life come about? Has it happened only once in the universe, against all odds? Or is its development as inevitable as the gravitational collapse of gas clouds and the ignition of the nuclear fires of stars? If the answer is, yes, it was inevitable, given the chemical and physical conditions of earth, and yes, the odds are that it has happened over and over again on thousands of planets out of the multitude that circle the billions of suns in the millions of galaxies, it is no less wonderful.

Chemical Evolution of Life

"And God said, Let the earth bring forth the living creature," wrote the ancient poet three thousand years ago (Genesis 1:24), and now we are beginning to figure out how it was done. The scene is the shore of a warm, shallow, moderately salty sea. The sun beams down its ultraviolet rays in unfiltered intensity through the primitive atmosphere. The high-energy radiation breaks some bonds, joining fragments together in new configurations. The larger molecules drift out of the atmosphere and into the sea. Thunderstorms rage, and lightning provides the energy for still further chemical reaction. The reaction products of both of these processes are, among other things, amino acids, the building blocks of protein. The nitrogen bases that compose the genetic code are similar compounds. It may have taken an eon of trial and error to develop the efficient mechanism for replication that is DNA, but there was plenty of time for this kind of chemical evolution to proceed, and to culminate in the formation of highly organized, self-replicating systems that we would call "alive." From that crucial point to the present there have been perhaps 3 or 3.5 more eons in which biological evolution could proceed.

The greatest nonideologic difficulty in the way of accepting both geological and biological evolution was this matter of time. Until the late eighteenth century the age of the earth arrived at by Bishop Ussher in 1658 by adding up the ages of the Biblical patriarchs—a method that dated the Creation of the Earth on October 22, 4004 B.C. at 8 P.M. (Greenwich Mean Time?)—seemed quite reasonable to nearly everybody in Christendom, at least. Because much had to have happened in that short time, any large-scale change had to be attributed to a major catastrophe. Opposing the prevailing theory of Catastrophism was the Uniformitarianism of James Hutton, who in the 1790's proposed that all the vast changes we can trace through earth's history came about by action of the very same forces and agents we see at work today: wind, water, ice, earthquakes, volcanoes, and the rivers running down to the sea (Plates 25–29). But that sort of thing takes time. So does the slow development and definition of species by natural selection that Darwin proposed in 1859. And it was not until radioactive dating of rocks was developed in the twentieth century that anyone had any clear idea that enough time was available. Now that we can place the age of the earth at no less than 4.5 billion years with considerable confidence, we can with equal confidence tell its story in terms of Uniformitarianism and evolution.

The oldest method of dating geologic events is relative, based on fossil remains in rocks. A fossil is an evidence of ancient life. It may be actual remains—a tooth or shell unchanged through millions of years. It may be a mold or cast or mineral replacement of shell or bone or wood. It may be a footprint or an impression of leaf or bark left in mud that hardened into rock undisturbed. Whatever their form, fossils are the best clues we have to earth's history. We can make two generalizations as to their interpretation: (1) in layers that are undisturbed over long periods (Plate 30), the age of a fossil is proportional to its depth; and (2) the oldest fossils are the simplest, and a regular progression in time from the simple to the complex can be clearly observed.

It is customary, for convenience, to divide the study of the earth into physical (structural) geology and historical geology. Today, however, the realization that the earth is a unit, an intradependent whole, is forcing itself upon us. So our story will be woven first with one thread and then the other and it doesn't really matter whether or not we recognize which is which.

We assume that warm, shallow seas were the setting for the chemical evolution because we can produce the required reactions from the raw materials believed to have been available by ultraviolet irradiation or electric discharge in solutions similar to those seas. Also, much later in time, when recognizable organisms appear in the fossil record, they are of the kind that must have lived in tidal waters. And there is no evidence of drastic change in between.

Landscape Evolution

It is possible that during this Azoic (no life) Era the predominant process was the wearing down of the original igneous (fire-formed) rocks, especially the light granites that formed most of the dry land. Undoubtedly vulcanism (volcanic action) produced basaltic lavas that are more resistant to weathering than the granites. Basalts are also denser than granites, and form the bedrock under the sea. However, when rock erodes and is deposited as sediment somewhere else, the balance of mass on various parts of the crust is upset, and a new equilibrium must be reached—the principle of *isostasy* (equal balance). So as accumulated sediments push down one place, the crust must buckle upward in another place. This is the process by which folded mountains are made.

Vulcanism

There are two main orogenic (mountain-building) processes. One is *vulcanism*—the building of mountains by the outpouring of subcrustal material (magma) as lava and cinder in volcanic eruption. A mountain of volcanic origin is characterized by a symmetrical cone shape—steep-sided if cinder predominates, gently sloping if formed from very fluid lavas. Probably most volcanoes have erupted many times, but eventually, the vent becomes plugged with hardened lava and subsequent disturbances erupt through weaker spots elsewhere in the crust. The volcano is then extinct. History records several disasters of great magnitude connected with the reviving of a volcano believed to be extinct, such as the famous destruction of the Roman city of Pompeii in A.D. 67 by an eruption of Mt. Vesuvius, which had been quiescent for so many centuries that no one then living knew it to be a volcano.

Mountain-building by vulcanism is a continual process. Parícutin in Mexico was born in a cornfield one February morning in 1943. Surtsey rose from the sea near Iceland in 1963. The Hawaiian Islands, Japan, the Azores, and most island chains of the Pacific have volcanic origins. In fact, the Pacific Ocean is sometimes said to be surrounded by a "ring of fire" because of the extensive volcanic (and earthquake) activity presently and in the recent past along the coasts of the American continents, Japan, the Philippines, Indonesia, and New Zealand. Another active zone runs along the northern shore of the Mediterranean through Italy, Sicily, Greece, Yugoslavia, and Turkey, and angles southward as a giant fault zone running from the Jordan River valley through the Red Sea and the Great Rift Valley of East Africa. The east-west branch circles India and joins the "ring of fire." A third active zone runs up the middle of the Atlantic Ocean and centers on the Mid-Atlantic Ridge, a submarine mountain range dotted with volcanic activity. This area is in fact one of the most interesting parts of the planet, geologically speaking, because the ocean floor is expanding there, pushing the American continents westward and Eurasia and Africa eastward, and incidentally, closing up the Pacific Ocean around on the other side of the world.

Diastrophism

The other mountain-building process is called *diastrophism*, the movement of large solid portions of the earth's crust. A glance at a map of the world shows that most mountain ranges are along the borders of continents, and most of them run roughly north and south, the Himalayas of Asia being the most notable exception. Such mountains are characterized by layers of rocks of different appearance and composition, thrust upward, cracked, tilted, jumbled, even turned upside down with respect to their neighbors. These folded layers give evidence of both up and down and sideways forces. We can account for the up and down forces by the principle of isostasy. In valleys, along sea shores, areas called *geosynclines*, sediments carried down from higher land by water and wind build up into thicker and thicker layers. These layers are pressed and heated into *sedimentary* rock. If the heating and pressing become sufficiently intense the crystal structure of the minerals themselves may be changed, and certain minerals may be concentrated in particular formations. Gem and ore deposits occur in this way, and usually occur in the pressure-cooked rocks we refer to as *metamorphic*. Eventually the geosyncline that has been depressed and refilled with sediments over a long period is uplifted, sometimes quite abruptly, to form a block mountain.

The side-ways forces that fold mountains are harder to account for. Once it was simply assumed that the cooling earth was shrinking, which caused the crustal "skin" to wrinkle. Today the alternative process of expanding ocean floors makes it likely that some wrinkling is due to expansion. At any rate, in addition to the long-known fact that land and sea have alternately replaced each other over much of the earth's face, we now must acknowledge that huge plates of the crust drift slowly over the hot and not entirely rigid mantle.

Continental Drift

The theory of continental drift was first proposed by a German meteorologist named Alfred Wegener in the 1920's. No one took him seriously for it seemed he was only playing with jigsaw puzzles. His idea was based largely on the fact that the eastern coasts of the Americas appear to fit rather neatly against the western edges of Europe and Africa. It also rather looks as if India came loose from East Africa and floated up against Asia and stuck there, and that Australia and Antarctica drifted off southward. In Wegener's day this seemed so impossible as to be ridiculous. But in the past two decades, matching rock formations and fossils in West Africa and Brazil, finding dinosaur remains in Antarctica, matching the directions of magnetic fields on the two sides of the Atlantic, along with the spreading of the ocean floor along the Mid-Atlantic Ridge, make the actuality of continental drift a near certainty.

The story as we see it now begins with two closely associated land masses—Laurasia, in the north, and Godwanaland, in the south. As Plates 32–34 show, we can account for the shape and location of our present continents by postulating the breakup and drift of these two land masses, and not only residual heat from the earth's core but heat continually produced by radioactivity supplies the energy for the convection currents in the mantle that move the continental plates of the crust around on the surface.

It is calculated that North America is moving westward at about a centimeter a year. It is probable that the rate hasn't varied much over the millennia, so life must have gone on in its own way, relatively oblivious to the long-range changes in geography. Other changes, in climate, in land elevation and the advancing and receding of the seas, probably occurred on a considerably shorter time scale, but this scale, too, was longer than many lifetimes of the organisms living on the earth.

Dating the Earth's Changes

As climate and topography changed, organisms changed, too. In fact, we can best follow the changes in the earth's surface as we see them mirrored in the fossils each layer of the crust contains. The story the fossils tell is summarized in Fig. 14.5.

The oldest surface rocks so far dated are in east central Canada and in South Africa in great areas called *shields* because of their convex surfaces and roundish shape. Radioactive dating places the age of these rocks at more than 3 billion years. They contain no fossils. The absence of fossils, however, does not

338

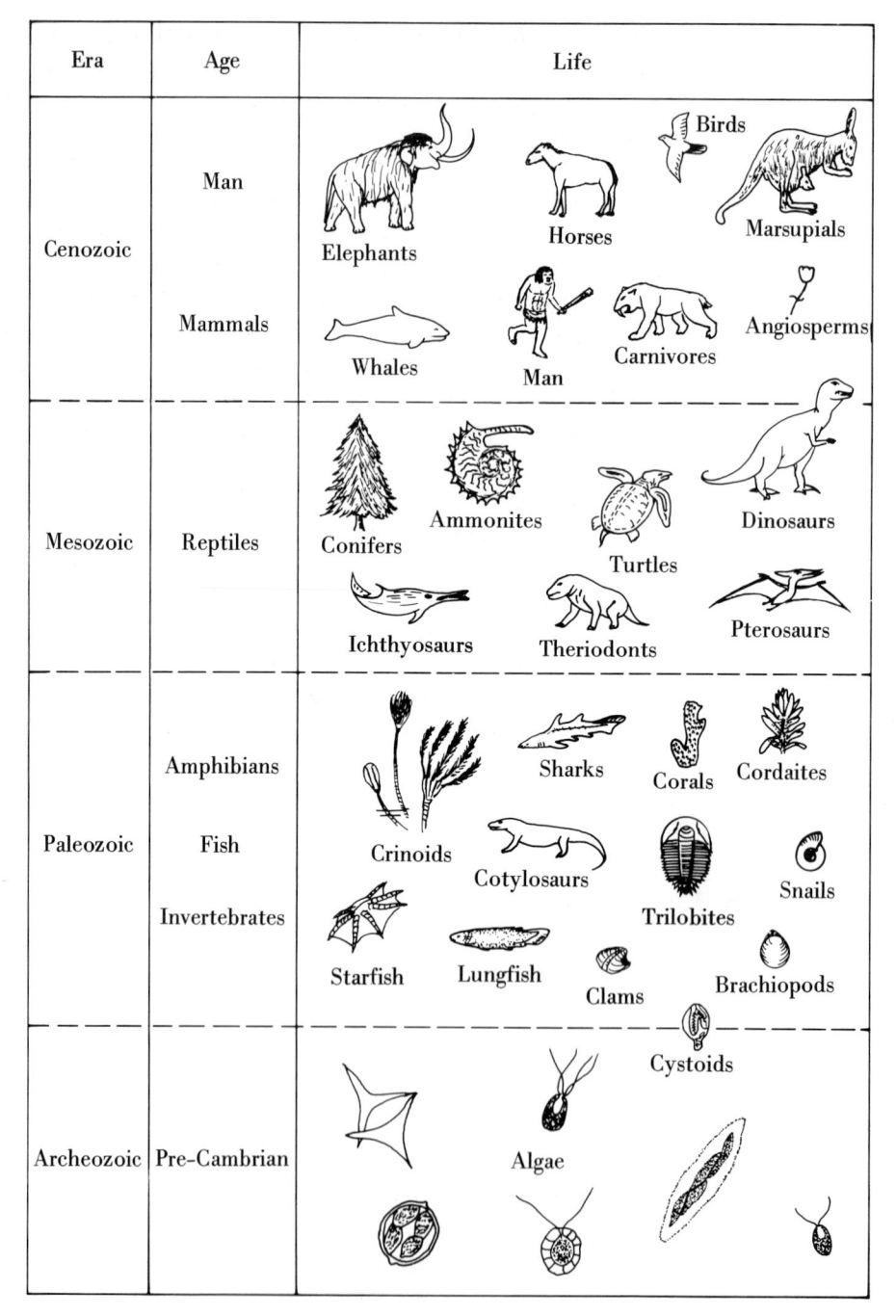

Era	Age	Life
Cenozoic	Man Mammals	Elephants, Horses, Birds, Marsupials, Whales, Man, Carnivores, Angiosperms
Mesozoic	Reptiles	Conifers, Ammonites, Turtles, Dinosaurs, Ichthyosaurs, Theriodonts, Pterosaurs
Paleozoic	Amphibians Fish Invertebrates	Crinoids, Sharks, Corals, Cordaites, Cotylosaurs, Trilobites, Snails, Starfish, Lungfish, Clams, Brachiopods
Archeozoic	Pre-Cambrian	Cystoids, Algae

Figure 14.5 The ages of the earth and the fossil record.

prove the absence of life, for the organisms of the first several million years of biological evolution may well have been too ephemeral to leave traces. It is likely that they had no "hard parts" such as shell or bone; they left no footprints. The oldest materials thought to be fossils are circular splotches in some 3-billion-year-old rocks found in the upper midwestern United States. The splotches are mineral, but they look a little like the mineral deposits left on rocks by lichens today. Perhaps the circles were formed by the chemical action of algae on the rock that supported their colony.

The oldest definitely recognized fossils were found in certain slaty strata (layers) in Wales. These strata were named *Cambrian* because Cambria had been the Roman name for Wales. Similar deposits were traced across Europe, but the name was applied to them all, thus establishing the system still in use. A geological period is named for the area in which its characteristic fossils are first identified, although the name may take various forms: geographic (Cambrian); historic (Ordovician, for ancient tribes in Wales); or structural (Cretaceous, for the chalk cliffs along the English Channel). Every geologic period has its characteristic or "index" fossils that label it unequivocally (these index fossils are shown in Figure 14.5). From the characteristics of the fossil remains and the strata in which they are imbedded we can reconstruct with some confidence the world in which these creatures lived.

Biological Evolution

The Precambrian world, we have guessed, provided the setting for chemical evolution that shaded into biological evolution. Compounds of carbon, hydrogen, nitrogen, and oxygen polymerized until they reached a highly organized stage at which they could direct other chemical reactions. It could be hypothesized that "chemical communities" formed that somehow, eventually, reached that stage of self-sustaining organization and activity that we call a living cell. We are not in a position at present to make any more precise statements about the origin of life. The mystery remains.

The animals and plants of Cambrian times reflect a long, long history of evolutionary development. Long ago cells had developed mechanisms for utilizing the most abundant energy available, the radiant energy from the sun, to maintain their activities. Remember, the Second Law of Thermodynamics — order can be maintained only by the expenditure of energy — was in operation then, too. The dominant mechanism was *photosynthesis*, and green plants were the organisms that developed it. Blue-green algae, the simplest photosynthetic organisms on earth today, may not be much different from their first ancestors. The operation of photosynthesis wrought a great change in the earth. Its by-product is oxygen, a highly reactive element, which must have reacted vigorously with the carbon and hydrogen compounds, and to a lesser extent with nitrogen. Eventually, the atmosphere contained little but oxygen along with the inert nitrogen and traces of the noble gases. The accumulation of oxygen also changed the character of radiation reaching the earth's surface. Oxygen near the earth's surface exists as a diatomic molecule, O_2. High up at the top of the stratosphere, oxygen exists largely in the form of O_3, or ozone, with the ring structure

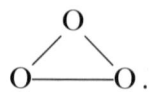

$$\begin{array}{c} O \\ \diagup \diagdown \\ O \!-\!\!-\! O \end{array}.$$

It is bluish and poisonous to breathe, but its presence in the upper atmosphere makes the earth habitable for modern organisms, because it absorbs and thus filters out nearly all the sun's ultraviolet radiation. This ultraviolet radiation that we consider essential for the beginning of chemical evolution is lethal to organisms. Once it was removed, other kinds of cells besides those that carried out photosynthesis could survive. We shall probably never know what kinds of organisms developed, or what fraction of them were successful. We do know that Cambrian strata display an amazing variety of "successes."

Geologic Time

The long span of Pre-Cambrian time includes some seven-eighths of the earth's history. Some paleontologists further divide it into the Azoic Era (up to 3 billion years ago), the Archaeozoic Era (3 to 2.4 billion years ago) and the Proterozoic Era (up to Cambrian times, which introduce the Paleozoic Era). Signs of life first occur in the Archaeozoic Era, though no proper fossils. These signs are in the types of sedimentary rocks that characterize the era, and include graphite and limestone deposits, which in later times are always the remains of organisms. These early forms had no substantial shells, no bones, nor feet with which to leave footprints, but the carbon and the carbonates concentrated from their teeming millions of cells over millions of years betray their presence.

No connection is yet apparent, but each period that has been characterized by rapid biological evolution has also been a time of drastic geological change. The Paleozoic Era dawned some half-billion years ago with the upheavals that built a mountain range at the southern border of the Canadian Shield in the region of the present Great Lakes, and another in the neighborhood of Ireland. No great climatic change seems to have occurred, however, for the creatures of the early Paleozoic were creatures of warm seas. Their variety, abundance, and widespread distribution are remarkable. The first bespeaks their long evolutionary history; the second and third, the uniformity and mildness of the worldwide climate of the time. The dominant life form of the Cambrian Period, the first of six divisions of the Paleozoic Era (era of ancient life), was the *trilobite*. These crustaceans had shells with three longitudinal lobes, a rigid section at head and tail, and a multijointed midriff that permitted them to roll up into a ball rather as their modern day cousins the "roly-polies" do. Most trilobites were small, but like some other species, they went out in a blaze of glory in which some specimens attained a length of 2 ft. A number of Cambrian animals looked more like plants, but by comparison with their descendants and by deduction as to their manner of life we must concede that crinoids, or stone lilies, and graptolites were actually members of the animal kingdom. None of these creatures survived the Paleozoic era (Fig. 14.6). Other contemporaries are still represented by present-day descendants: sea anemones, sponges, jellyfish, corals, starfish, sea urchins, and the minute diatoms called *foraminifera* and *radiolaria* whose siliceous shells still rain down on the ocean floor as they have since Cambrian times.

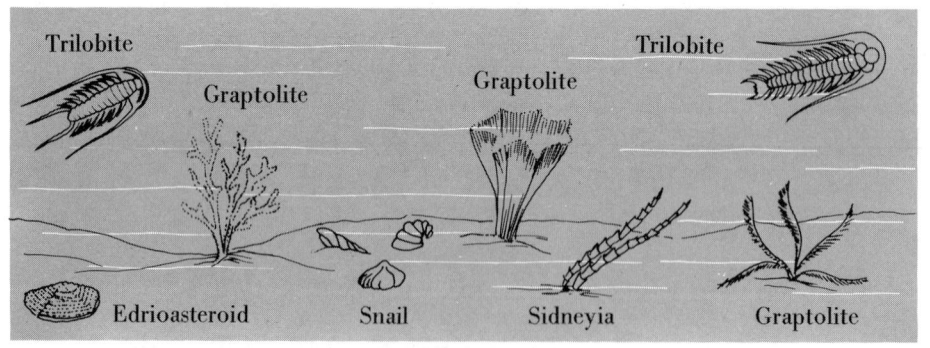

Figure 14.6 Cambrian life.

The next two periods, the Ordovician and Silurian (Fig. 14.7), were also first identified in Wales; both were named for ancient Celtic tribes of that region. Coelenterates such as jellyfish and corals reached their peak in the Silurian, as did the crinoids, which

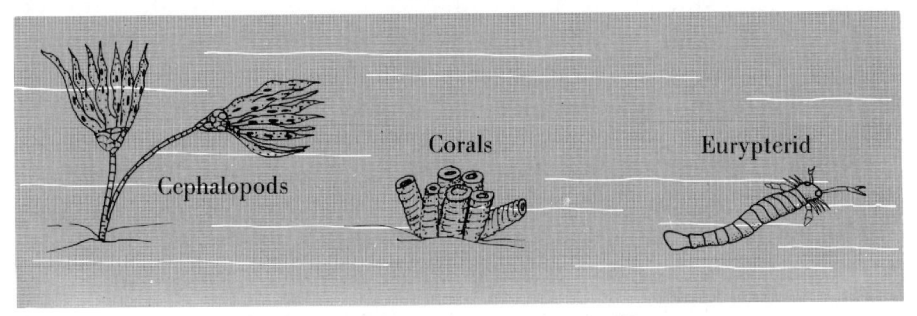

Figure 14.7 Ordovician and Silurian life.

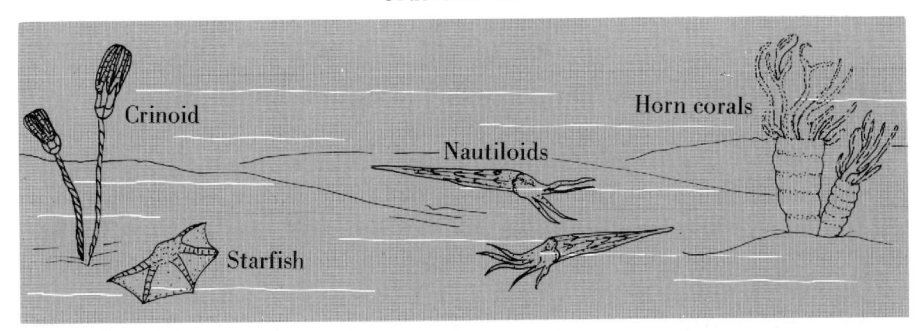

are echinoderms. The largest creature of Paleozoic times was a mollusk that flourished in Ordovician seas, *Endoceras proteiforme*, a nautilus-like animal that grew to a length of 15 ft. Its nearest living relative is the giant squid, the largest invertebrate animal of all time. Another creature that has become an index fossil for the Ordovician is the eurypterid, or sea scorpion. These flattened versions of the scorpions of today's warmer regions measured from a few inches to perhaps two feet in length.

The world climate of Ordovician times was quite similar to that of the Cambrian, although a drying trend began toward the end of the Silurian, which is evidenced today in salt deposits among Silurian rock formations. The mountain-building activities of the Ordovician period produced the Taconic Range in

Minnesota. In the Silurian, the Caledonian Range, which includes the mountains that run from Gibraltar through the Scottish Highlands and along the spine of Scandinavia, were thrust up.

The Devonian Period, named for Devon, in the south of England, was the Age of Fishes. Late in the Ordovician a new development appeared that portended great things: a spinal cord. As this first vestige of what was to become the most highly developed type of nervous system evolved, its protection evolved too: a series of hollow cylinders of carbonate and phosphate salts called vertebrae. During the Silurian, armored fishes evolved in freshwater streams and lakes. These were the first proper vertebrates. They had strange shapes and mere slits for mouths that enabled them to suck nourishment from the muddy bottoms of their watery habitats. But by Devonian times, there were not only the cartilage-skeletoned sharks but true fish, including a line called *crossopterygians*, or lobe-fins (Fig. 14.8). The discovery of a few members of this species in vigorous health off East Africa about twenty years ago was hailed as the

Figure 14.8 Crossopterygians, the first vertebrate colonists on the land.

discovery of a living fossil. As far as anybody had known until then, the *coelacanths,* as they were named, had been extinct for 300 million years!

The lobe-fins were important in the Devonian as the precursors of land animals, which first appeared in the Carboniferous Period. Many lobe-fins had swim bladders that could function for short periods as lungs. This permitted them to survive if the streams or lakes they inhabited filled up with sediment. The lobed fins served for land locomotion, too, enabling the animal to drag itself along. Apparently some lobe-fins preferred this way of life, for their descendants—the first amphibians—came to live all their adult lives on land, though they found it necessary to return to their native water to spawn.

The land was already inhabited, however, when the amphibians arrived. Algae and other nonvascular plants—that is, plants without specialized root, stem, and circulatory systems—had migrated onto the sea-shores as early as the Silurian. By Devonian times tree ferns and other vascular plants had spread well inland from the seashores, and among them lived various members of that still-numerous family, the Arthropods: scorpions and insects of various kinds.

The mild climate of earlier times persisted through the Devonian but rains tended to be seasonal instead of uniformly distributed throughout the year, which may have been crucial in directing the course of evolution toward animals that could live much of their lives out of water. Nevertheless, sea life still dominated the Devonian world. Trilobites, soon to be extinct, reached their largest size. Corals were giants in those days, too. Belemnites and ammonites, small mollusks with chambered shells, straight or curled like the modern nautilus, were so numerous as to become index fossils for the period, but none survived the Paleozoic.

The next period of the Paleozoic Era is named for its best-known product: coal. This is the Carboniferous Period. American geologists subdivide it into the Mississippian and the Pennsylvanian Periods, named for the areas in which the older and younger coal deposits are found. We have a clearer picture of the Carboniferous Period than of any preceding time, for the coal itself is the fossil remains of great forests. Strange forests, those, to a modern observer (Fig. 14.9). The trees were closely related to ferns, not to the trees of today. We know them by their leaf and bark prints, found in shales (rock hardened from clayey mud) associated with coal deposits, and from occasional fossilized stumps.

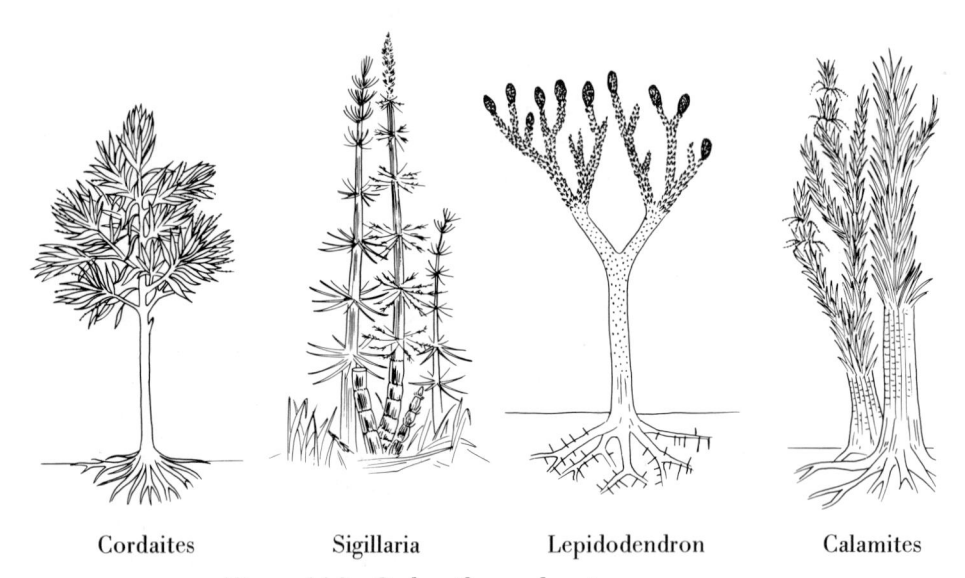

Cordaites   Sigillaria   Lepidodendron   Calamites

Figure 14.9 Carboniferous forests.

These plants thrived in warm, swampy places — almost everywhere on earth in those days. The great continents were still close together. The trees fell and were covered with water, layer upon layer through 50 million years or more. Bacteria and fungi aided their decay through various stages from peat to soft coal, and metamorphism in some places produced hard coal, or *anthracite*. Most of the plants that produced the world's coal beds thrived only in the Carboniferous Period, for the next period was one of great climatic change. The lycopods or scaly-barks and Cordaites, which were in some respects like modern pines, left no descendants. The 30-ft calamites and tree ferns are represented today by the modest horsetails, or scouring rushes, that appear in northern and temperate woods and waste places early in the spring, and by the lovely but seldom gigantic ferns that uncurl their fiddleheads through the leaf mulch of the forest floor.

The Carboniferous Period may be best remembered for its plants, but animal life was abundant, too. The largest species, in absolute terms, were amphibians, but insects grew to giant size in those days too: a fossil imprint of a dragonfly with a 29-in. wingspread dates from this time. The humble cockroach reached his final stage of evolution in the Carboniferous, and admire him or not, one has to admit he is a successful species; he's still around, unchanged.

The transition period with which the Paleozoic Era drew to a close, the Permian, was a time of great change. The name is derived from Perm, a region in eastern Russia, where the characteristic strata of the period were first found. In what was to be North America, the Appalachian mountains arose. Across the southern half of Gondwana stretched a great ice sheet, bringing to an end the Carboniferous swamps all over the world by reducing the available water supply. The drying land encouraged the development of organisms that could spend their whole lives out of water, and their rapid development in the Permian is called the Appalachian revolution, after the mountains that were formed concurrently.

The Age of Reptiles

The land animals that dominated the earth for 160 million years from the end of the Permian were the reptiles. This era, the Mesozoic, is divided into three periods, according to the predominant geological strata and life forms. The period names are not very systematic: "Triassic" comes from the Greek word for three, because the landforms in Germany from that time fall into three layers—a purely local phenomenon, but the name stuck; Jurassic strata were first studied in the Jura Mountains; Cretaceous means "chalky" and is the period in which Dover's famous white cliffs were formed—along with other more abundant kinds of rocks. However well or poorly the names of the periods fit, the Mesozoic Era is the Age of Reptiles.

The small, unspecialized reptiles that inherited the Permian earth from the amphibians, many of which became extinct as their breeding places dried up, developed in two main lines during the Triassic. These two orders were named *Ornithischia* (bird hips) and *Saurischia* (reptile hips), respectively. We commonly use the term *dinosaur* (terrible lizard) to include both the birdlike ornithischians, most of which walked upright on two legs with a sturdy tail for balance and had small grasping forelimbs, and the heavier four-footed saurischians. Triassic dinosaurs were not really big. Their maximum length was 10 to 15 ft, and most of them walked in a semierect posture. The Jurassic was the heyday of the reptilian monsters, when saurischians like Brontosaurus, Brachiosaurus, and Diplodocus grew to 65 to 85 ft and spent their whole lives in their native swamps where the water's buoyancy helped support their forty tons of body weight (Fig. 14.10). These were plant eaters, and doubtless of mild temperament in contrast to their overwhelming appearance. They had two "brains," both minute fractions of their total size: one at the back of the skull operated jaw muscles and enabled the creatures to receive and respond to stimuli of food and danger; the other, at the base of the vertebral column, operated the leg and tail muscles. Most ornithischians were predators, ranging

Tyrannosaurus   Brontosaurus

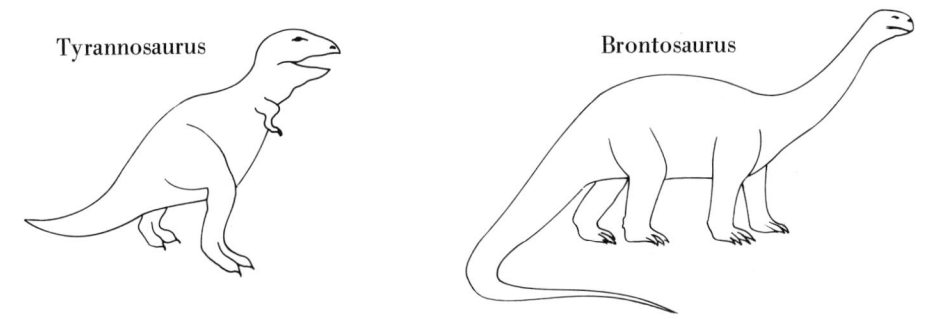

Figure 14.10 Jurassic dinosaurs.

from relatively small birdlike creatures to the 34-ft Allosaurus of the early Jurassic and the 47-ft Tyrannosaurus Rex, that most fearsome tyrant-king of the early Cretaceous.

Not all the dinosaurs were land dwellers, however. Early in the Triassic some reptiles returned to the sea. There was Elasmosaurus, with 23 feet of neck alone (his overall length was perhaps 35 ft), Kronosaurus, with nearly 3 yards of teeth in each side of his jaws, above and below (that makes 12 yards in all!), and the terror of the Cretaceous seas, the 40-ft mosasaurs.

Some dinosaurs took to the air, but they were gliders who apparently never mastered the art of powered flight. Birds evolved from reptiles, but the pterosaurs were probably not in the direct line of descent because they were too specialized in their own way to evolve into something quite different. True birds, with feathers, did develop in the Jurassic period, but they may not have been much more adept at flying than their reptilian contemporaries. Some fossils of early birds reveal webbed feet so far back on the bird's body that swimming must have been its only practical means of locomotion (Fig. 14.11).

The dinosaurs of the Cretaceous displayed some fantastic decorations in the way of horns and plates of armor, but their larger heads, four-legged gait, and

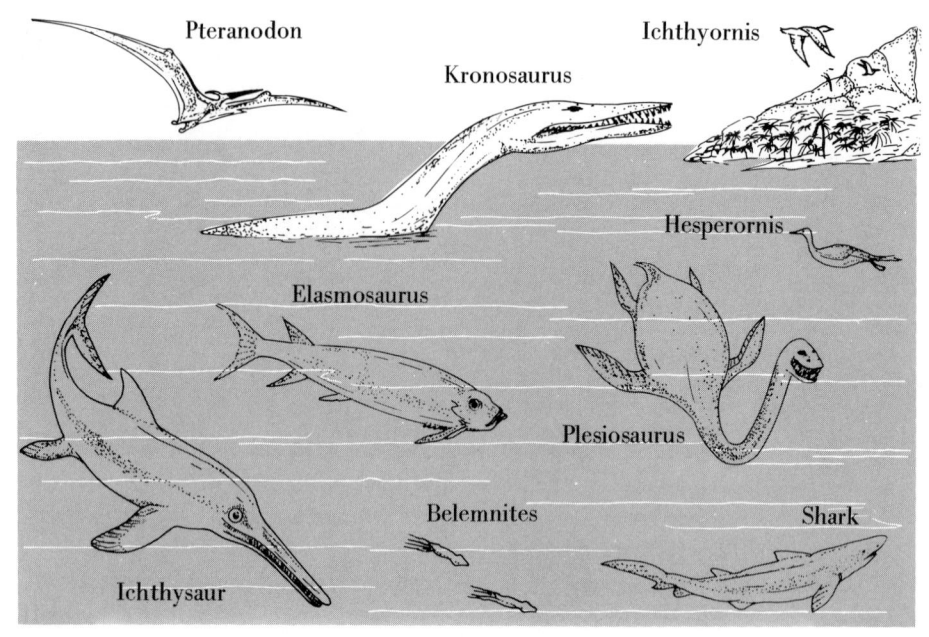

Figure 14.11 Flying and swimming dinosaurs.

smaller tails made them less strange to our eyes. But their days were as numbered as those of Tyrannosaurus Rex. Some of their unspecialized small cousins had been developing in quite a different direction to produce the order that would inherit the earth from the dinosaurs.

The dinosaurs are all gone. Their direct descendants, the lizards of today, are small and not very significant citizens of earth. Some species are highly localized in range and threatened with extinction. Nevertheless it's not quite fair to label them failures. They dominated the earth for 160 million years. Will man succeed that long?

During the era of middle life, the Mesozoic, vast changes occurred in the earth's geography. Orogenic processes were in nearly continuous operation, lifting up the palisades of New York and New Jersey, the Sierra Nevada, and the Eastern Alps in succession. It was also in the early Mesozoic that the primeval land mass, Pangaea, broke first into the northern continent, Laurasia, and the southern Gondwana. By the end of the Triassic, Gondwana had further split, with the crustal plate carrying India heading north and the Australia-Antarctica mass moving southward. By the end of the Jurassic, rifts were apparent that would mark the separation of the Americas, and the movements begun earlier continued apace. A stretching out of the southern mass foreshadowed the separation of Australia and Antarctica. By the end of the Cretaceous, there was an Atlantic Ocean, although Greenland still joined North America to Europe at its northern end. It is assumed that edges of westward-moving plates must have been "subducted"[1] into the mantle in a great north-south trench in the Pacific. The continents had a long way to go to reach their present positions, but the world was certainly different at the end of the Mesozoic Era from anything it had been before.

Geographic and climatic changes doubtless had much to do with the extinction of the dinosaurs. We ought not to suppose that these rulers of the earth died out overnight. They waned throughout the later Cretaceous, as other better adapted creatures displaced them, and perhaps also preyed upon their eggs and young. These "superior" creatures, evolved from some little saurischian, most likely, didn't lay eggs; they brought forth live young, which they fed by secretions of their own bodies and protected during infancy. They were less at the mercy of the weather than the reptiles, too, for their warm blood created their own constant internal climate, and their fur provided thermal insulation. These were the mammals, small, omnivorous, and unspecialized at

---

[1] This term is borrowed from R. S. Dietz and J. C. Holden, "The Breakup of Pangaea," *Sci. Am.* (October 1970), pp. 30–41, whose chronology of continental drift is followed here.

first, but destined to dominate the earth in the Cenozoic Era, the era of recent life.

The Era of Recent Life

The Cenozoic was a time of cooler climates, extensive mountain building, and relatively rapid movements of the drifting crustal plates. The Rocky Mountains were formed in two great bursts of orogenic activity. In between these bursts the Pyrenees and the Swiss Alps rose. At the end of this so-called Tertiary Period, the Indian subcontinent reached Asia, and the encounter thrust up the Himalayas.

The Cenozoic Era is divided into two periods, called Tertiary and Quaternary according to a naming system no longer used otherwise. The Tertiary is divided into five epochs: Paleocene, "the ancient recent"; Eocene, the "dawn of the recent"; Oligocene, "a few of the recent"; Miocene, "less recent"; Pliocene, "more recent." The Quaternary opened with the Pleistocene, "most recent"; we live in the Holocene, the "wholly recent."

Plant life changed as drastically at the end of the Mesozoic as animal life. The greatest change came in their means of reproduction, from the essentially asexual spore method to the bisexual seed. As in animals, this provided a mechanism for diversification, and soon the relatively few species of club mosses, horsetails, and ferns were reduced to even fewer small survivors, and first conifers and ginkos, then the flowering plants covered the land. The characteristic of flowers is that they provide a receptacle for the development of seeds. One of the most ancient flowering species, the lovely southern magnolia, offers a good example of the development of seeds in pods or "boxes" that marks the most abundant and varied order of plants, the angiosperms (Fig. 14.12).

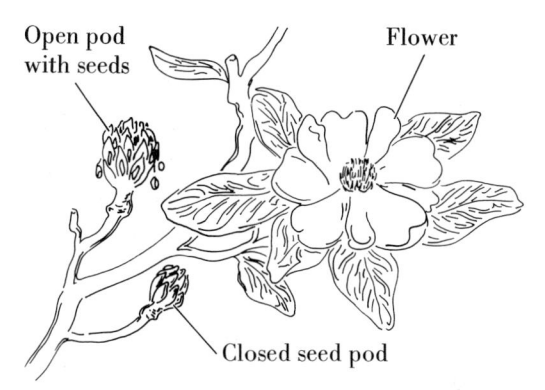

Figure 14.12 Magnolia: flower, seed pod, and seeds.

Natural Selection

So the small, "generalized" mammals of the Paleocene developed in habitats like those still prevailing: steaming jungles of tall trees, climbing plants, and sprawling undergrowth, temperate forests of soft and hard woods, and vast grassy savannahs. The mammals began to specialize to fit their habitats and food supplies—not by conscious choice, but by the long, slow process of natural selection in which the fittest survive to produce offspring, which inherit and pass on those characteristics that made them fit. Some specializations paid off better in the long run than others. Such giant mammals as the Baluchitherium, an Oligocene monster roughly the size of a double-decker bus, went the way of the giant reptiles. Other species evolved in even more successful forms. The development of the horse from the collie-sized eohippus is a classic example (Fig. 14.13).

Developing alongside the ground-dwelling mammals of all sizes throughout the Tertiary were small tree-dwelling omnivorous species whose forelimbs and brains developed even further, thus keeping them in the competition for survival. Bipedal locomotion—which simply means walking upright on one's hind legs—frees the hands not only to do things

Eohippus   Mesohippus   Merychippus

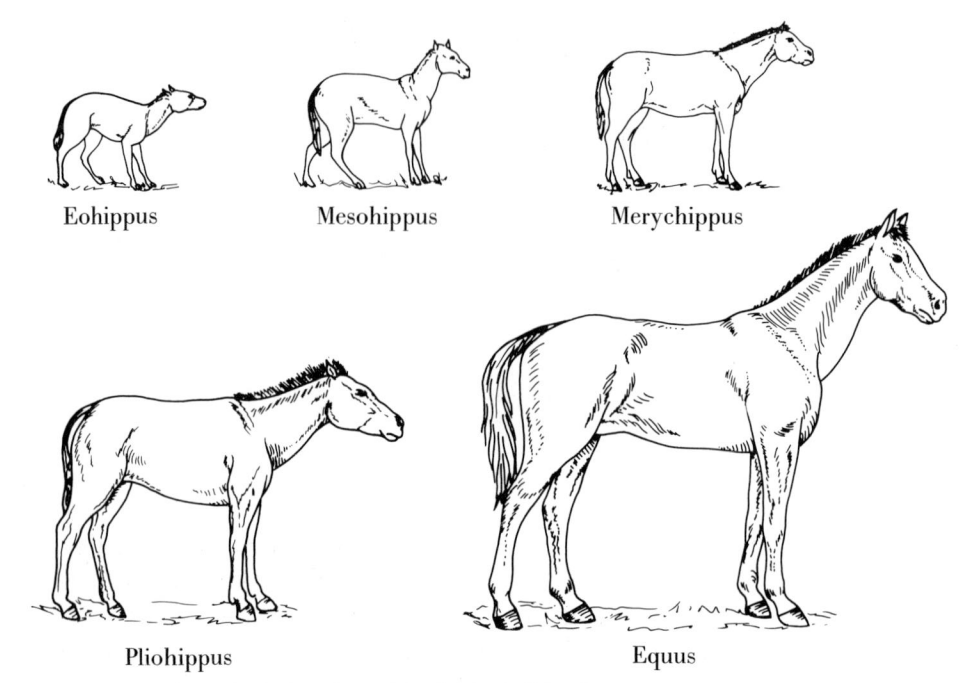

Pliohippus   Equus

Figure 14.13 Evolution of the horse.

themselves, but to extend their abilities with tools, the manufacture and use of which an improved brain can figure out. Some 2 to 3 million years ago at various sites in East Africa, remains of pebble tools and charcoal along with the bones of these mammals justify our calling them "man-like."

Tracing man's history back beyond the proverbial cave man to his probable ancestry among the man-apes of the Pliocene or beyond to the Miocene Proconsul, an anthropoid type which may be the common forebear of both man and the present-day great apes, is a fascinating, but lengthy story. Suffice it to say that by the time the great ice ages began in the early Pleistocene, creatures were indeed living in caves in southern Europe, North Africa, and probably parts of Asia, who by the vividness of perception reflected in their art, the craftsmanship of their stone tools, and the spiritual consciousness shown in their burial sites deserve full recognition as human.

Ice Age Mystery

The cause of the great ice ages of the Pleistocene remained a mystery to generations of geologists. Did some change in earth's atmosphere produce the drastic cooling of the northern latitudes, or did that steady and dependable star—the sun—actually diminish the intensity of its warming rays on three lengthy occasions? It seems now the theory of continental drift, which has solved so many mysteries, may solve the great ice age mystery, too. Columbia University geologists Maurice Ewing and William L. Donn explain it this way: Until the Pleistocene, the geographic north pole, well within that ocean depression called the *Arctic Basin*, was in open water. Then the drifting continents closed it in, permitting little exchange between its icy waters and those of warmer seas except east of Greenland, where an undersea ridge, of which Iceland is a peak, is only 2000 ft below the surface. The oceans are high now, for there is very little continental ice. Exchange of water between the Arctic and Atlantic basins goes on apace, and the warm water is melting the Arctic ice floes. The polar air masses, normally dry as they sweep over the Arctic ice cap, become wetter as the ice cap melts, and melting ocean ice means greater snowfall and accumulation of ice on the land, growing glaciers in northern mountains, and perhaps eventually continental ice sheets such as those that reached down into central Europe, Asia, and North America three times in the Pleistocene. Accumulation of ice on land means falling ocean levels, isolation of the Arctic basin by land barriers, and the freezing of the Arctic Ocean and the thawing of the land. If Ewing and Bonn are right we must expect a continued cycle of ice ages and interglacial periods until the drifting continents have swung far enough apart to break the cycle of isolation of the polar basin again (see Plate 32).

When one reads its history in a few pages, it is hard

to realize that all the vast and cataclysmic changes that have shaped the earth through four eons are still going on. The rivers run down to the sea. The rains fall in the highlands and flow downward. The gravitational potential energy of the water is changed into kinetic energy as it flows more and more rapidly, and into the work of excavating the valley, sharp and V-shaped at first, then in slow, sweeping curves, filling in here and digging out there. At last after many miles and many millennia, the ancient stream meanders among the eroded hills until or unless a great uplifting rejuvenates it and it begins cutting its own version of the Grand Canyon (Fig. 14.14).

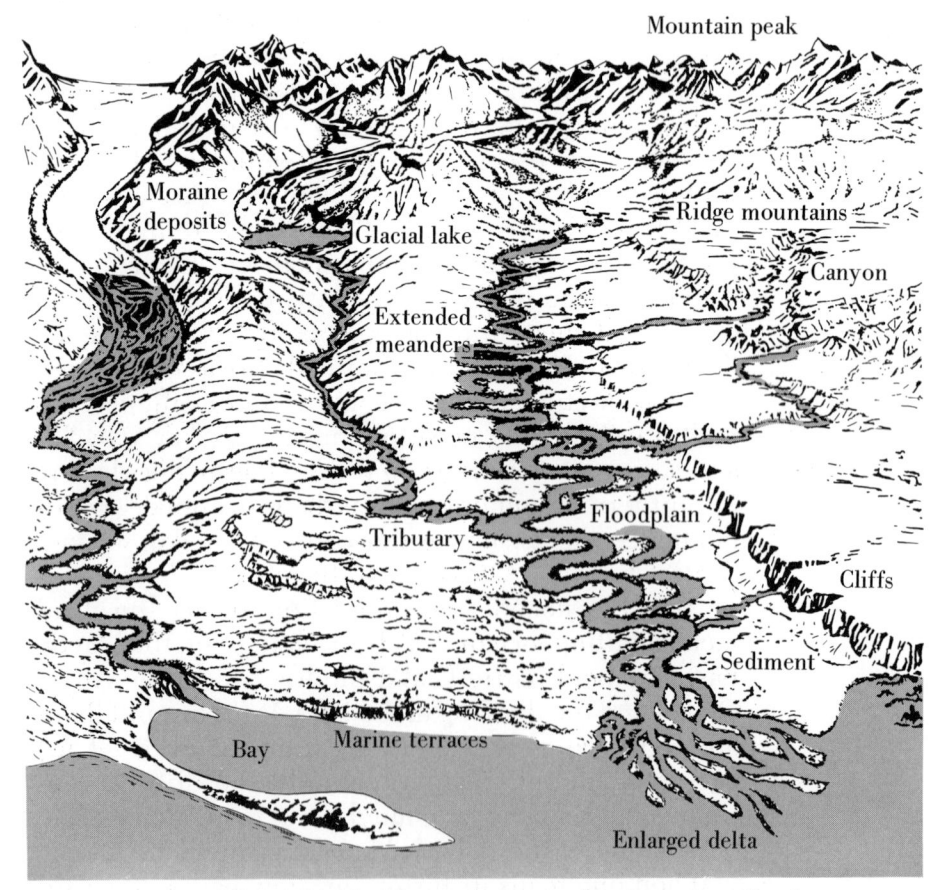

Figure 14.14 The evolutionary development of landscape.

When a river of ice, a glacier, is the excavator, the work goes more slowly, and the valley's cross-section is a U instead of a V, and at its head may be a cirque, as if the mountain had been excavated by a giant ice cream scoop. When the glacier reaches the warmer lowlands and melts, it drops its load of rock and soil and builds a moraine or strews boulders across the landscape that clearly come from somewhere else because they do not match the local rock.

The earth's present is the key to its past. In the shape of its valleys, the unconformities of its strata, the fossils neatly filed in the sediments in which the organisms fell at their death, and in the work of wind, waves, and running water we observe day by day, we read its story.

Weather

Only these daily changes we call *weather* remain. Weather is the product of the action of heat on air. The patterns of weather (or climates) are the product of the effects of latitude, land forms, ocean currents, and the rotation of the earth (Fig. 14.15).

We can understand weather phenomena by one main principle: warm air rises and absorbs moisture. The implications of this simple statement are far-reaching. The development of a thunderstorm on a hot afternoon illustrates the whole process on a small scale (Fig. 14.16).

Clouds form when warm moist air rises and cools so that the water vapor must condense out of it. The high wispy cirrus clouds are actually made of ice crystals instead of water droplets, and so are the top layers of towering cumulonimbus "thunderheads." Layers of stratus clouds may occur at almost any elevation from the ground (where they are called *fog*) up into the stratosphere. They portend the steady fall of rain or snow, sooner or later. The fleecy fairweather cumulus ride the tops of convection columns, but

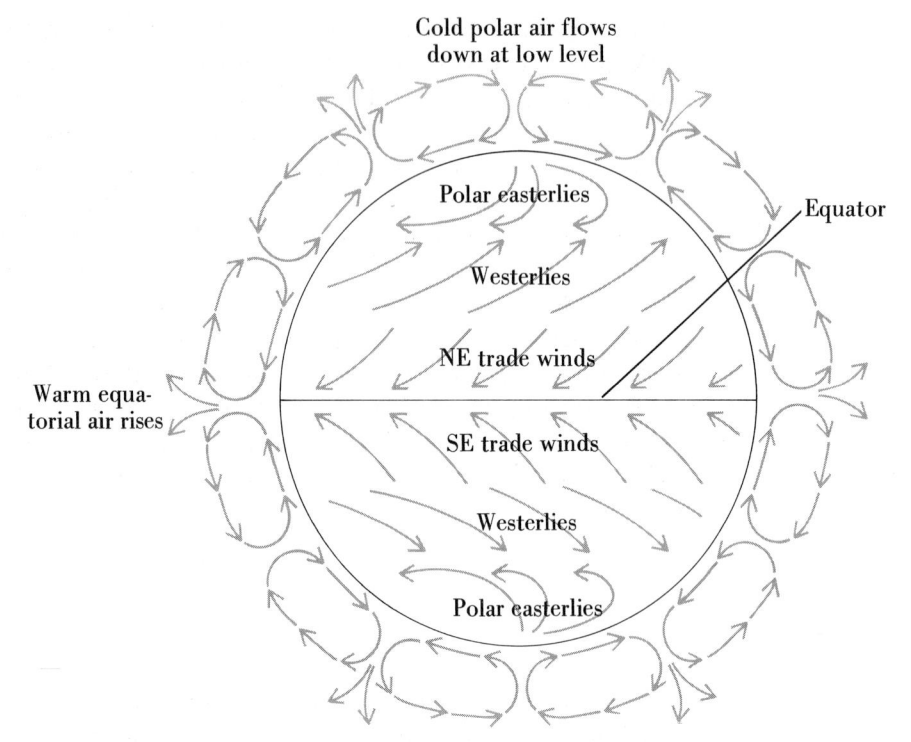

Figure 14.15 Global patterns of air circulation.

may grow into thunderheads six miles tall, under the proper conditions (see Plate 33).

Every newspaper reader and television weather watcher is familiar with "fronts," cold, warm, stationary, and occluded. A weather front is simply the leading edge of a moving mass of air. Because this mass will overtake or bump into other masses of different temperature and moisture content, "weather" tends to happen along fronts, where masses meet. Sometimes the signs of approaching fronts are so clear, once you know what to look for, that you can stand back and watch the events on the weather map happen before your eyes (Fig. 14.17).

The main weather fronts in the United States move across the continent from west to east in a week or

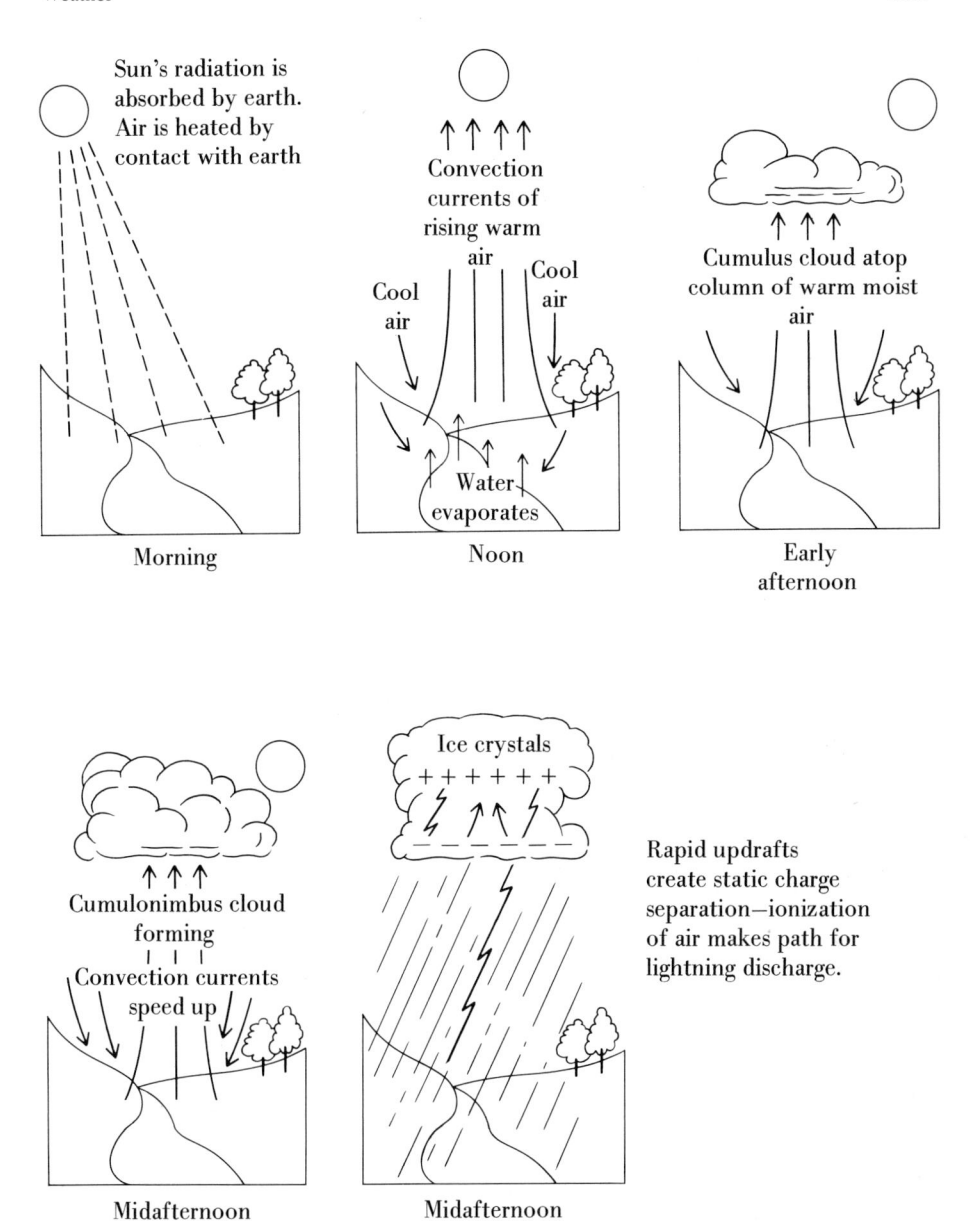

Figure 14.16 The development of a summer afternoon thunderstorm.

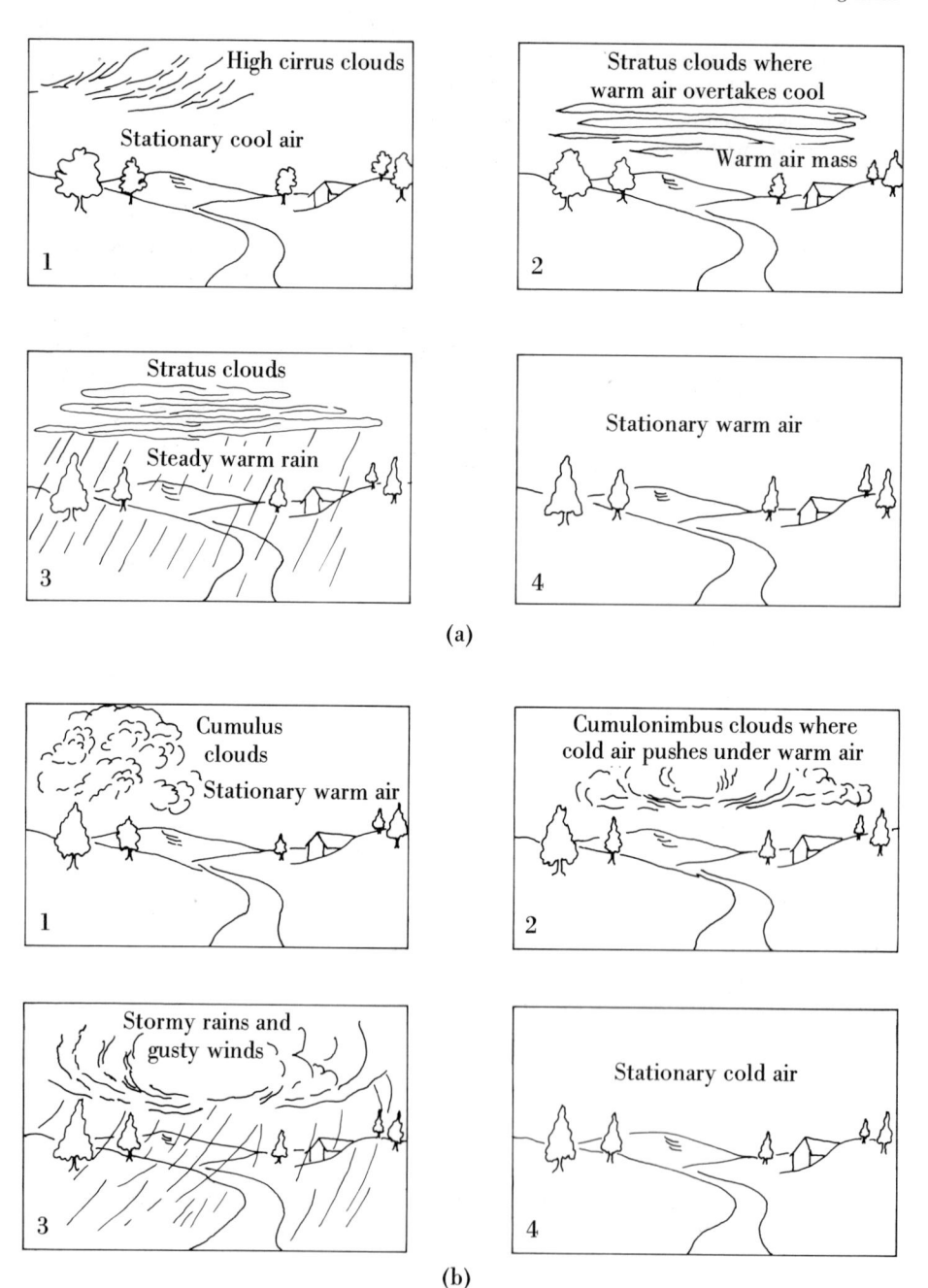

Figure 14.17 Weather fronts: (a) warm front; (b) cold front.

less. A stationary front is one that simply stops for a day or so. Occluded fronts are a warm front and a cold front meeting from opposite directions and staying together in one place, usually producing the worst weather features of both. We can understand the general pattern of frontal movement in terms of the prevailing winds and the earth's rotation.

Two other common features of the weather map are "highs" and "lows." A high is a mass of cool dry air that, having greater density (as shown by high barometric pressure) than the air around it, rotates in a clockwise direction in the northern hemisphere. A low is a mass of warm moist air of low barometric pressure rotating counterclockwise in the northern hemisphere. A high is properly called an *anticyclone*, a low, a *cyclone*. Although lows are generally associated with "bad weather," it is incorrect to equate "cyclone" to "tornado." Cyclone simply means a mass of air rotating counterclockwise. A tornado is a small but very violent storm developing out of a thunderstorm when extremely strong updrafts create very low pressure areas at their center. It is here that the characteristic funnel forms, sucking up debris and causing buildings to explode because their internal air pressure is so much greater than the pressure in the funnel outside.

Giant cyclonic systems forming over warm seas in late summer may become the destructive heat engines called *hurricanes* (Plate 31). The summer sun saturates the warm air with water evaporated from the sea. Water molecules are less massive than the oxygen and nitrogen molecules of dry air; the moist air mass becomes a "low," and begins to rotate. Condensation of the water vapor into clouds releases heat energy to drive the system landward. Then torrential rains fall, and the rapidly rotating air mass around the central "eye" or calm zone causes the devastating wind damage of a hurricane. Over land, the heat engine runs down because its supplies of water vapor to condense and produce heat run out.

The topography of the land also influences weather patterns. As warm air rises, it cools, and its moisture condenses out. So rain forests occur on the windward side of mountains, and deserts in their lee. Water has a higher heat capacity than rocks or soil. It heats up and cools off more slowly than the land and its presence moderates the climate of coastal regions. To say "continental climate" implies hot summers and cold winters, untempered by the presence of the sea.

"Great, wide, beautiful, wonderful world,
With the wonderful water around you curled. . . ."[2]

Are there any more like it? We may never know. It behooves us to exercise wise and responsible stewardship of what we have.

SUMMARY STATEMENTS

1. The best current theory of formation of the earth explains its occurrence along with the other planets of the solar system as an event associated with the formation of the sun itself. Not all materials in the original cloud of gas and heavier elements of supernova origin condensed into the central star, but the remainder formed nine major planetary bodies and many smaller ones. None of these, not even Jupiter, was massive enough to initiate nuclear fusion.

2. The original makeup of the earth was determined by its mass and distance from the sun.

3. The primitive, cooling earth formed into a spheroid with a partly liquid core primarily of iron, a mantle of dense minerals, a crust of lighter minerals that rose to the top and hardened, and an atmosphere of water, methane, ammonia, carbon dioxide, and other small molecules.

4. Life apparently arose through the formation of proteins from the amino acids produced by the action of ultraviolet radiation and lightning discharges on the primitive atmosphere, although the steps from chemical to biological evolution are not clear.

---

[2] A children's poem by William Brighty Rands, reprinted in *The Book of Knowledge*, vol. 1 (1953), p. 98.

5. We understand the changes in the earth itself in terms of the principle of Uniformitarianism: the same processes that we see at work today have been at work in the past, and account for presently existing features.

6. Among the continuing processes molding the earth are vulcanism, sedimentation, weathering, and erosion by water, wind, and ice, and continental drift.

7. Life forms that have evolved through the last half-billion years of earth's 4.5-billion-year existence have left behind fossil remains that tell of successful adaptations that have endured and of overspecialized species that died out.

8. Evolutionary development proceeds very slowly by mutations that introduce useful features and by natural selection that favors the survival of the best adapted individuals and their offspring.

9. The shortest range changes on earth are weather phenomena. The interaction of heat from the sun, air, water, and the earth's rotation produce weather fronts that move across the continents in the seasonal patterns that constitute the various climates of the planet.

REVIEW QUESTIONS

1. Trace the formation of the solar system according to the theory of C. F. von Weiszäcker.

2. How did the earth's original atmosphere differ from today's atmosphere? What was probably the most significant result of that difference?

3. Give some evidence for believing the earth has a core of molten iron.

4. How do earthquakes help us understand the earth's interior structure?

5. Outline two ways in which mountains can be formed. How could you tell which method accounted for a given mountain?

6. Trace the possible life history of a river.

7. Define the following; give examples where possible.

igneous	Catastrophism
sedimentary	Uniformitarianism
metamorphic	isostasy
geosyncline	evolution

8. What is a fossil? Mention some forms fossils take.

9. Cite some evidence for the theory of continental drift.

10. How can continental drift account for ice ages?

11. Trace the development of plant and animal life on earth from its possible beginnings to modern times.

12. Why do species become extinct?

13. What does "chemical evolution" mean?

14. What main factors are involved in producing weather phenomena?

15. Trace the development of a thunderstorm on a hot summer afternoon.

16. Why do most hurricanes occur in August and September?

17. Define: cyclone; tornado; warm front; cold front; "high"; "low."

THOUGHT QUESTIONS

1. What factors are involved in dating strata by their fossil content?

2. What common characteristics do all extinct creatures appear to have, if any?

3. Why is Uniformitarianism preferable to Catastrophism for accounting for changes in the earth?

4. Considering the probable course of development of life on earth, evaluate the possibilities of development of life on the other planets of our solar system.

5. What are some problems involved in trying to get in touch with intelligent creatures in other planetary systems, if they exist?

6. How would you attempt to detect life on Mars if you could land some detection device on its surface? Several such devices are being designed. After you have tried to think of one on your own, read about the "Wolf Trap," "Gulliver," and the "Multivator" being developed under NASA auspices.

7. Why does a "falling barometer" usually mean rain?

8. How can you square the spontaneous development of life with the Second Law of Thermodynamics?

READING LIST

Eiseley, Loren, *The Immense Journey*, Random House (Vintage), New York, 1957. A beautifully written account of human evolution—literature, not biology!

———, *Darwin's Century*, Doubleday (Anchor), Garden City, N.Y., 1961. The history of the theory of evolution and the men who discovered it.

Carson, Rachel, *The Sea Around Us*, New American Library (Mentor), New York, 1955. The classic story of the sea, its origin, its movements, and its life, beautifully told. *The Edge of the Sea* and *Under the Sea Wind* are companion volumes and equally interesting.

Carrington, Richard, *A Guide to Earth History*, New American Library (Mentor), New York, 1961. A brief, readable, well-illustrated history of the evolution of our planet and its life.

Edinger, James G., *Watching for the Wind*, Doubleday (Anchor), Garden City, N.Y., 1967. A clear, interesting study of seen and unseen influences on local weather.

Moore, Ruth E., *The Earth We Live On*, Alfred A. Knopf, New York, 1956. A fascinating book about the internal and external features of the earth and the men who discovered and explained them.

———, *Man, Time, and Fossils*, Alfred A. Knopf, New York, 1961. An equally fascinating account of the development of the theory of evolution and the science of genetics, with emphasis on people as well as events and ideas.

Ovenden, Michael W., *Life in the Universe*, Doubleday (Anchor), Garden City, N.Y., 1962. A scientific discussion of the possibility of life on planets other than earth and in other planetary systems.

15

Frontiers of Science: Problems and Possibilities

Perhaps the greatest fascination of science for the scientist is that it's never done. Every time a question of fundamental importance gets answered it opens the way for asking a whole new set of questions. A few years ago, a few people—not scientists, themselves—were suggesting that we were approaching a "golden age" when everything would be known. For a scientist, such a situation would be no golden age; it would be more comparable to the little room from which Sartre saw "No Exit." However, scientists today in every field have far too many questions clamoring for answers to worry about the day they might run out of questions.

What are some of the questions to which answers are being sought today? Most frontier areas of science are "interdisciplinary": biochemistry, biophysics, astrophysics, geochemistry, etc. Pushing back these frontiers requires the skills and knowledge of scientists of many specialties; and in fact, overspecialization at the expense of broad knowledge is becoming increasingly unproductive.

Perhaps the one realm still reserved for the relatively few with a narrow, highly theoretical, highly esoteric knowledge and methodology is that known as *particle* or *high-energy physics.* This is the realm of the entities that are fleetingly observed in cloud chambers and nuclear emulsions when nuclei are torn apart by high energy missiles, themselves neutrons, protons, and such. These entities have charge and mass like the more familiar particles, but they have lifetimes of nanoseconds (billionths of a second) and strange quantized properties with exotic names like "isospin" and "strangeness." No one now knows their relationship to each other or to ordinary matter, or even whether they are separate entities or different states of a few really fundamental particles. They can be talked about only in a complicated mathematical symmetry language that few speak. There must be much we don't know here, because most scientists share Einstein's feeling that the universe is "comprehensible," even simple, once we really understand the truth of it.

Related to the study of particles is the study of the structure of the nucleus itself. The shell model of Maria Goeppert-Mayer and others, based on "magic numbers" somewhat like the numbers of electrons in atomic orbitals, seems to be a step in the right direction, but there is a long way to go to full elucidation of the nucleus.

We know as little about the very large as we do the very small, and it is even less amenable to our experimentation. We still learn all we know about the stars and galaxies from their radiation, and as our instruments become more sensitive to wider ranges of radiation, we are presented with more and more puzzling phenomena to interpret. The decade of the 1960's presented astronomers with "quasars," "pulsars," X ray stars, and radio emissions revealing not only hydrogen atoms but OH and CN radicals and other molecules in interstellar space. Pulsars seem to fit into our scheme of evolution of stars as it is presently understood; quasars do not. Hence they open up broad new questions for investigation. Sometimes newly detected radiations come from long-known sources, causing our ideas about them to change, shedding new light on old mysteries. The discovery of small organic radicals and molecules in space increases the plausibility of the mechanisms for chemical evolution proposed for earth.

After a decade and a half of satellite observations of the earth and the moon, as well as the exploration of some of the moon's surface by man, we know a great deal about the surfaces of both these bodies. But we have only minimal knowledge of the interior of our own earth or even of the depths of its seas. And everybody still talks about the weather without understanding it well enough to do anything about it.

Mysteries of Life

The most fascinating questions for most nonscientists as well as for many scientists probe the mysteries of life. We still do not know, and indeed have few useful hypotheses, about how chemical evolution crossed over to biological evolution. In fact, faced with the simplest viruses, we are hard put to define a living organism. Even the simplest one-celled creature is enormously complicated, compared to a virus, and contains in its single cell *organelles*, whose functions remain mysterious.

We know at least in broad outline how cells make proteins, but we know the actual structure of the merest handful of proteins. We can synthesize very few of them. We cannot yet synthesize carbohydrates at all. That is still the prerogative of cells containing chlorophyll. We do not know how a cell "knows" when to synthesize a sugar or a protein, to duplicate its chromosomes, or to divide.

If we know little about the detailed functioning of our own bodies, we know orders of magnitude less about the functioning of our minds.

The solution of many practical problems in medicine—the prevention and cure of cancer, muscular dystrophy, multiple sclerosis, heart disease, and birth defects including mental retardation, and the problem of the acceptance of transplanted organs by the body—all await our fuller understanding of the chemistry and physics of life processes.

In some areas, the basic science has been "done," but the technology to put it into practical use has yet to be developed. The laser is an example. Around 1960 it was discovered that in several kinds of atoms electrons could be promoted to "metastable" states (levels that they would not ordinarily occupy), from which they would all fall down together when primed by energy of the appropriate frequency, giving a coherent (all in phase) beam of radiation of nearly a single frequency. This opened up tremendous possibilities for communications applications, because these finely tuned visible and microwave frequencies would be much more efficient and longer ranged than wires or radio waves. But so far our laser communication is mainly with the moon. It is now possible to measure the earth-moon distance by laser signal within a margin of error of 2 in. By making this measurement from various continents from time to time, it will be possible to measure continental drift with comparable precision.

The worst problems connected with the production of nuclear power, the storage and disposal of radioactive fission products and the danger of widespread contamination if a breakdown should occur, would be removed if we could use fusion reactions, rather than fission, to release nuclear energy. Fusion reactors are "clean"; they produce no radioactive byproducts. But the problem of producing and maintaining a plasma in which controlled (rather than

H-bomb type fusion can take place has not been solved, although scientists and engineers around the world are working on it.

Science and Moral Values

The answer to the questions of whether an omega-minus particle is truly fundamental, or how far away a quasar really is may not be of life-and-death importance to any living person, not even the graduate student writing his dissertation, but questions relating to organ transplants, to mutations caused by radiation damage to DNA, to deliberate changing of heredity, do have ultimate significance to persons. Questions like these do not have strictly scientific answers, and they bring us face-to-face with the question of the relation between science and moral values. The traditional "scientific" point of view has been that there is none. Certainly scientific truth itself is not good or evil in the moral sense, but clearly it may be applied for good or evil purposes, and more and more scientists are becoming convinced that they must assume some responsibility as human beings, as members of society, for the use of their discoveries, for they are the ones who understand best the potentials of those discoveries.

Three people will die without a transplanted heart. One heart is available. Who gets it? Science can evaluate the chances of a successful operation in each case, but "who gets it?" is a moral question. It involves the ultimate human value judgment: which life is most worth saving. And who is to decide?

If it becomes possible, as it may, to bear "made-to-order" children, who has the right to place the order? This is not a scientific question.

As we become more aware of the limitations of space and resources on "Spaceship Earth," how will the decisions about using that space and those resources be made? Not strictly on a "scientific" basis, for there can be none. Science can, and must, remind us that the laws of nature are inexorable. Energy is conserved. Moreover, it is degraded. It becomes less and less available, and more and more expensive. There is no recycling without loss. So the time has come to recognize that ability to do a thing is not justification to do it. We, members of a society inextricably based in scientific technology, must learn to count the long-range costs ahead of the immediate profits. It will obviously be easier for manufacturers to think in these terms if consumers set the example by being selective about the products they buy. Science can tell me the effects of phosphates on the lakes and streams that receive my dishwater, but I have to decide which detergent to buy.

As the dangers of nuclear war have become understood over the past quarter of a century voices have been raised from time to time crying for a moratorium on science. Such a thing would be neither possible nor helpful. A scientist cannot be stopped from thinking about his world. Then, too, scientific technology must provide some solutions to problems of pollution, use and recycling of metals and other resources, and to overpopulation, which many see as the most basic problem.

But as soon as the problems are listed we are aware that although science can provide techniques, methods, materials, and "know-how," the decision to apply science to solving the problem reaches into realms far beyond science, particularly into the realms of human value judgments and responsibility.

In this last third of the twentieth century we must see our world as one world, not just in a political sense as Wendell Willkie urged in the 1940's, but in an ecological sense, as one community of life. "Ask not for whom the bell tolls," said John Donne; "It tolls for thee." Or, as Thomas Merton wrote, also quoting Donne, "No man is an island." Now scientists are saying what the poets and philosophers said before them. And we had all better believe it—enough to act accordingly.

SUMMARY STATEMENT

Science is a human activity carried out to find answers to the questions humans ask about their world, and as such is never finished. Science has no magic instant solutions to human problems, but is a useful tool in finding solutions that also involve human choices and value judgments. The basic laws of science describe the world as it is and so provide the framework within which solutions may be found.

REVIEW QUESTIONS

1. List a dozen still unsolved scientific problems or unanswered questions.
2. Are scientists likely to "work themselves out of a job"? Why or why not?
3. To what sorts of things do these terms refer?
 (a) quasar (b) pulsar (c) fundamental particle
4. Do you think that because a thing *can* be done it *must* be done? Is it necessary or possible to ask if a thing *ought* to be done?
5. List some areas in which the considerations suggested in question 4 would be important.

THOUGHT QUESTIONS

1. A city of 20,000 people has determined that automobiles are its primary source of air pollution. Suggest some solutions to the problem. The only rule of the game is that the problem must be *solved*, not *transferred*. (If you prohibit cars, what will you do about unemployment in Detroit, for example?)
2. A city of 100,000 people is offering for sale a 2000-acre tract of land. You have been employed, as an expert in community planning, to advise the city council on which of five competing bids to accept: a dairy farmer; a chemical company; a firm that exploits mineral resources; a realtor who plans a new housing development; or a group of wealthy citizens who plan to develop a park and zoo for public use. Consider the map on page 373 and the fact that the city wishes to maintain its present population in making your decision.

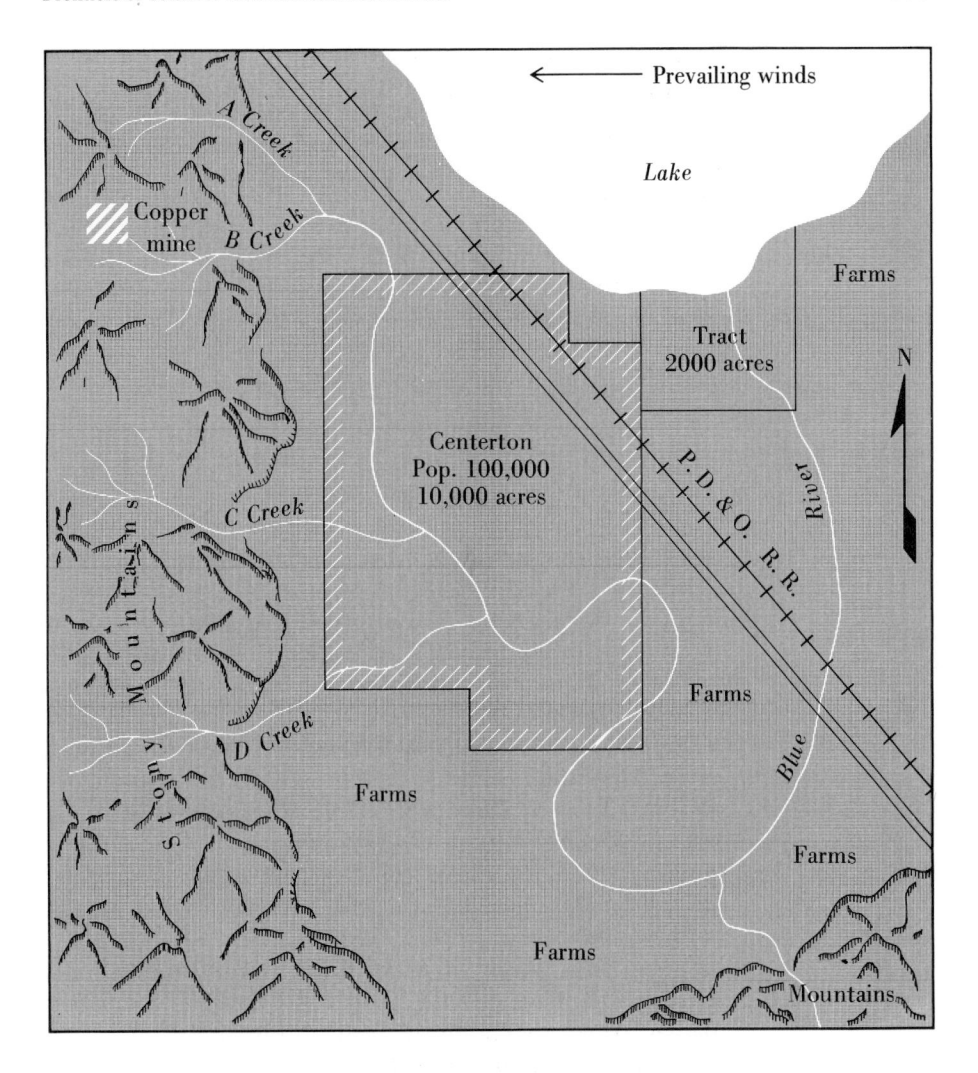

3. Your city government is holding public hearings on the advisability of letting the local electric power company build a nuclear power plant in your area. What questions will you want answered at these hearings?

4. Judge the "scientific" advertising claims below in the light of your scientific knowledge:

    a. Your hair is protein; so is our hair dressing, so it will be better for your hair.

    b. Our shampoo is made only of pure organic materials.

c. Keep the air pure; use our gasoline.
d. Does your wife nag? Take two tablets. . . .

5. Should scientists be encouraged to discover and develop techniques of controlling heredity?

6. How would you arrange the following in order of priority?

>Undersea exploration
>Manned flights to Mars
>All-out attack on cancer
>Made-to-order genetic material
>Fusion reactors for energy
>Development of artificial organs

READING LIST

Science and Society

Dubos, René, *So Human an Animal*, Scribner's, New York, 1968.

———, *Reason Awake*, Columbia University Press, New York and London, 1970. Two books by a concerned biologist about man's relation to science and nature. Readable and important.

Commoner, Barry, *Science and Survival*, The Viking Press, New York, 1963. Another concerned biologist points out the imminent dangers of dehumanized technology.

Carson, Rachel, *Silent Spring*, Fawcett Publications, Greenwich, Conn., 1962. The first, now famous, printed plea by a biologist that we stop and think what we are doing to our world.

Graham, Frank, *Since Silent Spring*, Fawcett Publications, Greenwich, Conn., 1970. A second look at ecological damage done by chemicals used carelessly.

Mesthene, Emmanuel G., *Technological Change*, New American Library (Mentor), New York, 1970. The director of the Harvard Program on Technology and Society writes about the impact of technology on society and how to cope with it.

Harrison, George Russell, *The Role of Science in the Modern World*, William Morrow, New York, 1956. A discussion of what man may become in a technological society.

Fischer, Robert B., *Science, Man and Society*, Saunders, Philadelphia, 1971. A brief, up-to-date discussion of science as a human activity, the people who practice it, and its relation to technology, education, and public policy.

Storer, John H., *The Web of Life*, New American Library (Signet), New York, 1953. The interdependence of all life on earth and man's responsibility for its preservation is simply and movingly explained.

Ruzic, Neil P., *Where the Winds Sleep*, Doubleday, Garden City, N.Y., 1970. A projected history of man's future on the moon. Intriguing.

Ehrlich, Paul R. and Anne H. Ehrlich, *Population, Resources, and Environment*, W. H. Freeman, San Francisco, 1970. Two population scientists' prescription for salvaging Planet Earth. It is somewhat more hopeful than Paul Ehrlich's earlier book, *The Population Bomb*.

For Further Information: Science in General for the Nonscientist Reader.

Asimov, Isaac, *New Intelligent Man's Guide to Science*, Basic Books, New York, 1965. Two volumes on the physical and biological sciences written with Asimov's usual charm and precision.

Weisskopf, Victor, *Knowledge and Wonder*, Doubleday (Anchor), Garden City, N.Y., 1966. A physicist writes about the natural world for students, comprehensively but simply.

Alfven, Hannes, *Atom, Man, and the Universe*, W. H. Freeman, San Francisco, 1969. A recent Nobel laureate writes about man's place in the universe.

Personal Recommendations

The reader of these lists will by now be aware of the author's admiration of the writings of Isaac Asimov, George Gamow, René Dubos, and Loren Eiseley, all of whom have written many good things not listed here. Two more names should be added. These men are naturalists, one amateur, the other professional, and their writings about the world of nature can deepen anyone's appreciation of his world. They and some of their books are: Joseph Wood Krutch, *The Desert Year* and *The Twelve Seasons;* and Edwin Way Teale, *The American Seasons* (four volumes).

APPENDIX A

Mathematics

Concepts

1. *Cartesian coordinates:* Two lines at right angles to each other used to locate and show the relationship between two or more points in space. The vertical line is designated the *y*-axis or *ordinate;* the horizontal line is the *x*-axis or *abscissa*. Their intersection is the origin, and each of the four line segments beginning at the origin is a number line infinitely extended. The left segment of the *x*-axis and the lower segment of the *y*-axis are negative number lines; the other two segments contain the positive numbers. The scale along each number line is arbitrarily chosen. In Figure A.1, if the marks along the axis are whole numbers, then the coordinates of point P are $x = 3$, $y = 2$, always written in the order (x, y). This coordinate system is convenient for representing the variation of any two quantities with each other. When successive points are connected, the result is a *curve*, and the whole representation is called a *graph*. A *z*-axis perpendicular to the other two may be added to provide a three-dimensional coordinate system.

Appendix A: Mathematics

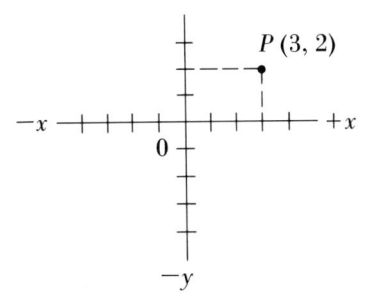

```
              | P (3, 2)
           ---•
              |
-x ++++++|++++++ +x
           0
              |
              |
              -y
```

Figure A.1 Cartesian coordinates.

2. *Constant:* A number that always remains the same regardless of how other quantities vary. A constant may be a ratio between two quantities measured in the same units, e.g., the ratio between the circumference and the diameter of a circle, π, which is a pure number, having no dimensions. A constant such as G, the universal gravitational constant, or h, Planck's constant, is the ratio between quantities having different dimensions; in this case the constant has dimensions consistent with those of the quantities it relates, and its numerical value will depend on the system of units being used.

For example, in the Law of Gravitation, $F = Gm_1m_2/d^2$, if force is measured in newtons, mass in kilograms, and distance in meters, i.e., in the mks system, G will have the value 6.67×10^{-11} newton-meters^2/kilogram^2. If force is in pounds, mass in slugs, and distance in feet, i.e., in the British Engineering system, $G = 3.41 \times 10^{-9}$ pound feet^2/slug^2. Within a given system of units the numerical value of a constant is always the same.

3. *Integers:* Whole numbers, 1, 2, 3, etc.

4. *Plane:* A two dimensional flat space. The study of plane figures is called *plane geometry*. Plane figures always have certain properties; for example the three angles of any *plane triangle* will add up to 180° (Fig. A.2). For plane triangles containing one right angle, that is, an angle measuring 90°, the *Pythagorean*

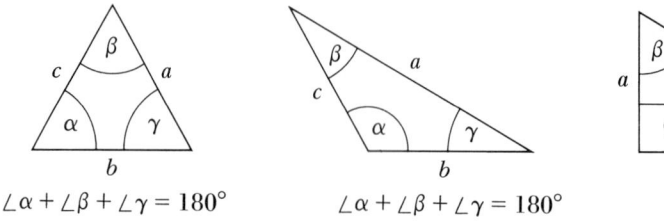

$\angle\alpha + \angle\beta + \angle\gamma = 180°$ $\angle\alpha + \angle\beta + \angle\gamma = 180°$ $c^2 = a^2 + b^2$

Figure A.2 Triangles.

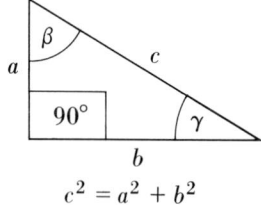

Theorem relates the sides thus: $c^2 = a^2 + b^2$, where c is the hypotenuse, or side opposite the right angle, and a and b are the mutually perpendicular sides.

5. *Powers of ten (orders of magnitude):* Scientists often compare quantities in terms of successive multiplications by ten; if we say a certain star is more massive than the sun by two orders of magnitude we mean it is 10 × 10 or 100 times as massive as the sun. When we deal with numbers involving many factors of 10 times some number, for example, astronomical distances, it is convenient to write these numbers in exponential form, or *scientific notation*. Exponents, the superscript numbers that express powers of ten, work like this: $10 \times 10 = 10^2$; $10 \times 10 \times 10 = 10^3$; 10 multipled by itself 100 times would be written 10^{100}. Any number can be written as a one-digit number plus a decimal fraction times a power of ten: 93,000,000 mi = 9.3×10^7 mi; 0.0000000000667 newton meters²/kilogram² (big "G") = 6.67×10^{-11} newton meters²/kilogram². A negative power means multiplication by 1/10 rather than by 10.

6. *Proportionality:* Most mathematical relationships in the sciences are expressed in terms of proportions. If two quantities A and B vary together in the same direction, that is, A gets larger when B gets larger and A gets smaller when B gets smaller, then A is *directly proportional* to B. The volume and temperature of a gas at constant pressure exhibit a direct proportionality. If values of V and T are plotted in Cartesian coordinates, the curve will look like Figure A.3. The curve is a straight line with a positive *slope*.

Appendix A: Mathematics

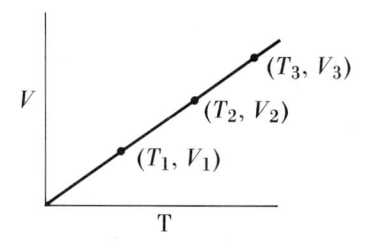

Figure A.3 Graph of a linear relationship such as Charles' Law.

"Slope" denotes the ratio of the change in the y values between two points to the change in x values between the same two points, and is often a useful quantity in relating the plotted quantities to each other. A straight line has a constant slope.

If two quantities A and B vary in opposite directions; that is, A decreases when B increases, and conversely, then A is *inversely proportional* to B. An example of inverse proportionality is the change in the volume of a gas as its pressure changes at constant temperature. If volume and pressure are plotted in Cartesian coordinates, the curve looks like Figure A.4. This curve is called a *hyperbola*. Any curve may be described by an equation. *Analytic geometry* is the study of the relation of equations to curves. Any straight line may be expressed by an equation of the form $y = mx + b$, where m is the slope of the line, and b is its y-intercept, the point at which $x = 0$ and the curve crosses the y-axis. For the curve in Figure A.3,

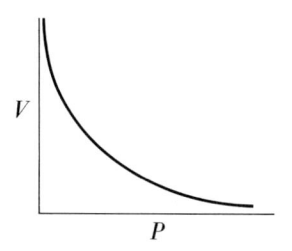

Figure A.4 Graph of a hyperbolic relationship such as Boyle's Law.

the equation becomes $V = kT$ (Charles' Law, Chap. 3) as V is plotted on the y-axis, T is plotted on the x-axis, we have previously called the slope k, and the y-intercept is zero. The curve in Figure A.4 is given by the equation for a hyperbola: $xy = k$. In our case it becomes $PV = k$ (Boyle's Law, Chap. 3). Since dividing each side by P gives $V = k/P$, we can say that if one quantity is inversely proportional to another, it is directly proportional to its reciprocal.

If one quantity is proportional to two or more quantities independently, it is proportional to the product of all those quantities. In Chapter 2 we found that the force we must exert to accelerate a mass is proportional to both mass and acceleration, so force is proportional to mass times acceleration.

As we have seen above, if two quantities are directly proportional their ratio is constant: e.g., $V/T = k$, while if they are inversely proportional, their product is constant: e.g., $PV = k$. Since an equation is more useful for making calculations than a proportionality expression, we introduce appropriate proportionality constants to make proportionalities into equations. Many "fundamental constants of the universe" were introduced into science in this manner, e.g., the universal gravitational constant, Planck's constant, and the constants in Coulomb's equations for electric and magnetic forces.

7. *Rational numbers:* Numbers that can be expressed as ratios of small integers. They can be expressed as finite decimal fractions: $1/4 = 0.25$ (exactly) or as repeating decimals: $1/3 = 0.333\ldots$; $1/7 = 0.14285714\ldots$. All numbers which cannot be expressed as ratios of integers are *irrational;* their decimal fractions do not repeat.

8. *Reciprocal:* The inverse of a number. Any number is itself divided by one; its reciprocal is one divided by the number. The reciprocal of 2 is 1/2; of 6×10^{23} is $1/(6 \times 10^{23})$.

9. (a) *Scalar:* A quantity that has only magnitude.

Mass, length, and time as dimensions of objects or duration of events have only magnitude; they are scalars. Energy is also a scalar.

(b) *Vector:* A quantity that has both magnitude and direction. Most kinematic and dynamical quantities: acceleration, velocity, force, momentum, displacement, are vectors.

10. *Trigonometric functions:* Some properties of plane triangles in addition to those mentioned in Section 4 are extremely useful in scientific calculations. Euclidean (plane) geometry shows that all triangles with the same three angles are *similar,* and their corresponding sides are in the same ratio to each other, whatever their absolute lengths. In Figure A.5, triangles ABC, $AB'C'$, and $AB''C''$ are all similar, and the ratios of the corresponding sides may be written equal:

$$\frac{AB}{AC} = \frac{AB'}{AC'} = \frac{AB''}{AC''}$$

and

$$\frac{BC}{AC} = \frac{B'C'}{AC'} = \frac{B''C''}{AC''}$$

and

$$\frac{AB}{BC} = \frac{AB'}{B'C'} = \frac{AB''}{B''C''}.$$

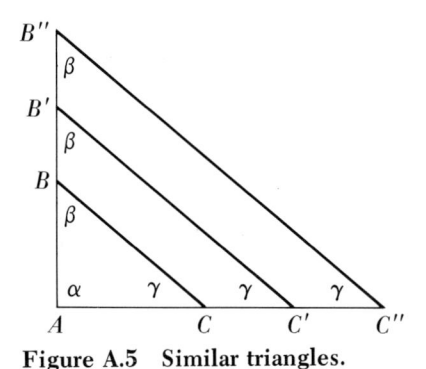

Figure A.5 Similar triangles.

This is true for any set of angles α, β, and γ; the triangles do not have to be right triangles. However, we define these ratios in a particularly useful way for right triangles in a Cartesian coordinate system, and call these ratios *trigonometric functions*. Let the arrow r, called the radius vector, form any angle α with the positive x-axis. Then r is the hypotenuse of the right triangle containing the angle α and its complement β. (Complementary angles are two angles whose sum is 90°.) The ratio of the y projection of r, line segment y, to r, y/r, is called the *sine* of angle α; the ratio x/r is called the *cosine* of α; the ratio y/x is called the *tangent* of α, and is also the slope of line segment r (Fig. A.6). Because of the ratio properties of sides of similar triangles, the sine, cosine, and tangent of a given angle are always the same, no matter in what triangle the angle occurs.

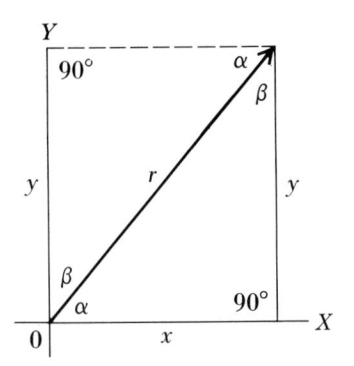

Figure A.6 Sine and cosine relationships.

Summarizing:

$$\text{sine of } \alpha \equiv \sin \alpha = y/r = \frac{\text{side opposite}}{\text{hypotenuse}};$$

$$\text{cosine of } \alpha \equiv \cos \alpha = x/r = \frac{\text{side adjacent}}{\text{hypotenuse}};$$

$$\text{tangent of } \alpha \equiv \tan \alpha = y/x = \frac{\text{side opposite}}{\text{side adjacent}}.$$

We can now state two laws by means of which, given three bits of information about a triangle—two sides and the included angle, two angles and the side opposite one, or three sides—we can find out all the other angles and sides.

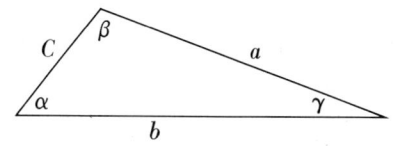

Figure A.7 Side and angle relationships for any triangle.

Law of Sines (Fig. A.7):

$$\frac{a}{\sin \alpha} = \frac{b}{\sin \beta} = \frac{c}{\sin \gamma}.$$

Law of Cosines: $c^2 = a^2 + b^2 - 2\,ab\,\cos \gamma.$

To use these laws for any triangle, it is necessary only to get the correct algebraic sign for the cosine (Fig. A.8). For any angle up to 180°, the sine is positive, but for an angle between 90° and 180°, the cosine is negative, since its x-projection lies on the negative x-axis. The sign of r is always positive.

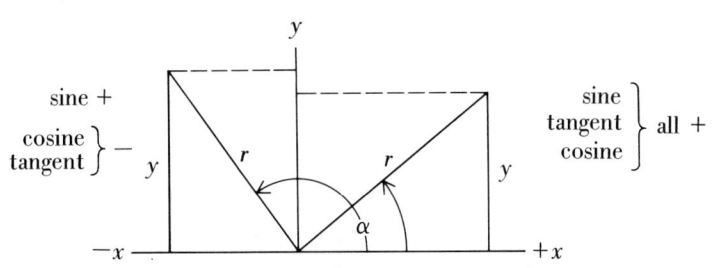

Figure A.8 Sign convention for trigonometric functions.

Algebraic Manipulations Used in the Text

1. *Law of Gravitation* (Chapter 2): Newton found that the force of gravity varies directly with the masses

of the two attracting bodies, but that it diminishes much faster than d increases. If force and distance had been merely inversely proportional, we could say F is proportional to $1/d$, but this does not express the facts accurately. Of course it could have turned out that the force was inversely proportional to distance to the 10/3 power or the 9/5 power, but fortunately for scientists, nature does follow rather simple laws, on the whole, and the force was found to be inversely proportional to the distance squared. So we write the law in mathematical shorthand as

$$F = G \frac{m_1 m_2}{d^2}$$

which says that force is directly proportional to each of the two masses, hence to their product, and inversely to the square of the distance between them (so we multiply $m_1 \times m_2$ by $1/d^2$), and then we express the constant ratio of F to $m_1 m_2 / d^2$ as G, the proportionality constant that enables us to write an easy-to-use equation.

When Cavendish wished to solve for a numerical value for G, these algebraic manipulations were necessary. To solve for a quantity we first "isolate" it; that is, we get it all by itself on one side of the equation. Given

$$F = G \frac{m_1 m_2}{d^2}$$

to isolate G requires that we divide both sides of the equation by m_1 and by m_2, and that we multiply both sides by d^2. (Whatever you do to one side of an equation you must do to the other, or the two sides are no longer equal.) Thus we get:

$$\frac{F \times d^2}{m_1 \times m_2} = G \frac{\cancel{m_1} \times \cancel{m_2}}{\cancel{d^2}} \times \cancel{d^2} \times \frac{1}{\cancel{m_1} \times \cancel{m_2}}$$

Appendix A: Mathematics

Notice that dividing by $m_1 \times m_2$ is exactly the same as multiplying by its reciprocal, $1/(m_1 \times m_2)$. Cancelling the factors that appear both above and below the fraction line on the right, we see that we have achieved our purpose. We can now switch right and left sides to get the conventional form $G = Fd^2/m_1m_2$. We are now ready to put in numerical equivalents for F, d, m_1, and m_2 and to perform the appropriate arithmetic $(F \times d \times d) \div (m_1 \times m_2)$.

The Law of Gravitation also allowed us to demonstrate the validity of Kepler's Harmonic Law. Gravity is a "central force"; that is, acceleration due to gravity is directed toward the center of mass of the two attracting bodies. Because the sun is so much more massive than any planet, the center of mass of any sun-planet pair is essentially the center of the sun at the (approximate) center of the planet's orbit. So making the rather good approximation of circular orbits, we write the force of gravity as the centripetal force on the planet:

$$G \frac{m_p m_s}{R^2} = \frac{m_p v^2}{R}$$

where m_p and m_s are the masses of the planet and the sun respectively, and R, the mean radius of the planet's orbit, is the distance between them. As the planet goes around the sun in one "year," designated as T, its orbital period, its velocity ($v = s/t$) is given by $2\pi R/T$, since the distance it travels in time $t = T$ is the circumference of its orbit, $2\pi R$. Substituting for v its equivalent, $2\pi R/T$, we get

$$G \frac{m_p m_s}{R^2} = \frac{m_p (2\pi R/T)^2}{R}$$

Squaring the factors in parenthesis and dividing out terms occurring on both sides of the equation we get

$$G \frac{m_p m_s}{R^2} = \frac{m_p \, 4\pi^2 R^2/T^2}{R}$$

Then we solve for T^2 by multiplying both sides by T and dividing by Gm_s/R^2:

$$\frac{T^2 \times \cancel{Gm_s}}{\cancel{R^2}} \atop {\cancel{Gm_s} \over \cancel{R^2}} = \left\{ \begin{array}{c} \text{multiply} \\ \text{together to} \\ \text{get the denominator.} \end{array} \right. \left[\frac{\frac{4\pi^2 R}{\cancel{T^2}} \times \cancel{T^2}}{\frac{Gm_s}{R^2}} \cdot \right\} \begin{array}{c} \text{multiply} \\ \text{together to} \\ \text{get numerator.} \end{array}$$

This gives, when we clear fractions as shown

$$T^2 = \frac{4\pi^2 R^3}{Gm_s}.$$

The constant factors $4\pi^2/Gm_s$ constitute the proportionality constant for the Harmonic Law.

Velocity Ratio Argument — Michelson-Morley Experiment (Chap. 6)

We supposed that our oarsman would row always at a constant velocity v_r with respect to the water in any direction, and that the current has a constant velocity v_c downstream. The velocities with respect to the river bank can then be expressed thus:

velocity upstream $= v_r - v_c$
velocity downstream $= v_r + v_c$
velocity cross-current (either direction) $= \sqrt{(v_r^2 - v_c^2)}$

The cross-current velocity v was calculated using the Pythagorean theorem applied to the vector diagram by which we add v_c and v_r when they are not parallel to each other. We can see that v is the same for both directions across the river (Fig. A.9)

Since $v^2 + v_c^2 = v_r^2$, $v^2 = v_r^2 - v_c^2$, and taking the square root of both sides of the equation we get $v = \sqrt{(v_r^2 - v_c^2)}$.

But we have set equal distances cross-current and upstream for our oarsman to go and return. Let us call

Cross-current velocity (v)

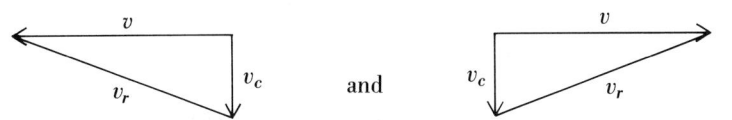

Figure A.9 Addition of velocity vectors for cross-current trip.

the one-way distance L; thus the round trip distance will be $2L$ in each case. Now we want to compare the times, $t_\perp$ for the trip cross-current and $t_\parallel$ for the upstream-downstream trip. Recalling that $v = s/t$, we can solve for $t = s/v$, by multiplying both sides of the equation by t and dividing by v. Let us next write down the expressions for $t_\perp$ and $t_\parallel$:

$$t_\perp = t_{\text{upstream}} + t_{\text{downstream}}$$

$$t_{\text{up}} = \frac{L}{v_r - v_c}; \quad t_{\text{down}} = \frac{L}{v_r + v_c};$$

hence

$$t_\perp = \frac{L}{v_r - v_c} + \frac{L}{v_r + v_c} = \frac{L(v_r + v_c) + L(v_r - v_c)}{(v_r + v_c)(v_r - v_c)} =$$

$$\frac{Lv_r + \cancel{Lv_c} + Lv_r - \cancel{Lv_c}}{v_r^2 - v_c^2} = \frac{2Lv_r}{v_r^2 - v_c^2}.$$

Also,

$$t_\parallel = \frac{2L}{\sqrt{(v_r^2 - v_c^2)}}$$

since $2L$ is the total distance and the velocity is the same both ways.

Let us finally show that for all possible velocities $t_\perp < t_\parallel$, or $t_\perp / t_\parallel < 1$:

$$\frac{t_\perp}{t_\parallel} = \frac{\frac{\cancel{2L}}{\sqrt{(v_r^2 - v_c^2)}}}{\frac{\cancel{2L}v_r}{v_r^2 - v_c^2}} = \frac{v_r^2 - v_c^2}{v_r \sqrt{(v_r^2 - v_c^2)}} = \frac{(\sqrt{(v_r^2 - v_c^2)})\cancel{^2}}{v_r(\cancel{\sqrt{(v_r^2 - v_c^2)}})}.$$

We have merely rewritten $(v_r^2 - v_c^2)$ as the square

root of itself, squared, to show how we divide out the factor $\sqrt{v_r^2 - v_c^2}$, leaving

$$\frac{t_\perp}{t_\parallel} = \frac{\sqrt{(v_r^2 - v_c^2)}}{v_r} \equiv \sqrt{\left(\frac{v_r^2 - v_c^2}{v_r^2}\right)} \equiv \sqrt{(1 - v_c^2/v_r^2)}.$$

For any possible velocities, v_c^2/v_r^2 is less than 1 and the square root of a number less than 1 is also less than 1. Note however, that a value of $v_c > v_r$ is not physically "possible." If the oarsman cannot row faster than the current he will never get upstream; $t_\parallel$ will be undefined, and the quantity $1 - v_c^2/v_r^2$ will be negative and its square root will be imaginary. So the velocity of the oarsman is the upper limit on velocities of current that can have physical meaning. Michelson and Morley chose light as their "oarsman." And its velocity also turned out to be a very important limit (see Chap. 7).

APPENDIX

B

Scientific Measurement

The metric system of units is used by scientists throughout the world. Its great advantage, in addition to its international usage, is that all the units are related to each other by powers of ten, and their names indicate this relation. Physicists and engineers, working on large-scale measurements, most often use the mks (meter-kilogram-second) units, while chemists and atomic and nuclear physicists find more use for the smaller cgs (centimeter-gram-second) units. But the two scales of units have a simple power of ten relationship to each other.

The fundamental measurable quantities of science are mass, length, and time. Time is measured in seconds in all systems of units. Tables B.1 and B.2 give the metric units of mass and length.

Mass and length units are related to each other by choosing water as the standard substance and defining the gram as the mass of one cubic centimeter of water at 15°C. The calorie was then defined as the amount of heat required to raise the temperature of this mass of water from 15°C to 16°C. The temperature must be specified because the volume of water changes with temperature.

Chemists and biological scientists make frequent use of metric volume measurements related to length units thus: one milliliter approximately equals one cubic centimeter (1 ml = 1.000027 cm^3). Thus one liter is very nearly 1000 cubic centimeters (1 liter = 1000.027 cm^3). The metric prefixes are applied to the liter to name its fractions and multiples as usual. The metric prefixes are also applied to seconds to name larger and smaller units of time.

In English-speaking nations, most practical engineers still use the British Engineering system of units, defined according to weight (rather than mass) and length. This system, made up of our commonly used units, grew up rather haphazardly, and the units bear no simple relation to each other. These are the commonly used British units:

Weight (force of gravity)
Ton = 2000 lb (T)
Pound = 1 lb (lb)
Ounce = 1/16 lb (oz)

Length
Mile = 5280 ft (mi)
Foot = 1 ft (ft)
Inch = 1/12 ft (in.)

The unit of mass is not used often in the British system, but the unit of mass consistent with pounds force and acceleration in ft/sec^2 is the *slug*. That is, according to Newton's second law, one pound of force will accelerate one slug of mass by one foot per second per second. Because the acceleration of gravity is 32 ft/sec^2, one slug weighs 32 lb by $F = mg$.

Appendix B: Scientific Measurement

Table B.1 Metric Units of Mass

Name	Number of Basic Units	Power of 10 × Basic Unit	Abbreviation
megagram°	1,000,000 grams	10^6	Mg
kilogram	1000	10^3	kg
hectogram°	100	10^2	hg
dekagram°	10	10^1	dkg
gram	1	10^0	g
decigram°	0.1	10^{-1}	dg
centigram°	0.01	10^{-2}	cg
milligram	0.001	10^{-3}	mg
microgram	0.000001	10^{-6}	μg
nanogram°	0.000000001	10^{-9}	ng
picagram°	0.000000000001	10^{-12}	pg

° Seldom used.

Table B.2 Metric Units of Length

Name	Number of Basic Units	Power of 10 × Basic Unit	Abbreviation
megameter°	1,000,000 meters	10^6	Mm
kilometer	1000	10^3	km
hectometer°	100	10^2	hm
dekameter°	10	10^1	dkm
meter	1	10^0	m
decimeter°	0.1	10^{-1}	dm
centimeter	0.01	10^{-2}	cm
millimeter	0.001	10^{-3}	mm
micrometer (micron)	0.000001	10^{-6}	μm
nanometer	0.000000001	10^{-9}	nm
picameter	0.000000000001	10^{-12}	pm

° Seldom used.

Table B.3 Derived Units — Mechanics and Heat

Quantity	mks name	mks mlt* equivalent	cgs name	cgs mlt equivalent	British name	British mlt equivalent
area		m²		cm²		ft²
volume		m³	milli-liter	cm³		ft³
density**		kg/m³		g/cm³		lb/ft³
velocity		m/sec		cm/sec		ft/sec
acceleration		m/sec²		cm/sec²		ft/sec²
force	newton (new)	kg m/sec²	dyne	g cm/sec²	pound	slug ft/sec²
momentum		kg m²/sec		g cm/sec		slug ft/sec
energy	joule	kg m²/sec²	erg	g cm²/sec²	foot pound	slug ft²/sec²
heat	calorie = 4185 joules		calorie = 4.185 joules		Btu = 778 ft lb	
power	watt = 1 joule/sec	kg m²/sec³		g cm²/sec³	horsepower = 550 ft lb/sec	

* mlt = mass, length, time
** Density in metric units is mass per unit volume. In British units it is weight per unit volume.

Table B.4 Electrical Units

A number of systems of electromagnetic units are in use, all based on metric units, but having no simple relationships among each other. This text, like most elementary physics textbooks, confines itself to using the "practical" or mks units, since this system involves the fewest and simplest conversion factors among its member units.

Symbol	Quantity	Name of Unit	Relation to others	Abbreviation
q	charge	coulomb	(defined)	coul
W	energy, work	joule	(see Table B.3)	j
V	potential	volt	joule/coulomb	V
I	current	ampere	coulomb/sec	amp
R	resistance	ohm	volt/amp	Ω
E	electric field strength		newton/coul	
p	magnetic pole strength	ampere-meter	(defined)	amp-m
B	magnetic field strength	webers*/m² or newton/amp-m		w/m²

* Weber is the unit in which "magnetic lines of force" were once counted. It is little used alone, but persists in the unit for B. A new name for webers/m², *tesla*, is coming into use.

Appendix B: Scientific Measurement   393

Table B.5 Some Useful Equivalents

1 meter	=	39.37 inches
2.54 centimeters	=	1 inch
1.6 kilometers	=	1 mile
453.6 grams	weighs	1 pound
1 kilogram	weighs	2.2 pounds
4.45 newtons	=	1 pound
252 calories	=	1 Btu
1.36 joules	=	1 ft lb

How to convert from one set of units to another:

Problem: Show that a mass density of 1 g/cm³ is equivalent to a weight density of 62.4 lb/ft³.

$$\frac{1 \cancel{g}}{\cancel{cm^3}} \times \underbrace{\frac{(2.54 \cancel{cm})^3}{\cancel{in^3}}}_{(1)} \times \underbrace{\frac{(12 \cancel{in})^3}{ft^3}}_{(2)} \times \underbrace{\frac{1 \text{ lb}}{453.6 \cancel{g}}}_{(3)} = 62.4 \text{ lb/ft}^3$$

Factor 1 changes cubic centimeters to cubic inches; factor 2 changes cubic inches into cubic feet, the desired unit in the denominator. Factor 3 changes grams to pounds, the desired unit of the numerator. Note that each factor is a fraction of the value of one, in that the numerator and denominator are equal quantities, but expressed in different units. Each time you are multiplying by "one," so the final result is equivalent to the quantity with which you started. To convert 62.4 lb/ft³ to 1 g/cm³, you would multiply 62.4 lb/ft³ by the reciprocals of each of the three factors above.

Glossary

This glossary is not comprehensive; it contains primarily terms which are important to the reader's understanding but which are used only a few times. Frequently used terms are defined in the text.

Acid: a chemical compound characterized by a sour taste, by the presence of hydrogen ions in its solution, and frequently by corrosive properties.

Adiabatic process: a process such as the expansion or compression of a gas in which no heat is added or removed from the system. Thus when a gas is compressed adiabatically its temperature will rise; when it is expanded adiabatically, its temperature will fall.

Analog computer: a computing device in which numerical data are represented by analogous, i.e., comparative, physical magnitudes or arrangements, in contrast to a digital computer that operates directly on numbers, usually in a binary system.

Analytical geometry: the branch of mathematics that relates algebra to geometry by expressing geometrical curves in terms of algebraic equations.

Base: a chemical compound characterized by the presence of hydroxide ions (OH^-) in its solution. Basic solutions frequently have a bitter taste and feel soapy or slippery to the fingers.

Chain reaction: A set of happenings in which each one triggers successive ones. A nuclear chain reaction occurs when neutrons released by the splitting of one nucleus cause the splitting of a few more nuclei, which release more neutrons that split other nuclei, and so on.

Chemical change: a process in which one or more new substances are formed from those originally present. Indicators of chemical change may be color change, heating or cooling effects, change in smell or taste.

Cirque: a scooped-out space on a mountainside produced by the grinding of grit carried by a glacier. The vertical sides of a cirque are often very straight and smooth with lateral curvature. Cirques frequently contain lakes left behind by the melting glaciers.

Cone cells: the cells in the retina of the eye that are associated primarily with color vision. The name indicates their shape, which distinguishes them from the rod cells associated with light and dark perception.

Conservation law: a fundamental statement about a physical quantity, stating that it cannot be created or destroyed. Energy, including mass, follows such a law, as do momentum, electric charge, baryons (heavy particles like neutrons and protons), and some less familiar quantities.

Corpuscular: pertaining to "little bodies." The corpuscular theory of light considered light to be made up of tiny particles, or corpuscles.

Covalent: the kind of chemical bond formed by the sharing of electrons, usually in pairs, between two atoms.

Density: the mass of a unit volume of a substance; briefly, mass per unit volume. When we speak of "light" substances we mean substances of low density; substances that are dense, i.e., of high density, we describe as "heavy."

Dielectric constant: the ratio of the electric field strength in a medium to its strength in vacuum. It is the ratio of the permittivity ϵ in the medium to ϵ_0, the permittivity of vacuum, and is also called *relative permittivity*. In general, materials of high dielectric constant are not good conductors, and the terms "dielectric," "insulator," and "nonconductor" are used somewhat synonymously.

Dipole: anything with a center of positive charge and a

center of negative charge that are not coincident. A water molecule is a dipole. A piece of copper wire moving perpendicular to a magnetic field would become a dipole. A magnet is a dipole in the magnetic sense of having a magnetic field in a definite direction associated with it.

Eccentric: off-center. To explain why the sun appeared to vary in brightness in the course of a year, the Ptolemaic system specified a circular path for the sun that was centered about a point called the *equant* point but that was eccentric with respect to the earth. These correcting devices, the equant and the eccentric, were applied to planets, also.

Endoenergetic process: one in which energy must be supplied from the environment to make it go. Boiling water by adding heat is an example.

Epicycle: a circle rotating about a point which itself revolves around some other point. The path of the point about which the epicycle rotates is called the *deferrent*. Both Ptolemy and Copernicus used epicycles and deferents to describe planetary orbits to explain retrograde motion.

Equilibrium: a state of balance; a state of no net change; a state of minimum potential energy of a system; the state of a system on which no unbalanced forces are acting.

Exoenergetic process: one in which energy is liberated; the release of heat when a fuel is burned is an example.

Extrapolation: extending a principle or explanation to cover cases not investigated experimentally, but reasonably expected to be a logical extension of those studied.

Field: a space in which forces act. Thus a gravitational field is the space in which a body exerts gravitational forces; an electric field is the space in which a charge exerts electrical forces; a magnetic field is the space in which moving charges exert magnetic forces.

Fission: the splitting of an atomic nucleus after it has absorbed a slow neutron. U^{235} and Pu^{239} are the most common fissionable nuclei. The fragments, in addition to some neutrons, are two middleweight nuclei, one usually of atomic weight about 140 and the other about 90.

Fusion: the building of larger nuclei out of smaller ones. This process goes on in stars, but has been accomplished on earth only in hydrogen bombs to date.

Gravitational mass: mass, or quantity of matter, as defined in terms of the effect of gravitational forces on it.

Half-life: the period of time during which, by completely random decay processes, one-half the total original number of radioactive nuclei will have disintegrated.

Historical geology: a branch of the study of the earth that traces the earth's history in terms of the fossil record and dates successive processes.

Homologous series: a series of organic compounds differing from each other by successive increments of —CH_2—. For example, the alkanes, CH_4, C_2H_6, C_3H_8, ... constitute a homologous series.

Hybridization: combination of two dissimilar things to produce a third having some features of each; hybrid orbitals in molecules are formed from two different atomic orbitals.

Igneous: fire formed. Igneous rocks such as granites and basalts were formed by hardening of melted minerals either as the earth cooled originally or from lava flows or other volcanic action.

Inductance: a quantity used to specify the magnetic properties of a coil of wire that becomes an electromagnet when current passes through it. The process by which a changing magnetic field causes a current to flow in a conductor is called *electromagnetic induction.*

Inertial mass: mass, or quantity of matter, as defined by Newton's second law, $F = ma$. Inertial mass and gravitational mass are identical; the theoretical reason for this is not fully understood.

Inorganic chemistry: the branch of chemistry that deals with elements and compounds of nonliving origin. Originally these were minerals and substances made from minerals in laboratories. The boundary between inorganic and organic chemistry has become very indefinite because it has become possible to synthesize not only compounds traditionally labeled one or the other, but also such "hybrids" as metal-organic compounds, unlike any occurring in nature.

Isomers: compounds having the same molecular formula, i.e., the same numbers and kinds of atoms, but different structures. For example, the empirical formula C_2H_6O represents both ethyl alcohol, C_2H_5OH, and dimethyl ether, CH_3OCH_3. There are many kinds of isomerism including optical, structural, and the kind of incidental isomerism just illustrated.

Isothermal process: a process that takes place at a constant temperature. In an isothermal expansion, heat will have to be supplied to keep the temperature constant, as the work of expansion uses up heat; in an isothermal compression, the heat produced when work is done on the gas to compress it must be removed to maintain a constant temperature.

Kinematic: having to do with the description of motion.

Kinetic: having to do with motion. *Kinetic energy* is the kind of energy a moving body possesses.

Luminosity: amount of light produced by a star; brightness. Apparent luminosity refers to how bright a star looks; intrinsic luminosity refers to the actual total amount of light radiated by the star. Luminosity is measured on a scale of magnitudes; the lower the magnitude number, the brighter the star. Rigel is a first magnitude star; Polaris is second magnitude; Sirius, the brightest star to our eyes, has an apparent magnitude of -1.4.

Macroscopic: large scale. The macroscopic world includes all those objects and phenomena easily observed with the naked eye, and also the large, if distant, objects of the universe.

Mean life: average length of life. *Mean* simply means *average*.

Mensuration: measurement, or the process of measurement.

Metamorphism: change or the process of change. Metamorphic rocks are those that have been changed in structure or content by extremely high pressures and temperatures within the earth's crust.

Microscopic: very small scale. The microscopic world includes all those objects and phenomena too small to be seen by the naked eye. Atomic phenomena cannot even be seen by microscopes.

Model: a representation of what reality is thought to be like. The physical sciences use many models. Some, like Faraday's model of electric and magnetic fields, may be pictured in drawings; others, like Heisenberg's model of the atom, are purely mathematical. A model does not claim to picture reality, but to provide an analogy in terms of which some aspect of reality may be understood.

Molecular biology: the study of life in terms of the physical and chemical processes within living cells.

Momentum: the product of the mass of a body and its velocity. Einstein chose to consider momentum as the fundamentally conserved physical quantity, and out of that choice grew the concept of mass as a function of velocity (i.e., mass changes with velocity) and the concept of the equivalency of mass and energy.

Monomer: a small molecule that can combine with itself or other small molecules to make the large molecules we call *polymers*. For example $H_2C=CH_2$, ethylene, is the monomer from which the giant molecule polyethylene is made.

Moraine: a low ridge or hill made of rock and other debris left behind at the sides and forward edge of a melting glacier. It is made of materials gathered up and carried along by the moving ice.

Nitrogen base: a nitrogen-containing organic compound that acts chemically as a base by absorbing hydrogen ions. The most famous nitrogen bases are the four that make up the genetic code. See Chapter 11.

Noble gas: a term applied to the gaseous elements of Group VIII A of the Periodic Table. "Noble" is an old-fashioned term for *unreactive* or *inert*. Gold, silver, and platinum were called "noble metals" because they do not rust or tarnish easily; that is, they are to some extent chemically inert.

Nuclear reaction: a change or interaction involving the nuclei of atoms rather than their electrons. A nuclear reaction commonly involves the absorption of a particle by a nucleus and emission of a different particle, leaving a new nucleus, e.g., $_7N^{15} + {_1}H^1 \rightarrow {_6}C^{12} + {_2}He^4$, which may be more simply written as $_7N^{15}$ (p, α) $_6C^{12}$.

Organelle: "little organ"; a term applied to structures within living cells that have particular functions. Some organelles may have evolved from free-living organisms that took up symbiotic (cooperative) residence in other simple cells.

Organic chemistry: originally, the chemistry of substances found in living things, or made by them; now, the chemistry of carbon compounds other than mineral carbonates. There are so many borderline cases that the boundary between organic and inorganic chemistry is very indefinite.

Oscillator: a device that vibrates back and forth about some average or equilibrium position. A diatomic molecule like O_2 has a vibrational motion somewhat like that of two weights on the opposite ends of a spring. In fact, oscillation occurs along any chemical bond in any molecule. If a charged particle oscillates, it emits electromagnetic waves.

Periodic motion: motion that repeats itself in a definite period of time. The time for one cycle or round trip is the period; the reciprocal of the period is the frequency of the motion. Examples of periodic motion are a swinging pendulum, a vibrating spring, and a vibrating molecule.

Perpetual motion: applied to machines, the term implies a device or process that goes on forever without further input of energy once it is started. A perpetual motion machine of the "first kind" requires the creation of energy, and so violates the First Law of Thermodynamics. A perpetual motion machine of the "second kind" depends on the 100% efficiency of a cyclic process, or otherwise violates the Second Law of Thermodynamics. Thus it is impossible to make a device that does work without consuming energy. On an atomic scale, however, there is perpetual motion of orbiting electrons, but it cannot be harnessed to do work.

Photon: a quantum of radiant energy. The term emphasizes the particle nature of radiant energy, and describes a particle of zero rest mass and velocity c.

Photosynthesis: the many-stepped process by which green plants utilize the sun's energy to transform carbon dioxide and water into carbohydrates. Chlorophyll is the essential catalyst. The reaction may be summarized: $6CO_2 + 6H_2O \rightarrow C_6H_{12}O_6 + 6O_2$, but many complex steps are involved, and the process has never been carried out synthetically by man.

Physical change: a change in size, shape, state, or other physical property of matter that leaves its chemical constitution unchanged.

Physical geology: also, structural geology; the study of the formations and mineral composition of the earth's surface and the processes that produced them.

Plane geometry: that branch of mathematics that deals with the properties of lines and curves in two dimensions.

Polymer: a large molecule made by the chemical combination of many smaller molecules. Polyethylene and teflon are polymers each made of many thousands of the same monomeric units, C_2H_4 and C_2F_4, respectively. Proteins are polymers of up to twenty different monomeric units, the amino acids, each of which may recur many times and in many orders.

Postulate: a statement given as true and used as the base on which a theory is built. Postulates are usually not provable, but if they are assumed to be true, then the rest of the theoretical structure follows.

Potential energy: energy due to place or position; stored energy. Examples are the energy of a pendulum at one end of its swing, the energy of a rock on the edge of a cliff, the energy in a dry cell not in use, and the energy in a lump of coal.

Pressure: the force on a unit area of a surface. It is usually measured in pounds per square inch, newtons per square meter, or atmospheres, taking the standard sea level pressure of the earth's atmosphere to be 14.7 lb/in^2. Pressure may be due to gravitational forces, to collisions of rapidly moving molecules, or to radiation, i.e., the pressure of photons.

Prism: a three-dimensional figure normally having one axis longer than the other two. The cross-section perpendicular to the long axis can be any polygon, but triangular prisms are most common. Light entering one face of a triangular prism will be refracted both at its entrance point and as it leaves the adjacent face, with the result that each color in the light beam will be bent through some characteristic angle. Thus a prism disperses a beam of white light that passes through it into a spectrum of colors.

Quantization: the property or process of being "quantized," that is of being restricted to only certain locations or values. To say electron orbitals are quantized is to say that there is a very high probability of the electron being in a certain location and a near zero probability of it being elsewhere. To say an electron's angular momentum is quantized is to say that the product of its mass, velocity, and orbital radius may have only specified values.

Quantum number: one of a set of numbers, defined by quantum theory, that describes the state of an electron in an atom. Each electron in an atom is described by a set of four quantum numbers, which specify (1) its orbit, (2) its subshell within the orbit, (3) the orientation of its subshell in space, and (4) its spin.

Radiant energy: a term applied to electromagnetic energy because its transmission can be described in terms of waves moving out in all directions from a point source, i.e., radiating.

Science: as used primarily in this text, the study of the world in terms of causes and effects; the seeking, by a combination of experimentation and reasoning, of a rational understanding of phenomena. See also *technology.*

Sedimentary rocks: rocks formed from mineral grains and chunks deposited under water and later pressed and heated into rock by the accumulation of more and more layers of similar materials on top.

Selection rule: a term used in quantum mechanics to specify what values certain quantum numbers may have in relation to the others. For example, since an electron can spin only clockwise or counterclockwise, represented by $s = +\frac{1}{2}$ or $s = -\frac{1}{2}$, respectively, the selection rule for the spin quantum number s is $s = \pm\frac{1}{2}$. Spectroscopists also use the term to specify what the "allowed transitions" of electrons in an atom or the allowed modes of motion of atoms in a molecule are.

Solution: a homogeneous mixture of molecules. We are most familiar with the solutions of substances in water. The dissolved material (solute) is dispersed uniformly in the dissolving medium (solvent) in molecule- or ion-sized particles.

Spectrum: a spread-out array. In physics it usually refers to all the wavelengths of light spread out in order, i.e., the visible spectrum, or all wavelengths of electromagnetic radiation, i.e., the electromagnetic spectrum.

Symmetry: balance; a condition of proportion, fitness, or "oughtness." There are many kinds of symmetry of form, shape, and design, and also the kind of symmetry of behavior of natural phenomena that made Faraday expect to be able to produce electric current from magnetism, as current produces a magnetic field.

Technology: applied science; the inventions, machines, materials, and processes through which science touches most people's lives are the products of technology. Technological advances rest on the discoveries of "science-for-its-own-sake," but basic research in science also benefits from the inventions of technology. Neither can get along without the other.

Theoretical science: science for its own sake; "pure" science, without consideration of its applications; basic or fundamental research; search for understanding of phenomena for its own sake. More narrowly, each science has its theoretical aspects in contrast to its experimental aspects. It is not recorded that Einstein ever did an original experiment; his research was done with pencil and paper and mathematics. He was a theoretical physicist. Linus Pauling is one of the greatest living theoretical chemists. The thinking of men like these must be tested by the experimentalists; they rarely do the experiments themselves.

Time dilation: the stretching out of a time interval or slowing down of the passage of time for systems traveling at velocities near the velocity of light.

Trigonometry: the branch of mathematics that deals with triangles and angles and their functions (sine, cosine, tangent, and their reciprocals). It is extremely useful in making indirect measurements of distances, since the Law of Sines and the Law of Cosines may be applied to the measurement of astronomical distances from a base line within the solar system. (See Mathematical Appendix.)

Vibrator: something that stretches and contracts with a regular frequency, i.e., undergoes a back-and-forth periodic motion. An atom or molecule moving back and forth about its equilibrium position in its crystal lattice is a vibrator; so is the prong of a tuning fork or a spring with a weight on it, moving up and down. To start vibrating, a vibrator must acquire energy; it gives up energy, most often as heat, as it slows to a stop.

Vulcanism: any manifestation of igneous (fire-related) activity at the earth's surface; lava flows, volcanic eruptions, upwellings of molten materials through faults.

Wave: a means of transmitting energy by the periodic motion of a medium. The disturbance carrying the energy is propagated in some (or many) directions, but the medium itself oscillates or vibrates only around an equilibrium position. Thus a wave in a material medium, such as sound in air, moves forward, while the air molecules are compressed and expand about their "rest" position. For example, a vibrating guitar string pushes adjacent air molecules together; they spring apart, crowding the molecules on either side of them together. These also spring apart, compressing their next neighbors. Thus the disturbance is transmitted across the room, although no single air molecule is likely to go that far. For electromagnetic waves, the medium is the electric and magnetic field pervading space, and varying according to the frequency of the wave it is transmitting.

Work: in physics, the product of a force and the distance through which it acts. Work is a form of energy. Since when work is done one form of energy is changed into another, work has been described as energy in transit.

INDEX

Absolute zero:
  Carnot cycle, 69
  gas volume, 62
  temperature scale, 65, 69
Acceleration:
  centripetal, 39
  definition, 30
  due to gravity, 36
  general relativity, 149, 151
  law of, 31
  Newton's Second Law, 32
  periodic motion, 41
Accelerators, 270–274
Acid amide, 237
Acids, carboxylic, 212–214
  ionization, 214
Acids, mineral:
  binary, 194
  formation, 193
  ionization, 193
Adenine, 243, 250
Adiabatic processes, 68
Air resistance, 21, 22
Alchemy, 171
Alcohols, 209–210
Aldehydes, 211
Alexander the Great, 8
Algebra, 14
Alicyclic compounds, 218
Aliphatic compounds, 218
Alkanes, 204–207
Alkenes, 207–208
Alkyl groups, 206
Alkynes, 207

Alpha rays, 131, 155, 258–259, 261, 267, 276
Alternating current, 88–91
Amines, 215–217
Amino acids, 216–217
  essential, 238
  origin, 332
  proteins, 237
  recognition by RNA, 247–248
Ampère, 94
Amphibians, 345
Analytical geometry, 14
Anaximander, 8
Andromeda, 295
Angiosperms, 352–353
Angle of rotation, of polarized light, 233
Angular momentum, quantization of, 158
Annihilation, positron and electron, 269
Anticyclone, 361
Antiparticles, 269
Arabs, 5, 14
Archimedes, 13
Argon dating, 261
Aristarchus of Samos, 9, 10
Aristotle, 7, 12, 15
  ideas of motion, 20, 21
Aromatic compounds, 219–223
Arthropods, 345
Asteroids (planetoids), 293
Aston, F. J., 266
Asymmetric carbon atoms, 231–233
Atmosphere:
  changed by photosynthesis, 340

405

Atmosphere (*Continued*):
  planetary, 326
  primitive, 324-326, 329-330
Atom:
  chemistry, 175
  Democritus, 13, 51
  parts, 258
  structure, 154-156
Atomic models:
  Thomson, 154-155
  Rutherford, 156
  Bohr, 157-158
  mathematical, 163
Atomic number, 160, 177
Atomic pile, 278
Atomic weight, 175, 187, 197
ATP (adenosine triphosphate), 250, 251
Avogadro, 64
  hypothesis, 175
  number, 65
Azoic Era, 333

Babylonians, 4, 16
Bacon, Roger, 15
Baryons, 275
Bases, 195
Becquerel, 130-131
Benzene, 219-221
Berzelius, 202
Beta rays, 131, 154, 258-259, 261, 276
Bethe, Hans, 298-299
Big Bang theory, 297-298
Birds, 349
Blackbody radiation, 114, 124-128
  model of stars, 305
"Black hole," 314
Black, Joseph, 57
Blue shift, Doppler, 306-307
Bohr atom, 156-159, 167
Bohr, Niels, 156, 167, 258
Bondi, Hermann, 297
Born, Max, 164
Boyle, Robert, 60
  Law, 60
Brahe, Tycho, 25
Breeder reactors, 276, 280-281
Bruno, Giordano, 15
Brushes, a.c. generators, 90
Bubble chamber, 262-264

Bunsen, line spectra, 157
Buoyancy, Archimedes principle, 13
Buridan, 15
Burning, as a chemical reaction, 172

Calendars, 4, 286
Caloric fluid, 51
Calorie, 54
Cambrian Period:
  life in, 340, 342
  strata, 339
Camera, as astronomical tool, 300
Capacitor, 83
Carbohydrates, 230-236
Carbon bonds, 191, 203-204
Carbon-14 dating, 261
Carboniferous Period, 345-347
  giant insects, 347
Carboxylic acids, 212-214
Carnot, Sadi, 66
  cycle, 67-69
Cartesian coordinates, 137-138, 376-377
Catalyst, 236
  in body, 254
Catastrophism, 332
Causality, in classical physics, 118
Cause-and-effect, in science, 2, 3
Cavendish, Henry, 36, 37, 77
Cavendish Laboratory, 241
Celestial spheres, 9, 20
Cell, living:
  components, 241
  development, 331-332
  membrane, 240
Cellulose, 236
Celsius, 54-55
Cenozoic Era, 352-354
Centigrade, 54
Central Dogma, of molecular biology, 246
Central forces, 26, 29
Centrifugal force, 40
Cepheid variables, 304
Chadwick, James, 187
Chain reaction, 277-278
Charles, Jacques Alexander, 61
  Law, 61
Chemical bonds, 52, 180-182, 197
  covalent, 180

Index

ionic, 181
polar, 180
Chemical equations, 190–191
Chemical evolution, 331–332
Chemical formulas, 189–190
Chemical reactions, 197
  in voltaic cells, 86–88
Chemical stability and electron structure, 160, 178–180
Chemical symbols, 189–190
Chemistry:
  evolution from alchemy 171
  laws of, 174, 196
  reactions of inorganic type, 192–193
China, medieval, science in, 15
Chromosomes, 241
Church:
  opponent of science, 23
  repository of learning, 13
Circuits, 95
Classical mechanics, 29
Classical physics, failure of, 119, 132
Climate:
  effect on life, 337, 345
  topography, 357, 362
  water, properties of, 59
  weather, 357, 362
Cloud chamber, 262–263
Clouds of Magellan, 304
Coal, 345–346
Cockcroft-Walton generator, 270–271
Coelacanths, 345
Color:
  complementary, 108
  electron structure effect, 186–187
  primary, 107–108
  summary, 115
  vision, 107
Combining Volumes, Law of, 175
Comets, 290–292
  as meteorites, 294
Communication, in science, 4
Compass, 15, 76–77
Complementary bases, and pairing in DNA, 246
Completion, in chemical reactions, 196
Compounds, chemical, 52, 174
Condensation reaction:
  in ADP-ATP, 250–252
  in sugars, 235
Condenser, electric, 75

Conductors, of electricity, 75, 79
Conservation Law (*see also* Law of Conservation of Energy):
  applied to circuits, 84
  maximum temperature, 64
Conservation of matter, in chemistry, 190
Constant, definition, 377
Constellations, 14, 286–289, 319
Continental drift, 336–337, 351, 355
Continuity vs. discontinuity:
  energy, 127–128
  matter, 51
  motion, 19
Continuous creation, 297
Control mechanisms, for nuclear reactors, 278–279
Conventional current, 94
Coordinates, 147
Copenhagen Interpretation, of quantum mechanics, 167
Copernicus, 23
  world system, 13, 23
Cosmic rays, 275
Coulomb, Laws, 77
Counters, 265–266
Crab Nebula, 314
Creation, mechanism of, 227
Crick, Francis, 241
Crossopterygians, 344–345
Crusades, 15
Crustal plates, 337
Curie, Marie, 131
Cycle (vibration, oscillation), 41
Cyclone, 361
Cyclotron, 143, 272–273
Cytosine, 243

Dalton, atomic theory, 52, 132, 175
Dark Ages, 15, 16
Darwin, Charles, 332
Dating, radioactive, 261
Davisson and Germer, electron diffraction, 163
de Broglie, Louis, 161–162
Decomposition reactions, 192
Definite Proportions, Law of, 174
Democritus, 13, 51
De Revolutionibus, 23
Derived quantities, 30, 392
Detectors of radiation, 262–265

Determinism, in classical physics, 118
Devonian Period, 344–345
Diastrophism, 335
Dielectric constant, 79
Diffraction:
  electrons, 163
  light, 109–111
Digestion:
  carbohydrates, 235–236
  fats, 229–230
  protein, 238
Dinosaurs, 348–350
  extinction, 351
Dirac, P. A. M., 167, 269
Direct combination reactions, 192
Direct current, 88, 94
Dispersion, Newton's explanation, 100
Displacement, 30
Displacement reactions, 192–193
Distance, 29
DNA:
  origin, 332
  structure, 242–244
Döbereiner's triads, 173
Domains, magnetic, 186
Donn, William L., 355
Doppler effect, and star motions, 306
Double helix, of DNA, 245–246
Dry cell, 86, 89
Dualism, of matter and energy, 163–167

Earth:
  age, 332–333
  density, 326–327
  distance to sun and moon, 9–10
  geologic processes, 363
  interior, 327–329
  life development, 362–363
  magnetic field, 76, 329
  mass, 328
  shape, 8–9
  size, 11
  summary, 362–363
Earthquakes, 328–329
Eclipse, lunar, 8
Ecliptic, 287
Edison battery, 86
Efficiency, of heat engine, 66–69
Egyptians, early science of, 4, 16
Einstein, 128
  general relativity, 148
  photoelectric equation, 138
  special relativity, 140
Eiseley, Loren, 228
Electric charges, 75, 94
Electric current, 83, 84
Electricity:
  form of energy, 74
  relation to magnetism, 77
  static, 75
Electric permittivity, 79
Electrodes, 87
Electrolyte, 87
Electromagnetic spectrum, 115, 319
Electromagnetism:
  classical, 78
  relativistic view, 138–139
Electromotive force (emf), 88
Electronegativity, and periodic table, 182
Electronics, 93, 94, 95
Electron pumps, 92
Electrons, 74, 154
  diffraction, 163
Electron structure, of atoms,
  explanation of properties by, 197
Electron volt, as energy unit, 270
Electroscope, Lauritsen, 262
Elementary particles, 274–275
Elements, chemical, 52, 172–173, 176–188
  abundance, 324–326
  formation, 297–298, 314
Ellipse, as orbital shape, 25
Empirical laws, 71
Endoceras proteiforme, 343
Endoplasmic reticulum, 247
Energy:
  conservation, 46, 71
  definition, 45
  degradation, 46
  equivalence to mass, 144, 151
  forms, 45, 49, 71
  kinetic, 46, 49
  organisms, 249–250, 254
  potential, 45
  radiant, 125, 128, 130
Engineer, Archimedes, 13
Entropy, 70
Enzymes, 236
Eocene epoch, 352

Epicycles:
  Ptolemaic, 12
  Copernican, 13, 24
Equal areas, Kepler's Second Law, 25
Equation of state for gases, 64–65
Equilibrium:
  chemical, 195, 214–215
  definition, 42
  living systems, 253, 331
  mechanics, 31, 32
  stars, 311
Equivalence Principle, in General Theory of Relativity, 148–149, 151
Eratosthenes, 11
Esterification, 217
Esters, 217–218
Ether:
  luminiferous, 112, 120
  wind, 121
Ethers, 210–211
Euclid, 8
Evolution:
  biological, 332
  galactic, 315–320
  stellar, 310–315
Ewing, Maurice, 355
Exclusion Principle (Pauli), 161
Expanding universe, 306–307
Expansion, and molecular motion, 53
Experimental approach, beginnings, 20
Extrapolation, to ideal conditions, 22, 30, 56

Fahrenheit, 54
Falling, explained by gravitation, 26, 29
Fallout, 277–278
Faraday, M., field model, 81
Fats, 229–230
Faults, in earth, 328
Fermi, Enrico, 278
Field:
  electromagnetic, 80–81, 94
  gravitational, 80
First Law of Thermodynamics (conservation law), 70
Fission, 277–281
  fragments, 277–278
FitzGerald, contraction, 124
Fizeau, H. L., and determination of c, 99

Fluorescence, 130
Food, 228, 253
Force:
  action-reaction pairs, 34, 35
  Aristotle's view, 20
  at-a-distance, 80
  cause of acceleration, 31
  centripetal, 38
  definition, 30
  electric and magnetic, 75–78, 94
  equilibrium, 35
  nuclear, 259
  units, 33
Fossils:
  dating, 333
  oldest, 339
Fourth dimension (time), 147
Frames of reference:
  absolute, 120
  equivalence, 142
  inertial, 140
Franklin, Benjamin, discovery of electricity, 75
Frequency, of periodic motion, 41
Friction, 22, 31, 50
Fulton, Robert, 66
Functional groups, 208
Fundamental measurable quantities, 30
Fusion:
  processes in stars, 298–299, 317
  reactors, 369–370

G (universal gravitational constant), 28, 37
Galaxy, 295, 315–318
  elliptical, 315–316
  evolution, 315–318
  irregular, 315
  rotation, 305
  spiral, 315–316
  types, 315
Galileo, 21, 42
  periodic motion, 40–41
  telescope, 24, 98
  thermometer, 54
  velocity of light, 21, 97–98
Galle, Johann Gottfried, 119
Galvani, 85
Gamma rays, 131, 258–259, 282
Gamow, George, 297

Gases:
    ideal, and kinetic theory, 55
    properties, 59-65
    real, 56
Gas Laws, 59-65
Geiger-Müller counters, 265-266
Gell-Mann, Murray, 274
General Theory of Relativity, 148-150
Generators, 88-91, 94
Genes, 241, 248
Genetic code, 247, 254
Genetic damage, from radioactive materials, 278, 281-282
Geological change, and biological evolution, 341
Geology:
    historical, 333
    physical, 333
Geosynclines, 335
Gilbert, William, and magnetism, 76
Glaciers, 357
Glaser, Donald, 262
Glucose, as fuel, 249-250, 252
Goeppert-Mayer, Maria, 367
Gold, Thomas, 297
Gondwanaland, 337, 351
Gravitation:
    centripetal force, 40
    geometric property of space, 150
    law, origin of, 27-28
    third law of motion, 34
Gravitational collapse, 311, 323, 326
Gravitational potential energy, and heating of stars, 298
Gravitation, Law of:
    and Harmonic Law, 29, 40, 383-385
    mass of stars, 303
    origin, 27-28
    understanding the universe, 289-290
Greeks:
    discovery of electricity, 74
    early science, 6, 16
Grosseteste, Robert, 15
Guanine, 243
Gutenberg, 16

Half-life, 260-261
    from particle tracks, 263, 265
Halley, Edmund:
    comet, 291
    proper motion of stars, 305

Halogenated hydrocarbons, 208-209
Hansen, H. M., 156
Harmonic Law:
    gravitation, 29, 40, 383-386
    Kepler's Third Law, 25
Heat:
    direction of flow, 50
    effect on matter, 52-54
    from electricity, 92
    from friction, 50
    kinetic energy, 53, 71
    measurement, 53, 58
    mechanical work, 51
Heat engine:
    efficiency, 66-69
    ideal, 67
Heisenberg, Werner:
    atomic model, 163
    uncertainty principle, 164-167
Hertz, Heinrich, radio waves, 112
Hertzsprung, Ejnar, 308
Hexoses, isomeric, 231
High energy bonds, 250
High energy physics, 367
Highs (weather systems), 361
Hipparchus of Nicea, 11
Hippocrates, 8
Holocene epoch, 352
Homologous series, 205
Hooke, Robert, wave theory of light, 100
Hoyle, Fred, 297
H-R diagram, 308-309
Hurricane, 361
Hutton, James, 332
Huygens, Christian:
    principle, 110
    wave theory of light, 100
Hyades, 289
Hydrocarbons, 204-208
Hydrogen bond, 244-245
Hydrogen, spectrum, 157-158
Hydrolysis:
    fats, 238
    polysaccharides, 236
    protein, 238
Hyperbola, Boyle's Law, 61-62

Ice ages:
    Permian, 347
    Pleistocene, 355

Ideal gas:
  in Carnot cycle, 67
  model, 63
Igneous rocks, 333
Impetus, 45
Impulse, 33
Indeterminism, in atomic systems, 164–165
Index fossils, 339, 343, 345
Industrial Revolution, 118
Inertia, 22, 31
Insulator, 79
Interval, invariant in Special Theory of Relativity, 147
Invariants:
  classical, 138
  special relativity, 151
Inverse square law of light intensity, 304
Ionic reactions, 193
Ions, 181
Irrational numbers, 7
Islands, volcanic origins, 335
Isomers, 206
Isostasy, 334, 335
Isothermal processes, 67–68
Isotopes, 187
  dating methods, 261
  radioactive, 259

Joliot-Curie, Irène and Frédéric, 268
Joule, unit of work, 82, 85
Jupiter:
  atmosphere, 326
  moons, 24
  velocity of light, 98–99

K-capture, 259
Kekulé von Stradonitz, F. A., 220
Kelvin, Lord, temperature scale, 65, 68–69
Kepler, Johannes, 25
  laws of planetary motion, 25, 42, 289
Ketones, 211–212
Kinetic energy, and heat, 53
Kinetic Theory of Gases, 55
Kirchhoff, identification of elements by spectra, 157
Kolbe, A. H. H., 202, 227

Land, colonization of, 345

La Place, Pierre Simon, 119
Laser, 369
Latent heat, 57
Laurasia, 337, 351
Lava flow, 328
Lavoisier, Antoine:
  "father of chemistry," 72
  quantitative experiments, 172
Lawrence, E. O., 272
Laws of impotence, 72
Laws, scientific, nature of, 47
Lead storage battery, 86, 89
Leavitt, Henrietta, 304
Length, dependence on velocity, 141, 142
Lenses, 105–106
Leptons, 275
Leverrier, U. J. J., 119
Life:
  mysteries remaining, 368
  origin, 228
Life expectancy, of stars, 311
Light:
  absorption, 106
  colors, 106–109
  diffraction, 109
  rectilinear propagation, 100
  reflection, 101
  refraction, 101
  source of astronomical information, 300
  summary of behavior, 115
  theories, 97, 109, 129–130
Lightning:
  electricity, 75
  energy for chemical evolution, 332
Linac, 272
Linear accelerators, 272
Line spectra, 157–158, 167
Lippershey, H., 98
"Little g," 36
Lorentz-FitzGerald contraction, 142
Lorentz, H. A., 140
  transformations, 141
Lowell, Percival, 290
Lows (weather systems), 361
Lucretius, 13, 51

Mach, Ernst, principle and General Theory of Relativity, 149, 151

Magic, 2
Magma, 334
Magnetism:
 and electron structure, 185–186
 field, 77, 80–81
 properties, 76
 relation to electricity, 77, 92
Main sequence of stars, 309, 311
Mammals, evolution of, 351–352
Man, evolution of, 353–354
Mars, Brahe's observations of, 25
Mass:
 defect, 277
 definition, 29
 dependence on velocity, 141, 143
 equivalence to energy, 144, 151
 gravitational, 28
  compared to inertial, 148
 relation to stellar evolution, 317
Mass-energy transformation, 277
Mass-luminosity relationship, 303, 309
Mass spectrograph, 266–267
Mathematics, early history of, 5
Maxwell, J. C.:
 electromagnetic spectrum, 111
 electromagnetic theory, 138–140
 field concept, 80
 future of physics, 119
Mechanical energy, 71
Mechanical work and heat, 51
Mechanics, beginnings of, 20
Mendeleev, Dmitri:
 periodic table, 159, 176
 properties of ekasilicon, 177
Mensuration (ancient), 5
Mercaptans, 215
Mesons, 275
 muon, 269
Metamorphic rock, 336
Meteorites, 294
Meteors (shooting stars), 294
Metric system, units in, 389–393
Michelson, A. A., and determination of c, 99–100
Michelson interferometer, 123
Michelson-Morley experiment, 114, 120–124, 132
 mathematical argument, 386–388
Mid-Atlantic Ridge, 335, 337
Middle Ages, science in, 15
Milky Way, 295, 296

Millikan, Robert A., and electronic charge, 154
Minerals, in food, 238
Miocene epoch, 352, 354
Mirrors, image formation by, 102–104
Mitochondria, 241, 249
Models of gases, 56
Moderator, 278–279
Mohole, project, 327
Mohorovicic, 327
Mole, gram molecular weight, 64
Molecular orbitals, 204
Molecules, 52, 71–72, 175
 in space, 367
Momentum, 33
 conservation of, 144
Monitoring, of radiation levels, 281
Monomers, 223
Monosaccharides, 231, 233–234
Moon:
 as seen by Galileo, 24
 exploration of, 368
Morgan, Thomas Hunt, 241
Morley, E. W., 120
Moseley, H. G. J., 160
Moslem science, 14
Motion:
 cause, 20
 circular, 38
 early theories, 19
 falling:
  Aristotle, 20
  Galileo, 21
 planets and moons, 29
 solar system, 23
Motor, 94
Multiple Proportions, Law of, 174
Muons, mean life and time dilation, 144
Mutagens, 282
Mutations, radiation induced, 281–282

Natural selection, 332, 353
Neptune, discovery of, 119, 290
Neptunium, 276
Nervous system, evolution of, 344
Neutralization, of acids by bases, 196
Neutrinos, 268, 275
Neutron, 188
Neutron stars, 313
Newcomen, Thomas, 66

Newlands, John A. R., octaves of elements, 173
Newton, Isaac:
  calculus, 5
  central forces, 26
  motion, 23
    laws, 29, 42
  theory of light, 100
Nitrogen bases:
  origin, 332
  structure, 242-243
Normal distribution, 55
Nova, 312-313
Nuclear power:
  dangers, 282-283
  future, 369-370
Nuclear technology, summary of, 283-284
Nucleons, 188
Nucleotides, 246
Nucleus, atomic, 258
  stability, 188
  structure, 367
Nucleus, cell, 240-241

Observation, of natural phenomena, 4
  and Ptolemaic system, 12
Occam's Razor, 12
Oersted, H. C., 76
Ohm, G. S., Law, 84, 95
Ohm, unit, 84
Oligocene epoch, 352, 353
Operational definition, 32, 167
Optical activity, 232-233
Optics, 97
  geometric, 100-106
Ordovician Period, 339, 342-344
Oresme, Nicholas, 15
Organic compounds:
  sources, 201
  summary, 223-224
  synthesis, 201-203
Organism:
  improbability, 253
  needs, 228
Origin of life, mechanism of, 331-333
Orogenic processes, 334-336, 350-351
Oscillation, 41
Osiander, 23

Oxidation, in biological systems, 252-253
Ozone, 340

Paleocene epoch, 352, 353
Paleozoic Era, 341-347
Pangaea, 351
Parallel circuit, 92
Particle physics, 367
Pasteur, 227
  optically active substances, 232
Pauli, Wolfgang, exclusion principle of, 161
Pauling, Linus, 239, 241
Penicillin, mechanism, 240
Pentoses, 231
Peptide linkage, 237
Periodic Law of the elements, 197
Periodic motion, 41-42
  energy conservation, 47-48
  period, 41
Periodic table, 159-160
  and electron structure, 178
Period-Luminosity relation, 304
Perkin, William Henry, 202
Permian Period, 347
Perpetual motion:
  kinds, 67
  machines, 67
Phlogiston, 172
Photoelectric effect, 114
  and quantum theory, 128-130
  use in counters, 266
Photographic emulsions, as radiation detectors, 263
Photosynthesis, 253, 340
Planck, Max, 114, 127
Planetoid (asteroid), 293
Plasma, 297
Plato, 7
Pleiades, 288
Pleistocene epoch, 352, 354, 355
Pliocene epoch, 352, 354
Pluto, discovery of, 290
Plutonium, 276, 277, 280-281
Poincaré, Jules Henri, 140
Polarimeter, 232
Polaris, 287, 289
Polarization of light:
  methods, 112-113, 115-116
  rotation of plane, 231-233

Polar molecules, 180, 194
Pole:
  direction of field, 186
  magnetic, 76
Pollution, reversibility, 71
Polonium, discovery of, 131
Polymerization, 208
Polymers, 208
  fibers, 223
  plastics, 223
  sugars, 235–236
Polysaccharides, 235–236
Polysomes, 247
Population I stars, 295
Population II stars, 296
Positron, 269
Potential, 81–83
  difference, 83
Pre-Cambrian era, 339–340, 341
Pressure of gases, 56, 60
  relation to temperature, 61
Priestley, Joseph, 172
Primary cell, 86
Probability, 70
Proper motion of stars, 305
Proportionality, 28, 378–380
Proteins:
  amino acids, 217
  enzymes, 236
  formation, 246–248
  natural polymers, 208
  structure, 237
Protoearth, 326
Protons, 74, 268
Protoplasm, 59, 240
Prout's hypothesis, 159, 175
Ptolemy, 12
Pulsars, 312, 313, 367–368
Pulsating universe, 298
Pulsation, in stars, 311
Punctuation, in genetic code, 248
Pure substances, identification of, 171–173
Purines, 243
Pyrimidines, 243
Pythagoras, 5, 7

Quantitative basis for chemistry, 172
Quantization of energy emission, 128, 132
  and line spectra, 158
Quantum mechanics, 164, 167–168
Quantum numbers, 161
Quantum theory, 114
  atomic structure, 167
Quark, 275
Quasars, 318–319, 367–378
Quaternary Period, 352

Radiation, effects on living cells, 281–282
Radiationless states of atoms, 156, 158
Radiation pressure, in stars, 304–311
Radicals, 194
Radioactivity:
  artificial, 268, 283
  decay chains, 275–276
  discovery, 131–132
  emissions, 154
  natural, 259, 283
  source of earth's heat, 337
Radioisotopes:
  artificial, 268
  uses, 282
Radio waves, 112
Radium, discovery of, 131
Range, of radioactive particles, 261
Rates, of chemical reaction, 214–215
Rational numbers, 7
Rayleigh-Jeans, radiation law, 126
Reaction, Law of, 31
Reactors:
  breeder, 280–281
  graphite pile, 278
  nuclear power, 279–280
Reality:
  inaccessibility, 167
  Platonic, 7, 20
Real substances, 63
Red giants, 309
Red shift:
  cosmological, 318
  Doppler, 306–307
Reducing sugar, 234
Reflection, Laws of, 101
Refraction Law, 104
  index, 104–105
Relativism, moral, 136
Relativity principle:
  alternatives to violation, 139
  classical, 136
  general relativity, 148

Religion, 3
Renaissance, 16
Replication, of DNA, 245–246
Reptiles, Age of, 347–350
Resistance, 83, 95
Resonance, in benzene, 220
Respiration, 250, 253
  chart, 251
Retrograde motion, of Mars, 12
Ribosomes, 241, 247
Richards, T. W., determination of atomic weights, 187
"Ring of fire," 335
Rivers, and landscape, 356
RNA, 247–248
Rockets, 34
Rocks:
  densities, 326–327
  elementary composition, 326
Roemer, Olaus, velocity of light, 98
Roentgen, Wilhelm, discovery of X rays, 130
Rumford, Count (Benjamin Thompson), 50–51, 66
Russell, Henry Norris, 308
Rutherford, Ernest:
  α-scattering experiment, 155
  atomic model, 156, 167, 258
  nuclear physics, 132, 267

Safety, in nuclear installations, 279–281
Satellites, natural, 293–294
Scalar, definition of, 30
Scheele, Karl Wilhelm, 173
Schrödinger, Erwin, atomic model, 163–164
Science, 1, 3, 16
  dawn of age, 16
  frontiers, 366–371
  moral values, 370–372
  nature, 366
  progress, 19
Scientific approach, 2
Scientist, 1, 3
Scintillation counters, 266
Secondary cell, 86
Second-generation stars, 314–315
Second genetic code, 248
Second Law of Thermodynamics, 69–71
  apparent exceptions, 71
  living systems, 249, 253, 340
Second transition series (of elements), 184
Sedimentary rock, 336
Selection rules, for electron structure, 161
Sequence of N-bases, and genetic code, 247
Series circuit, 88, 92
Shields, 337
Silurian Period, 342–343
Sirius, 4
Slip rings, 90
Slope, 379
Snell, Willebrod, law of refraction, 104
Solar system:
  Aristarchus' idea, 10
  elements, 324
  formation, 323–324
Space, curvature of, 150
Spaceship Earth, 370–371
Spacetime, 151
Spallanzani, 227
Special Theory of Relativity, postulates of, 140, 150–151
Specific heat, 53–54
Spectra:
  electromagnetic, 111–112
  of stars, 305
  visible, 107
Spontaneous processes:
  heat flow, 69
  order to disorder, 70
  probability, 183
Star:
  chemical composition, 305
  distance, 301–302, 304
  life line, 310
  luminosity, 303–304
  mass, 303–304
  origin, 298
  parallax, 301–302
  spectra, 305
  surface temperature, 305
Starch, 236
States (phases), of matter, 57
Steady State theory, 297
Stefan-Boltzmann Law, 125
Stonehenge (analog computer), 4
Strata, 339

Structural formulas, 205
Sugars, 231, 235
Sun, fusion reactions in, 299
Supergiants, 310
Supernova, element formation in, 313–314
Superposition, principle of, 110
Sutton, Walter S., 241
Symmetry, 140
  in solar system, 25
  in twin paradox, 146
Synchrotron principle, 273–274
System of the World:
  Copernican, 13, 23
  Greek, 6, 12
  Modern, 25–26

Technology, 1, 3, 370–371
Telescope, 24, 300
Temperature:
  definitions, 53, 55, 72
  maximum and minimum, 63
  relation to pressure, 61
  scales, 54, 65
Template function, of DNA, 246
Terminal velocity, 21
Tertiary Period, 352–353
Tetrahedron, bonds in carbon, 204
Thales, 6
Theoretical science, 1, 4
Thermodynamics, Laws of, (*see also* Conservation Law; Second Law of Thermodynamics) 69–71
  and life, 253, 331
Thompson, Benjamin (Count Rumford), 50–51, 66
Thomson, J. J.:
  atomic model, 154
  discovery of electrons, 93
  mass spectrograph, 266
  nuclear physics, 132
Thomson, William (Lord Kelvin), 68
Thought experiments:
  in General Relativity, 149
  in Uncertainty, 164–166
Thunderstorms, 357, 359
Thymine, 243
Time, 30, 144–147
  looking back, 307
Time dilation, 144–146
  evidence from muon lifetimes, 275

Tombaugh, Clyde, 290
Topography, effect on life, 337
Tornado, 361
Tracers, radioactive, 282
Transformation equations:
  Galilean-Newtonian, 137, 151
  Lorentz, 141, 151
Transistors, 93, 95
Transition elements, 182–184
  electronic structure, 183
  variable valence, 184
Transmutation, of elements:
  via alchemy, 171
  via nuclear reactions, 259, 267
Transuranium elements, 276
Trilobites, 342, 345
Twin paradox, 145–146

Ultraviolet catastrophe, 127
Ultraviolet radiation, and life, 331, 340–341
Uncertainty Principle, 164, 168
Uniformitarianism, 332
Uniform linear motion, 22
Universe:
  current model, 150, 164, 307–315
  evolution, 295–298, 315–318
  expansion, 296–297
  size, 295
Units:
  electrical (table), 85
  mechanical (table), 33
  summary (tables), 391–393
Unsaturated compounds, 207–208
Uracil, 247
Uranium:
  dating, 261
  ore, 131
  radioactivity, 131
  reactor fuel, 278–291
Ussher, Bishop, 332

Vacuum tubes, 93, 95
Valence:
  and electron structure, 160
  as combining power, 175–176
Valley:
  river, 356
  glacier, 357
Van de Graaff generator, 270–272
Van der Waals equation, 65

Variable stars, 312
Vector, definition of, 30
Velocity:
  definition, 30
  dependence of mass, length, and time, 141
  molecular, 53
  relative to observer, 142
Velocity of light (c):
  as limiting velocity of masses, 146-147, 151
  constancy, 124
  determination of:
    Fizeau, 99
    Galileo, 21
    Michelson, 99-100
    Roemer, 98
  same for all electromagnetic waves, 120
  value, 100
Vibration, 41
Vitalism, 227
Vitalist-mechanist argument, 228
Vitamin C, 239
Vitamins, 238-239
Volcano, 328-329, 334-335
Volta, A., and voltaic cells, 86
Voltaic cell, 84-88, 94
von Weiszäcker, Carl F., 324
Vulcanism, 334-335

Water, unusual properties, 58-59
Watson, James D., 241
Watt, James, 66
Wave properties of matter, 162-163
Wave theory of light, 110-111, 115
Weather, 357-362
  fronts, 358, 360-361
Wegener, Alfred, 336
"Weighing the world" (Cavendish experiment), 38
Weight, 36
Weightlessness, 37
Weisskopf, Victor, 274
White dwarfs, 310
Wien:
  Displacement Law, 305
  Radiation Law, 126
Wilkins, Maurice, 241
William of Occam (Ockham), 12, 15
Wilson, C. T. R., 262
Wöhler, Friedrich, 202, 227
Work and energy, 45, 46

X rays:
  discovery, 130
  spectra, 160

Ylem, 297
Young, Thomas:
  color vision, 107
  double slit experiment, 109, 111, 115
Yukawa, 269

Zeno, paradoxes, 19
Zodiac, 5

Credits

The color illustrations following pages 18, 226, and 322 are reproduced from the Life Nature Library and the Life Science Library, published by Time-Life Books. Individual photographers and artists are listed here.

PLATE 1. Fritz Henle from Monkmeyer Press Photo Service. PLATE 2. Lewis C. Thomas and Joseph F. Ossanna Jr. from The Bell Telephone Laboratories, Inc., Murray Hill, New Jersey. PLATE 3. Don Moss. PLATE 4. Leo and Diane Dillon. PLATE 5. Ken Kay. PLATE 6. Gordon Parks. PLATE 7. Mark A. Binn. PLATE 8. Fritz Goro. PLATE 9. Fritz Goro. PLATE 10. Ken Kay. PLATE 11. Ken Kay. PLATE 12. Ken Kay. PLATE 13. Fritz Goro. PLATE 14. Fritz Goro. PLATE 15. Fritz Goro. PLATE 16. Fritz Goro. PLATE 17. Fritz Goro. PLATE 18. Mel Hunter. PLATE 19. George V. Kelvin. PLATE 20. George V. Kelvin. PLATE 21. Charles W. Halgren, courtesy Caru Studios. PLATE 22. Copyright by the California Institute of Technology and the Carnegie Institute of Washington. Photograph courtesy of the Hale Observatories. PLATE 23. Copyright © 1959 by the California Institute of Technology and the Carnegie Institute of Washington. Photograph by William C. Miller, courtesy of the Hale Observatories. PLATE 24. Copyright © 1959 by the California Institute of Technology and the Carnegie Institute of Washington. Photograph by William C. Miller, courtesy of the Hale Observatories. PLATE 25. Copyright 1956 by William A. Garnett. PLATE 26. Josef Muench. PLATE 27. Andreas Feininger. PLATE 28. Camera Hawaii from Alpha Photo Associates, Inc. PLATE 29. Ned Haines from Rapho Guillumette. PLATE 30. Ansel Adams from Magnum. PLATE 31. NASA. PLATE 32. David Klein. PLATE 33. David Klein. PLATE 34. David Klein. PLATE 35(a) The Reverend Frank B. Dinwiddie; (b) C. R. Rouillon © World Meteorological Organization; (c) Ruth Galaid from Photo Researchers, Inc. (d) Russ Kinne from Photo Researchers, Inc.

Symbols and Constants

Symbol	Quantity
a	acceleration (in equations and vector diagrams involving force and motion)
	a side of a triangle (in geometric examples)
α	radioactive rays that are helium nuclei
	used to label an angle
B	magnetic field strength
	force per unit pole
β	radioactive rays that are electrons
	used to label an angle
C	capacitance, the amount of charge stored per unit of potential difference across a capacitor
c	the velocity of light
	specific heat in heat equations
	a side of a triangle (in geometric examples)
γ	very short X rays
	used to label an angle
d	distance
Δ	an interval or change in a quantity
E	energy, in equations $E = h\nu$ and $E = mc^2$
	electric field strength
	force per unit charge
$\varepsilon, \varepsilon_0$	electric permittivity (the subscript zero refers to a vacuum)
F	force
f	force of friction
G	universal gravitational constant
g	acceleration due to gravity
h	Planck's constant
	height, in potential energy equation
I	electric current
i	angle of incidence of a light ray
k	generalized constant, in showing proportionalities
	In the equation $\frac{1}{2}mv^2 = \frac{3}{2}kT$, k is the Boltzmann constant.
	In equations for electric forces and fields k is the Coulomb's Law constant.
	In Wien's Displacement Law, $T\lambda = k$, k is a specific but nameless constant.